T0298461

Multirate Signal Processing for Communication Systems
Second Edition

RIVER PUBLISHERS SERIES IN SIGNAL, IMAGE AND SPEECH PROCESSING

Series Editors:

MARCELO SAMPAIO DE ALENCAR
Universidade Federal da Bahia UFBA, Brasil

MONCEF GABBOUJ
Tampere University of Technology, Finland

THANOS STOURAITIS
University of Patras, Greece
and
Khalifa University, UAE

Indexing: all books published in this series are submitted to the Web of Science Book Citation Index (BkCI), to SCOPUS, to CrossRef and to Google Scholar for evaluation and indexing

The "River Publishers Series in Signal, Image and Speech Processing" is a series of comprehensive academic and professional books which focus on all aspects of the theory and practice of signal processing. Books published in the series include research monographs, edited volumes, handbooks and textbooks. The books provide professionals, researchers, educators, and advanced students in the field with an invaluable insight into the latest research and developments.

Topics covered in the series include, but are by no means restricted to the following:

- Signal Processing Systems
- Digital Signal Processing
- Image Processing
- Signal Theory
- Stochastic Processes
- Detection and Estimation
- Pattern Recognition
- Optical Signal Processing
- Multi-dimensional Signal Processing
- Communication Signal Processing
- Biomedical Signal Processing
- Acoustic and Vibration Signal Processing
- Data Processing
- Remote Sensing
- Signal Processing Technology
- Speech Processing
- Radar Signal Processing

For a list of other books in this series, visit www.riverpublishers.com

Multirate Signal Processing for Communication Systems
Second Edition

fredric j harris

University of California San Diego, USA

River Publishers

Published, sold and distributed by:
River Publishers
Alsbjergvej 10
9260 Gistrup
Denmark

www.riverpublishers.com

ISBN: 978-87-7022-210-5 (Hardback)
       978-87-7022-209-9 (Ebook)

©2021 River Publishers

To the memory of my parents,
Edith and Seymour Harris,
My Wife, Penelope,
Our children, Danielle and Robyn,
Our grandson, Justin,
And to all my students and colleagues
with whom I share the joy of learning.

*"A little learning is a dangerous thing;
drink deep, or taste not the Pierian spring:
there shallow draughts intoxicate the brain,
and drinking largely sobers us again."*

*Alexander Pope (1688–1744)
An Essay on Criticism*

Contents

Preface   **xiii**

List of Figures   **xxi**

List of Tables   **lv**

List of Abbreviations   **lvii**

1 Why Multirate Filters?   **1**
   1.1 Compact Disc 4-to-1 Oversample 2
   1.2 Anti-Alias Filtering 7

2 The Resampling Process   **13**
   2.1 The Sampling Sequence 15
       2.1.1 Modulation Description of Resampled Sequence . . 19
   2.2 What Is a Multirate Filter? 20
       2.2.1 Properties of Resamplers 23
       2.2.2 Examples of Resampling Filters 26
   2.3 Useful Perspectives for Multirate Filters 29
   2.4 Nyquist and the Sampling Process 33

3 Digital Filters   **41**
   3.1 Filter Specifications 42
   3.2 Windowing 45
   3.3 The Remez-Firpm Algorithm 55
       3.3.1 Equiripple vs. $1/f$ Ripple Designs 64
       3.3.2 Acceptable In-Band Ripple Levels 71

4 Useful Classes of Filters   **87**
   4.1 Nyquist Filter and Square-Root Nyquist Filter 88
   4.2 The Communication Path 91

    4.3   The Sampled Cosine Taper   95
         4.3.1   Root Raised Cosine Side-Lobe Levels 97
         4.3.2   Improving the Stop-Band Attenuation 99
    4.4   Half-Band Filters 106

5   Systems That Use Resampling Filters   117
    5.1   Filtering With Large Ratio of Sample Rate to Bandwidth . . 117
         5.1.1   Partial Sum Accumulator: The Dual Form 121
         5.1.2   Generate Baseband Narrowband Noise 126
         5.1.3   Generate Narrowband Noise at a Carrier Frequency . 129
    5.2   Workload of Multirate Filter 131

6   Polyphase FIR Filters   139
    6.1   Channelizer 140
         6.1.1   Transforming the Band-Pass Filter 146
    6.2   Separating the Aliases 155
    6.3   Problems 163

7   Resampling Filters   167
    7.1   Interpolators 168
         7.1.1   Simple 1-to-M Interpolator 168
    7.2   Interpolator Architecture 174
         7.2.1   Polyphase Partition 175
    7.3   Band-Pass Interpolator 178
    7.4   Rational Ratio Resampling 182
    7.5   Arbitrary Resampling Ratio 185
         7.5.1   Nearest Left Neighbor Interpolation 186
         7.5.2   Two-Neighbor Interpolation 196
    7.6   Farrow Filter 200
         7.6.1   Classical Interpolator 200
         7.6.2   Polynomial Approximation 205
         7.6.3   Farrow Structure 208

8   Half-Band Filters   219
    8.1   Half-Band Low-Pass Filters 220
    8.2   Half-Band High-Pass Filter 221
    8.3   Window Design of Half-Band Filter 223
    8.4   Firpm-Remez Algorithm Design of Half-Band Filters 224
         8.4.1   Half-Band Firpm Algorithm Design Trick 225

8.5 Hilbert Transform Band-Pass Filter 227
    8.5.1 Applying the Hilbert Transform Filter 228
8.6 Interpolating With Low-Pass Half-Band Filters 231
8.7 Dyadic Half-Band Filters 234

9 Polyphase Channelizers   **241**
9.1 Analysis Channel Bank 242
9.2 Arbitrary Output Sample Rates 245
    9.2.1 Noble Identity Based Analysis Filter Bank 257
9.3 Noble Identity Based Synthesis Filter Bank 268

10 Cascade Channelizers   **275**
10.1 Perfect Reconstruction Analysis–Synthesis Filter Banks . . . 276
10.2 Cascade Analysis and Synthesis Channelizers 282
    10.2.1 Cascade Channelizers with Channel Masks 284
    10.2.2 Compare Cascade Channelizers to Direct
        Implementation FIR 289
10.3 Enhanced Capabilities of Coupled Channelizers 293
    10.3.1 IFFT Centered Channelizer Enhancements 296
10.4 Multiple Bandwidths Arbitrary Frequency Center
    Channelizers 302
10.5 Channelizers with Even and Odd Indexed Bin Centers 309

11 Recursive Polyphase Filters   **323**
11.1 All-Pass Recursive Filters 324
    11.1.1 Properties of All-Pass Filters 326
    11.1.2 Implementing First-Order All-Pass Networks 332
11.2 Two-Path All-Pass Recursive Filters 337
    11.2.1 Two-Path Half-Band Filters: Non-Uniform Phase . . 338
    11.2.2 Two-Path Half-Band Filters: Linear Phase 347
11.3 Comparison of Non-Uniform and Equal Ripple Phase Two-
    Path Filters 350
11.4 Pass-Band and Stop-Band Response in Half-Band Filters . . 355
11.5 Transforming Half-Band to Arbitrary-Bandwidth 357
    11.5.1 Low-Pass to Low-Pass Transformation 357
    11.5.2 Low-Pass to Band-Pass Transformation 361
11.6 Multirate Considerations of Recursive Half-Band Filters . . 367
11.7 Hilbert-Transform Filter Variant of Two-Path All-Pass Filter 376
11.8 M-Path Recursive All-Pass Filters 380

11.9 Iterated Half-Band Filters   385
    11.9.1 Final Comparisons   386

12 Cascade Integrator Comb Filters   393
12.1 A Multiply-Free Filter   394
12.2 Binary Integers and Overflow   400
12.3 Multistage CIC   403
12.4 Hogenauer Filter   407
    12.4.1 Accumulator Bit Width   408
    12.4.2 Pruning Accumulator Width   410
        12.4.2.1 Up Sampling CIC   411
    12.4.3 Down Sampling CIC   417
12.5 CIC Interpolator Example   419
12.6 Coherent and Incoherent Gain in CIC Integrators   423
12.7 Equal Ripple Stopband Bifurcate Zeros   426
12.8 Compensation of CIC Main-Lobe Droop   431

13 Cascade and Multiple Stage Filter Structures   437
13.1 Interpolated FIR Filters   437
    13.1.1 Interpolated FIR Example   439
13.2 Spectral Masking Filters Based on Half-Band Filters   443
13.3 Spectral Masking Filters: Complementary Filters   447
13.4 Proportional Bandwidth Filter Banks   449
    13.4.1 Octave Partition   450
    13.4.2 Proportional Bandwidth Filters   452
        13.4.2.1 Example Proportional Bandwidth Design .   452
        13.4.2.2 Fractional Bandwidth Design Example . .   456
13.5 10-Channel Audiometric Filter Bank Example   458
    13.5.1 Signal Reconstruction in Synthesis Filter Bank . . .   460

14 Communication Systems Applications   471
14.1 Conventional Digital Down Converters   472
14.2 Aliasing Digital Down Converters   476
    14.2.1 IF Subsampling Example   477
14.3 Timing Recovery in a Digital Demodulator   484
    14.3.1 Background   484
    14.3.2 Modern Timing Recovery   488
14.4 Modem Carrier Recovery   492
    14.4.1 Background   493

14.4.2 Modern Carrier Recovery 494
      14.4.2.1 Design and Partition of Band-Edge Filter . 496
14.5 Digitally Controlled Sampled Data Delay 501
  14.5.1 Recursive All-Pass Filter Delay Lines 502
14.6 Interpolated Shaping Filter 509
14.7 Sigma-Delta Decimating Filter 523
  14.7.1 Sigma-Delta Filter 526
14.8 FM Receiver and Demodulator 534
  14.8.1 FM Band Channelizer 535
  14.8.2 FM Demodulator 539
  14.8.3 Stereo Decoding 541

Index   **549**

About the Author   **557**

MATLAB software available for this text:
  https://www.riverpublishers.com/book_details.php?book_id=788

Preface

Digital signal processing (DSP) has become a core body of material in undergraduate electrical engineering programs. Several threads branch from this core to enable related disciplines, such as communication systems, source coding, multimedia entertainment, radar, sonar, medical and laboratory instruments, and others. Multirate signal processing is one of these major threads. Multirate signal processing is the body of material that deals with concepts, algorithms, and architectures that embed sample rate changes at one or more sites in the signal flow path.

There are two reasons to include multirate signal processing in the solution of a particular signal processing task. The first is reduction in cost of the implementation. The second is enhanced performance of the implementation. We might also include a third, personal incentive, which is that it is fun to apply clever concepts to solve problems. We can hardly complete a multirate DSP design without a smile and the accompanying thought, "Boy, this is neat!"

Traditional concepts developed in the DSP world are the same as those developed in the analog-processing world. In both domains, we learn and use concepts such as convolution, Fourier transforms, transfer functions, poles and zeros, and others. When required to distinguish the two approaches, we use the qualifier "discrete" when discussing the DSP version of these fundamental concepts. The reason the two approaches are so similar is that they both emphasize linear time invariant (LTI) systems for which the tools of analysis and synthesis are well developed.

Multirate signal processing brings to the designer an important tool not available to the traditional DSP designer, who to the first order applies DSP techniques to emulate analog systems. We note that the interface between the two versions of the world, continuous and discrete, is the sampling process. In the traditional DSP perspective, the sample rate is selected to satisfy the Nyquist criterion but is otherwise incidental to the problem. In multirate signal processing, selection and modification of the sample rate are primary considerations and options in the signal processing chain. The option to change the sample rate is the additional tool offered to the DSP designer. Discrete systems with embedded sample rate changes are characterized as linear time varying (LTV) or as periodically time varying (PTV). Most of us have very little experience in the continuous world with LTV filters; thus, their unique properties come as a pleasant surprise as we learn how to design and use them.

The ability to change sample rate within the processing stream presents a remarkable list of processing tricks and performance enhancements. A consistent theme in this book is the presentation of perspectives that access these processing tricks. The first perspective we present is that a processing task should always be performed at the lowest rate commensurate with the signal bandwidth. This is the Nyquist rate of the signal component of interest.

We note that a common processing task is to reduce the bandwidth of a signal by filtering and then reduce the sample rate to match the reduced bandwidth. Our first processing trick interchanges the order of filtering and sample rate change so that the processing proceeds at the reduced output sample rate rather than at the high input sample rate. The condition under which this interchange is permitted is known as the noble identity. Reducing the sample rate prior to reducing the bandwidth causes aliasing of the input spectrum. Multirate signal processing permits and, in fact, supports this intentional aliasing, which can be unwrapped by subsequent processing. In fact, most of the tricks and enhancements associated with multirate signal processing are related to spectral aliasing due to the sample rate change. It seems counter-intuitive to use intentional aliasing as part of the signal processing scheme particularly when we have been told over and over not to alias the signal in the data collection process. It also seems a bit suspicious to claim that aliasing can be reversed. But, in fact, it can be when the aliasing occurs with a specified structure that we ensure in the multirate processing scheme. We can also use the change in sample rate to intentionally alias a signal at one center frequency to another. This option includes aliasing a signal from an intermediate center frequency to baseband by reducing the sample rate, as well as aliasing a signal from baseband to an intermediate center frequency by increasing the sample rate.

Purpose of the Book

The purpose of this book is to present a clear and intuitive description of the unique capabilities of multirate signal processing. This is accomplished by presenting the core material along with numerous examples supported by the liberal use of figures to illustrate the time and frequency representations of the multirate processing options. We also present a number of useful perspectives to facilitate development of insight of multirate systems. One such insight is that when describing a multirate system, since the sample rate is changing, we do not use the sample rate as our reference as is done in conventional DSP. Rather, it is useful to use the signal symbol rate or signal bandwidth as our reference since that is the single parameter that remains fixed in the process. The book includes many practical applications of multirate systems to help the reader see novel ways they can be applied to solve real problems. Commentary of traditional design techniques and alternate improved options are sprinkled throughout the text. Some of the material presented in the book, by necessity, must mimic similar expositions found in other texts on multirate processing, while other segments of the material reflect my unique perspective and experience. There are specific segments of material that are covered quite extensively here that are only lightly covered in other texts.

In particular, the chapter on recursive all-pass filters is a wealth of material deserving greater exposure than traditionally allocated to this topic. Much of the material presented in this textbook has been used in my graduate course, "Multirate Signal Processing." Significant segments of this material have found their way into my undergraduate course, "Modem Design," as well as a series of short courses and presentations dealing with synchronization techniques in modern communication systems. A light sprinkling of multirate filters can even be found in my undergraduate course in "Digital Signal Processing" as well as our undergraduate course in "Real-Time Digital Signal Processing." This book can be used in an undergraduate course in advanced DSP concepts or in a graduate course in multirate signal processing. Segments can also be used in support of various courses in communication system design and modem design, and can be used as a source of real-world applications in a DSP programming lab.

One of the pleasures of being a faculty member proficient in DSP skills as well as knowledgeable in modem communication systems is the ability to roam the halls of knowledge as well as the commercial centers of excellence that feed the economic engines of our society. I have had the good fortune of participating in the development of many systems that require the capabilities of high-performance, cost-effective DSP solutions. These systems include laboratory instruments, cable modems, satellite modems, sonar systems, radar systems, wireless systems, and consumer entertainment products. With a foot in each camp, one in academia and one in commercial, I am exposed to a rich and varied set of questions of interest from residents of the two areas.

Questions posed by commercial folks addressing focused problems are very different from those posed by the ones in academics. Some of my most creative work has been spawned by questions posed, and challenges offered, by perceptive folks in the industrial arena.

The academic environment provides me access to promising and talented students with whom I can share the pleasure of learning and understanding an established knowledge base in multirate digital processing while developing and expanding that same base. This text reflects much of that knowledge base tempered by the insight gathered from problem-solving tasks in the commercial sector and nurtured by scholarly interaction with curious, motivated students.

Organization of Book

This book is divided into 14 chapters. Chapter 1 is an introduction to multirate signal processing. The 1-to-4 up-sampler of the common consumer CD player is shown as an example of a ubiquitous application. Chapter 2 describes the process of sampling and resampling in time and frequency domains. Chapter 3 presents the relationship between the specifications of a finite impulse response (FIR) filter and the number of taps, or the length, of the filter. We also compare the window design and equal-ripple or Remez (FIRPM) design techniques. The effects of in-band ripple and of constant level stop-band side lobes are examined and various modifications to the design process are introduced to control pass-band ripple and stop-band side-lobe levels. Chapter 4 presents special filters such as square-root Nyquist filters and half-band filters. Discussions on standard and improved design techniques appropriate for these specific filters are also presented.

Chapter 5 presents examples of systems that use multirate filters and illustrates applications and demonstrates the wide range of applications. Chapter 6 presents resampling low-pass and band-pass FIR filters for which the noble identity has been applied to interchange the operations of resampling and filtering. Up-sampling, down-sampling, and cascade up–down sampling filters are examined. Chapter 7 describes polyphase interpolators and filters that perform arbitrary sample rate change. We also examine Farrow filters as well as filters that interpolate while performing alias-based translation. Chapter 8 covers quadrature mirror filters and dyadic half-band FIR filters. Chapter 9 covers M-path down-sampling analysis and M-path up-sampling synthesis channelizers. Also discussed are simultaneous interpolation and channel bank formation.

Chapter 10 looks at interesting variations of analysis and synthesis channelizers, in particular, their cascade operation for extreme performance and unusual capabilities. Chapter 11 covers recursive all-pass filters implemented as non-uniform phase and equiripple approximations to linear phase filters. A number of structures, including half-band, M-path filter banks, resampling structures, and arbitrary bandwidth non-resampling structures, are presented and illustrated. Chapter 12 presents the cascade integrator comb (CIC) filter and its resampling version, the Hogenauer filter as well as interesting variants. Chapter 13 describes cascades of low-order zero-packed up-sampled filters that exhibit periodic spectra with narrow transition band. Chapter 14 presents areas of applications in which multirate filters have significant presence.

At the end of each chapter are a number of problems designed to highlight key concepts presented in the chapter. These problems also serve to test the reader's knowledge and understanding of the material. Following each problem set is a list of references to guide the reader to related areas for further study.

Acknowledgements

We are able to see further than those who came before us because we stand on their shoulders. In addition, each day presents a fresh horizon because of the work we have performed as well as that performed by those around us. Interaction with our peers, our students, and our patrons enriches us by moving boundaries and lifting veils through the exchange of ideas. I cherish a group of people with whom I have had the great pleasure of sharing work and developing knowledge and with whom I have established long-standing friendships. I would like to cite some of them with the understanding that their ordering is a random process involving fleeting pleasant remembrances. Tony Constantinides of Imperial College in London has had a special influence on my work. I met Tony at a conference in Brisbane where we started a lifelong friendship and working relationship. It was through Tony that I gained an appreciation of the recursive polyphase half-band filter. I have yet to put down this valuable filter structure. It was Markku Renfors of Tempre University in Finland who first touched me through a classic paper on M-path recursive polyphase filters. That paper brought forth a torrent of ideas, papers, student projects, and commercial products. Many years later, meeting Markku at a conference in Santorini, I quickly warmed to him as a fellow traveler and I continue to enjoy our too infrequent academic and personal encounters. In a similar vein, Ron Crochiere through the classic book that he and Larry Rabiner authored, "Multirate Digital Signal Processing" helped form many of the ideas I nurtured into my concepts and perspectives on multirate filters. I thank Ron for reviewing the manuscript of this book and I am relieved that it passed muster. I am always pleased to see and spend time with Ron. From the other side of the playing field, I take great pleasure in my nearly daily interaction with Bernie Sklar, formerly with Aerospace Corporation and now with Communications Engineering Services. I hone my applied DSP skills on the communication system questions and problems that Bernie brings to me. Interacting with Bernie is like dancing with a skilled partner with both participants exhilarated and charmed by the encounter. Another fellow dancer who sways to the music of applied multirate DSP is Chis Dick of Xilinx. Chis and I met at a radar conference in Adelaide while he was at LaTrobe University. Our common interest and passion have been the core of a special friendship and our wives tell us that we are boring company when we climb into our common DSP shell to do our DSP magic. Michael Rice of BYU is another special person with whom I have been sharpening my applied DSP skills.

The list of people with whom I initially interacted through professional contacts that have warmed to friendships mingled with professional respect and admiration hold another special place for me. These include Don Steinbrecher, a very creative individual with a piece of his mind firmly planted in our future. Itzhak Gurantz, formerly of Comstream, Rockwell and Entropic, is a very smart person who many times has posed fascinating questions that guided me to places I would never have thought to tread. Ralph Hudson with whom I shared an office at Hughes Aircraft while I was on a leave of absence from SDSU, and David Lynch, who enticed me to visit Hughes, are clever individuals with enormous vision and ability to see beyond the view of most folks. I have not seen either of these fellows for many years, but they flooded me with great problems from which I was able to develop insight I could not have had without them. Others with whom I have worked and from whom I have learned are Pranish Sinha, Ragavan Sreen, Anton Monk, and Ron Porat. In more recent times, Bob McGwier of Virginia Tech brought me some fun to solve signal processing tasks.

The third group of people I feel compelled to acknowledge are those with whom I initially interacted through a student−mentor relationship and have continued to interact as friend and colleague. I will limit this list to current and former students who have worked with me in the area of multirate signal processing. Maximilien d'Oreye de Lantramange and Michael Orchard are the best of the best. Dragan Vuletic with whom I interact on various applied problems is a star following other exceptional former students, such as Bob Bernardi, Christian Bettwieser, and Benjamin Egg. The line continues with more recent and current graduates, Elettra Vanessa, Xiaofei Chen, and Alice Sokolova leading the pack. A very special bond connects me to my former students and now long-standing friends Joe Kiani, Eric, and Peggy Johnson. In many years of teaching, I have touched the minds of many students whom I hope have enjoyed the encounters as much as I have. I know that many names are missing from my acknowledgment list and if you feel offended by this, do not be. Please attribute it to a state of absentmindedness rather than to one of intentional omission.

fredric j. harris
University of California San Diego

List of Figures

Figure 1.1   Signal conditioning tasks required to convert a digital signal to its analog representation.   2

Figure 1.2   Three time signals: input and samples of input, DAC output, and filtered output.   3

Figure 1.3   Three spectra: sampled input, DAC output, and filtered output.   4

Figure 1.4   Modified signal conditioning tasks required to convert digital signal to analog representation. . . .   4

Figure 1.5   Four time signals: input samples zero-packed 1-to-4, 1-to-4 interpolated samples of input, DAC output of 1-to-4 interpolated samples, and analog filter output.   5

Figure 1.6   Four spectra: sampled input, 1-to-4 up-sampled output, DAC output, and filtered output.   6

Figure 1.7   Signal conditioning tasks required to convert an analog signal to its digital representation.   7

Figure 1.8   Spectra of input signal, analog filtered analog signal, and sampled input signal.   8

Figure 1.9   Modified signal conditioning to convert an analog signal to its digital representation.   9

Figure 1.10   Four spectra: filtered input, four-times oversampled sampled input, digitally filtered output, and 4-to-1 down-sampled output.   10

Figure 2.1   Resampling by zeroing sample values and by inserting zero-valued samples.   14

Figure 2.2   Two examples of 4-to-1 down-sampling of input series.   15

Figure 2.3   Sampling sequence $s_5(n)$ in time and frequency domain.   16

Figure 2.4   Sampling sequence $s_5(n-1)$ in time and frequency domain.   17

Figure 2.5   Time and frequency domain representation of sampled sequence $x(n)$.   18

Figure 2.6   Time and frequency domain representation of sampled sequence $x_0(5n)$.   18

Figure 2.7   Time and frequency domain representation of sampled sequence $x_1(5n)$.   19

Figure 2.8   Symbols representing down-sampling and up-sampling elements.   21

Figure 2.9   Down-sampling process filtering and sample rate reduction.   22

Figure 2.10   Time series and spectra for signal points in 3-to-1 down-sample process.   22

Figure 2.11   Up-sampling process filtering and sample rate increase.   23

Figure 2.12   Time series and spectra for signal points in 1-to-5 up-sample process.   23

Figure 2.13   Down-sampled sum of scaled sequences equivalent to sum of scaled down-sampled sequences.   24

Figure 2.14   Scaled up-sampled sequences equivalent to up-sampled scaled sequences.   24

Figure 2.15 Q-units of delay and Q-to-1 down-sampler is equivalent to Q-to-1 down-sampler and 1-unit of delay.   25

Figure 2.16   1-to-P up-sampler followed by P-units of delay is equivalent to 1-unit of delay and P-to-1 up-sampler.   25

Figure 2.17   Exchanging order of filtering and down-sampling. .   25

Figure 2.18   Exchanging order of up-sampling and filtering. . .   26

Figure 2.19   Reordering cascade up-samplers and down-samplers.   26

Figure 2.20   Interchange of relative prime input and output resamplers.   26

Figure 2.21   Standard down-sampling filter architectures: M-to-1 polyphase filter, 8-to-1 dyadic half-band filter chain, and M-to-1 Hogenauer (CIC with embedded resampler) filter.   27

Figure 2.22   Standard up-sampling filter architectures: 1-to-M polyphase filter, 1-to-8 dyadic half-band filter chain, and 1-to-M Hogenauer (CIC with embedded resampler) filter.   28

Figure 2.23 Filter impulse response and filter spectrum sampled at four times the bandwidth and at five times the signal bandwidth. 30

Figure 2.24 Time and spectral presentations of a continuous and sampled data sine wave for two different sample rates and two different scaling options. 31

Figure 2.25 Time and frequency description of four, fixed time interval, fixed sample rate = 1, with successively higher center frequencies f_0 = 0.1, 0.2, 0.4, and 0.8. 32

Figure 2.26 Time and frequency description of four, fixed time interval, fixed input frequency f_0 = 0.1, with successively lower sample rates f_s = 1, 1/2, 1/4, and 1/8. 33

Figure 2.27 Spectrum of input signal and of filter, of filtered input signal, and of sampled input signal. 35

Figure 3.1 Frequency response prototype low-pass filter. . . . 43

Figure 3.2 Impulse response of prototype low-pass filter. . . . 43

Figure 3.3 Sampled data impulse response of prototype filter. . 45

Figure 3.4 Spectrum of 100-point rectangle window with zoom to main lobe. 47

Figure 3.5 Spectrum of rectangle windowed prototype filter obtained as convolution between spectrum of prototype filter and spectrum of rectangle window. . 48

Figure 3.6 Log display of spectrum to emphasize high levels of out-of-band side-lobe response of rectangle windowed prototype filter. 49

Figure 3.7 Raised cosine window and transform illustrating sum of translated and scaled Dirichlet kernels. . . . 50

Figure 3.8 Side-lobe levels of Kaiser–Bessel windowed FIR filter and of the Kaiser–Bessel window as a function of window parameter β. 53

Figure 3.9 Time and frequency response of FIR filter windowed with Kaiser–Bessel: first designed for 6 dB band edge and second for pass-band and stop-band edges. 54

Figure 3.10 Parameters required to specify sampled data low-pass filter. 55

Figure 3.11 Frequency response of equiripple filter and error frequency profile. 56

Figure 3.12   Distribution of initial estimates of extrema for multiple exchange algorithm. 57

Figure 3.13   Polynomial passed through initial estimate sample points. 58

Figure 3.14   Exchanging sample points for iteration $k + 1$ with sample points located at extremal values obtained from estimate at iteration k. 59

Figure 3.15   Multiplier parameter $K(\delta_1, \delta_2, \Delta f)$ as function of pass-band ripple δ_1, stop-band ripple δ_2, and transition bandwidth $\Delta f/f_s$. 60

Figure 3.16   Input and output arrays in the Remez algorithm. . . 62

Figure 3.17   Time and frequency response of FIR filter designed with the firpm algorithm. 63

Figure 3.18   Remez impulse response showing details of end point, close-up of end point, and spectra of original filter and of filter with clipped end point. 65

Figure 3.19   Resampling low-pass filter. 66

Figure 3.20   Frequency response of reference filters with equiripple and $1/f$ side lobes and of 10-bit quantized version of same filters. 67

Figure 3.21   Modified weight functions for Remez algorithm to obtain increasing penalty function in stop band to obtain $1/f$ side lobes. 69

Figure 3.22   Remez filter time and frequency response designed with linear frequency and with staircase frequency weight function. 69

Figure 3.23   Spectral response of filters designed by the default remezfrf and by myfrf routines in Remez call. . . . 70

Figure 3.24   Input and output of a linear filter. 71

Figure 3.25   Frequency response of equiripple FIR filter and spectrum of input signal. 72

Figure 3.26   Time signals at input and output of filter with equiripple spectral response. 73

Figure 3.27   Impulse and frequency response of a FIR filter. Details of pass band show ± 1 dB in-band ripple. Spectrum of input signal is fully contained in filter pass band. 74

Figure 3.28   Time-aligned input and output signal waveforms and difference of two waveforms to show paired echoes. 75

Figure 3.29   Impulse, magnitude response, and group delay response of an IIR filter and of a phase equalized version of same filter. Details of equalized phase show ± 2.2 degree in-band ripple. 77

Figure 3.30   Time-aligned input and output waveforms from nonuniform phase IIR filter with the difference of two waveforms to show quadratic phase distortion. 78

Figure 3.31   Time-aligned input and output waveforms from phase equalized IIR filter with the difference of two waveforms to show paired echoes. 79

Figure 3.32   Block diagram of 16-QAM modulator, demodulator chain used to illustrate ISI due to in-band ripple of shaping filter and matched filter. 80

Figure 3.33   Constellation at output of matched filter, error profile of converging equalizer, constellation at output of equalizer, and spectra of shaping filter and compensating equalizing. 81

Figure 4.1   Various wave shapes satisfying condition for zero ISI. 89

Figure 4.2 $\mathrm{Sin}(x)/x$ waveform and uniform pass-band spectrum. 90

Figure 4.3   Spectral convolution of prototype Nyquist spectrum with even symmetric spectral mass, and equivalent time domain product of Nyquist pulse with window. 91

Figure 4.4   Simple model of signal flow in communication. . . 92

Figure 4.5   Spectrum and time response of the Nyquist filter and SQRT Nyquist filter. 94

Figure 4.6   Highest out-of-band side-lobe levels for SQRT raised cosine Nyquist filter for a range of roll-off bandwidths and filter lengths. 98

Figure 4.7   Spectrum of windowed SQRT raised cosine Nyquist filter, Kaiser(N,0) (-37 dB), Kaiser(N,2) (-50 dB), and Kaiser(N,3) (-60 dB), inserted details of spectrum near 3-dB bandwidth and details of ISI in matched filter outputs. 99

Figure 4.8   Top subplot: spectral cosine and harris tapers; center subplot: cosine and harris tapered Nyquist spectra; bottom subplot: cosine and harris tapered SQRT tapered spectra. 100

Figure 4.9   Top subplot: impulse responses of harris taper and cosine taper SQRT Nyquist shaping filters with 25% excess bandwidth; center subplot: spectra of shaping filters with same 25% transition bandwidth; bottom subplot: zoom to in-band ripple levels of the harris and cosine-tapered shaping filters. 102

Figure 4.10   Frequency and time responses of cosine and harris tapered SQRT Nyquist filters. Top subplots: in-band ripple frequency responses. Bottom subplots: low-level ISI, deviations from nominal time response zero crossings. 103

Figure 4.11   Spectrum of Remez algorithm output at iteration n matching band edges of Nyquist pulse and then increase of frequency f_1 for iteration $n + 1$ to raise $h_1(n)$ to $h_1(n + 1)$ and eventually to the desired -3 dB level. 104

Figure 4.12   Spectra of converged and SQRT cosine filters, details of same spectra, and finer details of same spectra. 105

Figure 4.13   Details of ISI levels matched filter responses of harris–Moerder and cosine taper SQRT Nyquist filters. 106

Figure 4.14   Impulse response and spectrum of ideal low-pass half-band filter. 107

Figure 4.15   Impulse response and spectrum of finite duration low-pass half-band filter. 108

Figure 4.16   Impulse response and spectrum of finite impulse response (FIR) high-pass half-band filter. 109

Figure 4.17   Polyphase partition of half-band filter pair. 110

Figure 4.18   Complementary partition of nonrealizable half-band filter pair. 111

Figure 4.19   Noncausal and causal forms of half-band low-pass filter. 111

Figure 4.20   Complementary partitions of realizable half-band filter pair. 111

Figure 5.1   Specifications of low-pass filter. 118
Figure 5.2   Initial implementation of a filtering task. 118
Figure 5.3   50-to-1 polyphase partition and down-sampling of low-pass filter. 119
Figure 5.4   Cascade 50-to-1 and 1-to-50 polyphase maintains constant sample rate. 120
Figure 5.5   Reduced resource cascade 50-to-1 and 1-to-50 polyphase filter. 121
Figure 5.6   Standard three-stage, three-MAC, polyphase up-sampling filter. 122
Figure 5.7   Shared registers of three-stage, three-MAC, polyphase up-sampling filter. 123
Figure 5.8   Minimum resource version of three-stage polyphase up-sampling filter. 123
Figure 5.9   Standard three-stage, three-MAC, polyphase down-sampling filter. 124
Figure 5.10   Dual form three-stage polyphase down-sampling filter. 125
Figure 5.11   Shared registers of three-stage, multiplier-accumulator polyphase up- sampling filter. 126
Figure 5.12   Minimum resource version of three-stage polyphase down-sampling filter. 126
Figure 5.13   Block diagram of simple brute force narrowband noise generator. 127
Figure 5.14   Input spectrum, filter spectral response, and output spectrum. 127
Figure 5.15   Block diagram of efficient narrowband noise generator. 128
Figure 5.16   400-Hz input spectrum, 1-to-50 zero-packed 20 kHz input spectrum, filter spectral response, and filtered output spectrum. 129
Figure 5.17   Translate baseband signal formed by polyphase filter. 129
Figure 5.18   Baseband spectrum and translated spectrum at 20 kHz sample rate. 130
Figure 5.19   Band pass signal formed by translated polyphase filter. 131
Figure 5.20   Multirate filter indicating input and output sample rates. 132

Figure 5.21   Spectra of low-pass signal at input and output sample rates.   132

Figure 5.22   Spectra of two low-pass filters of different bandwidth but with fixed ratio of bandwidth to transition bandwidth.   133

Figure 6.1   Spectrum of multichannel input signal, processing task: extract complex envelope of selected channel.   140

Figure 6.2   Standard single channel down-converter.   141

Figure 6.3   Spectra observed at various points in processing chain of standard down-converter.   142

Figure 6.4   Tapped delay line digital filter: coefficients and data registers, multipliers, and adders.   143

Figure 6.5   Band-pass filter version of down-converter.   143

Figure 6.6   Block diagrams illustrating equivalency between operations of heterodyne and base-band filter with band-pass filter and heterodyne.   144

Figure 6.7   Down sampled band-pass down-converter.   145

Figure 6.8   Band-pass down-converter aliased to base band by down sampler.   146

Figure 6.9   Spectral description of down-conversion realized by a complex band-pass filter at a multiple of output sample rate aliased to base band by output resampling.   147

Figure 6.10   Statement of noble identity: a filter processing every math input sample followed by an output M-to-1 down sampler is the same as an input M-to-1 down sampler followed by a filter processing every input sample.   148

Figure 6.11 M-path partition of prototype low-pass filter with output resampler.   149

Figure 6.12 M-path partition of prototype low-pass filter with input resamplers.   149

Figure 6.13 M-path partition of prototype low-pass filter with input path delays and M-to-1 resamplers replaced by input commutator.   150

Figure 6.14   Resampling M-path down-converter.   153

Figure 6.15   Time delay function performed by path filter of m-path polyphase partition.   157

Figure 6.16 Time alignment of input samples performed by M-path filter stages. 157

Figure 6.17 Phase profiles for 10 paths of 10-path polyphase partition. 158

Figure 6.18 Group delay profiles for 10 paths of 10-path polyphase partition. 159

Figure 6.19 Phase profiles of 10 separate paths of 10-path polyphase partition. 159

Figure 6.20 Details of phase profiles of 10 paths of 10-path polyphase partition. 160

Figure 6.21 Polyphase filter bank: M-path polyphase filter and M-point FFT. 161

Figure 7.1 Example of interpolator following initial time series generator. 168

Figure 7.2 Spectrum at output of shaping filter with embedded 1-to-4 up-sampler. 169

Figure 7.3 Periodic input spectrum and filter response of 1-to-32 interpolator filter. 169

Figure 7.4 Spectrum of interpolator and spectrum of zero-packed shaping filter. 171

Figure 7.5 Spectrum of 1-to-32 interpolated shaping filter. . . 172

Figure 7.6 Spectrum of alternate interpolator and periodic spectrum of zero-packed shaping filter. 173

Figure 7.7 Spectrum of alternate 1-to-32 interpolated shaping filter. 173

Figure 7.8 Impulse response of interpolator filter with single stop band and with multiple stop bands with do-not-care band. 174

Figure 7.9 Initial structure of 1-to-32 interpolator. 175

Figure 7.10 Initial structure of 1-to-M polyphase interpolator. . 176

Figure 7.11 Transition structure of 1-to-M polyphase interpolator. 177

Figure 7.12 Standard structure of 1-to-M polyphase interpolator. 177

Figure 7.13 Efficient implementation structure of 1-to-M polyphase interpolator. 178

Figure 7.14 Baseband modulator with dual analog components to be replaced by digital components in digital IF modulator. 179

Figure 7.15   Digital baseband and IF modulator with minimum
analog components.   179

Figure 7.16   Structure of 1-to-M polyphase interpolator with
phase rotators for selecting spectrum centered in
Mth Nyquist.   181

Figure 7.17   Spectrum of translated interpolator and periodic
spectrum of zero-packed shaping filter.   181

Figure 7.18   Spectrum of 1-to-32 interpolated and translated
shaping filter.   182

Figure 7.19   Straightforward approach to 5/3 interpolator.   183

Figure 7.20   Detailed presentation of 5/3 interpolator.   183

Figure 7.21   Time indices for input series, for output commutator
series, save and discard indicator for 3-to-1
resampler, saved commutator indices, and output
series.   184

Figure 7.22   Resampling an input sequence by accessing every
fifth output port of a 12-stage polyphase filter. . . .   185

Figure 7.23   Interpolating to a position between available output
points in a P-stage interpolator.   186

Figure 7.24   Replacing desired sample value with nearest
neighbor sample value.   187

Figure 7.25   Timing jitter model: sampling of virtual DAC output
for maximum interpolated sample values.   187

Figure 7.26   Spectrum of up-sampled signal at input and output
of DAC.   188

Figure 7.27   Frequency response in neighborhood of the DAC's
first spectral null.   189

Figure 7.28   Shaping filter: time and frequency response four
times oversampled.   191

Figure 7.29   Time and frequency response of 32/6.4 nearest
neighbor interpolator.   191

Figure 7.30   Time and frequency response of straight 1-to 5
polyphase interpolator.   192

Figure 7.31   Time and frequency response of 32/6.37 nearest
neighbor interpolator.   193

Figure 7.32   Block diagram of input, output, and filter weight set
index control.   194

Figure 7.33   Visualization of relationship between input clock,
output clock, and polyphase filter index pointer. . .   194

Figure 7.34 Example of indexing sequence formed by address control in resampling routine. 195

Figure 7.35 Linear interpolation between available left and right sample values. 196

Figure 7.36 Linear interpolation as convolution of sample points with triangle pulse. 197

Figure 7.37 Spectrum of up-sampled signal at input and output of triangle interpolator. 197

Figure 7.38 Frequency response in neighborhood of triangle first spectral null. 198

Figure 7.39 Shaping filter: time and frequency response four times oversampled. 199

Figure 7.40 Time and frequency response of 32/6.4 linear interpolator. 199

Figure 7.41 A polyphase up-sampling filter. 201

Figure 7.42 Input and output sample location for *P*-stage resampling filter. 201

Figure 7.43 Coefficient mapping of *P*-stage polyphase filter. . . 202

Figure 7.44 Time and frequency response of 300-tap, 1-to-50 interpolating filter. 203

Figure 7.45 Contents of five columns in polyphase partition of 250-tap interpolating filter. 204

Figure 7.46 First three columns of polyphase filter, fourth-order polynomial approximation, and errors between columns and approximation. 206

Figure 7.47 Polynomial form of arbitrary resampling filter. . . . 207

Figure 7.48 Time and frequency response of 50-stage interpolator formed by fourth-order polynomial approximations to five columns of polyphase partition. 208

Figure 7.49 Impulse and frequency response of farrow filters: filters form Taylor series coefficients of input time series. 210

Figure 7.50 Signal flow diagram of Horner's rule for evaluating polynomial. 212

Figure 7.51 Efficient hardware version of Horner's rule processor. 212

Figure 8.1 Spectral characteristics of half-band low-pass filter. 220

Figure 8.2 Impulse response of non-causal half-band low-pass filter. 221

Figure 8.3   Spectral characteristics of half-band high-pass filter. 222

Figure 8.4   Impulse response of non-causal half-band high-pass filter. 222

Figure 8.5   Impulse response and frequency response of half-band low-pass filter designed with Kaiser window. . 224

Figure 8.6   Impulse response and frequency response of half-band low-pass filter designed with firpm algorithm. 225

Figure 8.7   Impulse response and frequency response of trick one-band filter designed with Remez algorithm. . . 226

Figure 8.8   Impulse response and frequency response of trick one-band filter with inserted zero samples and symmetry point sample. 227

Figure 8.9   Spectral characteristics of Hilbert transform half-band filter. 228

Figure 8.10   Impulse response and frequency response of half-band filter designed by firpm algorithm and converted to Hilbert transform by complex heterodyne to $f_s/4$. 229

Figure 8.11   Hilbert transform filter as a spectrally translated version of low-pass half-band filter. 229

Figure 8.12   Two-path model of Hilbert transform filter with complex impulse response. 230

Figure 8.13   Noble identity applied to Hilbert transform filter. 230

Figure 8.14   Two-path commutator driven Hilbert transform filter. 230

Figure 8.15   One-to-two up-sampling process with half-band filter. 231

Figure 8.16   One-to-two up-sampling with polyphase half-band filter. 231

Figure 8.17   One-to-two up-sampling at output of polyphase half-band filter. 232

Figure 8.18   One-to-two up-sampling with commutated two-path half-band filter. 232

Figure 8.19   Spectral characteristics of half-band filter suppressing spectral replicate. 232

Figure 8.20   Filter length parameter $K(\alpha)$ as function of fractional bandwidth α. 233

Figure 8.21   Cascade of four half-band filters to raise the sample rate by 16 in steps of increase by 2 per stage. 234

Figure 8.22   Frequency plan of three successive half-band filters showing transition bandwidth and sample rate. . . . 235

Figure 8.23   Proportionality factor $K_1(\alpha)$ in ops/input and $K_2(\alpha)$ in ops/output for one to five cascade stages of half-band filters. 237

Figure 8.24   Proportionality factor $K_3(\alpha)$ in ops/input and $K_4(\alpha)$ in ops/output in M-path filter for values of $M = 2, 4, 8, 16,$ and 32. 238

Figure 9.1   Polyphase partition of M-to-1 down-sampled band-pass filter. 242

Figure 9.2   Polyphase analysis filter bank as a polyphase filter input and an IDFT output process. 244

Figure 9.3   Spectral response of two analysis filter banks with same channel spacing: one for spectral analysis and one for FDM channel separation. 245

Figure 9.4   Data memory loading for successive 32-point sequences in a 64-stage polyphase filter. 248

Figure 9.5   Cyclic shift of input data to FFT to absorb phase shift due to 32 sample time shift of data in the polyphase filter. 250

Figure 9.6   Alias output frequencies for successive Nyquist zones for M-to-1 and for M-to-2 down-sampling of M-path polyphase analysis filter bank. 251

Figure 9.7   Input and output spectra of 50-channel channelizer and resampler. 253

Figure 9.8   Maximally decimated filter bank structure and modified two-samples-per-symbol filter bank structure. 255

Figure 9.9   Memory contents for successive 48-point input data blocks into a 64-point prototype pre-polyphase partitioned filter and FFT. 256

Figure 9.10   Memory contents for successive 48-point input data blocks into a 64-point polyphase filter. 258

Figure 9.11   Cyclic shift schedule for input array to FFT. 258

Figure 9.12 $M/2$-to-1 down-sampler following filter polynomials in Z^M. 259

Figure 9.13 $M/2$-to-1 down-samplers preceding filter polynomials in Z^2. 260

Figure 9.14 $M/2$-to-1 down-samplers moved through delays in bottom half of delay chain. 261

Figure 9.15 Replace input delay chains and $M/2$-to-1 down-samplers with dual commutators. 262

Figure 9.16 Top $M/2$ filters use even indices and bottom $M/2$ filters use odd indices. 263

Figure 9.17 M-path, $M/2$-to-1 down-sample polyphase analysis filter architecture formed by noble identity transformations. 264

Figure 9.18 M-path, $M/2$-to-1 down-sample polyphase analysis filter architecture formed by interleaving top and bottom half data buffers. 265

Figure 9.19 M-path, 1-to-$M/2$ up-sample polyphase synthesis filter architecture formed by dual of noble identity transformations. 269

Figure 10.1 M-path analysis filter bank coupled to an M-path synthesis filter bank. 277

Figure 10.2 Top subplot: analysis Nyquist spectra spaced at 20 kHz centers with half-amplitude crossings at 10 kHz. Center subplot: single Nyquist spectrum with transition bandwidth of $5-15$ kHz sampled at 40 kHz. Bottom subplot: periodic Nyquist spectra with replica spectra at multiples of 40 kHz sample rate and overlay synthesis spectrum transition bandwidth of $15-25$ kHz. 279

Figure 10.3 Top subplot: three adjacent Nyquist magnitude spectra of synthesis filter bank. Center subplot: three adjacent Nyquist log magnitude spectra of synthesis filter bank. Bottom subplot: single half-band Nyquist spectra with adjacent spectral replicas and overlay spectrum of synthesis filter. 280

Figure 10.4 Top subplot: impulse response of full bandwidth cascade analysis and synthesis filter banks. Center subplot: zoom to low-level details of impulse response. Bottom subplot: spectrum ripple in-band ripple levels, perfect reconstruction errors. 283

Figure 10.5   Top subplot: spectrum ripple in-band ripple levels, for 239-tap filter with appended leading zero. Bottom subplot: spectrum ripple in-band ripple levels, for 240-tap filter. 285

Figure 10.6   Top subplot: spectrum of synthesized wide bandwidth filter from analysis channels (−9 to +9). Center subplot: spectrum of synthesized narrow bandwidth filter from analysis channels (−4 to 4). Bottom subplot: spectrum of synthesized Hilbert transform filter from channels (+1 to +13). 286

Figure 10.7   Top subplot: spectrum of multiple sinusoid input signal to coupled channelizers. Second subplot: spectrum of synthesized wide bandwidth filter response. Third subplot: spectrum of synthesized narrow bandwidth filter response. Bottom subplot: spectrum of synthesized Hilbert transform filter response. 288

Figure 10.8   Spectra of desired, same length channel filter, spectral span of polyphase Nyquist zones, mask enabled Nyquist channels, and synthesized filter bandwidth from selected Nyquist channel bands. . . 290

Figure 10.9   60-Path analysis and synthesis channelizer implementation of 600-tap FIR filter. 60-Path filters designed with 600 coefficients to obtain the same filter performance. 291

Figure 10.10   Time and frequency responses of 60-path coupled channelizers. Top subplot: impulse response. Middle subplot: frequency response with overlay of analysis filter spectra input to synthesizer. Bottom-left subplot: zoom to in-band spectral ripple. Bottom-right subplot: zoom to filter transition bandwidth. 292

Figure 10.11   Time and frequency responses of 599-tap direct implementation FIR filter. Top subplot: impulse response. Middle subplot: frequency response with overlay of analysis filter spectra input to synthesizer. Bottom-left subplot: zoom to in-band spectral ripple. Bottom-right subplot: zoom to filter transition bandwidth. 293

Figure 10.12 Efficient two-tier channelizer with spectral modification of edge channels to enable arbitrary variable bandwidth filter synthesizer. 294

Figure 10.13 Spectra of channelizer channel responses, inner tier filter applied to end channel, and modified end channel response to obtain the desired synthesized filter with reduced bandwidth. 295

Figure 10.14 Spectra of synthesized broadband channel with inner tier filters reducing end channel bandwidth to reduce synthesized bandwidth. 295

Figure 10.15 Perfect reconstruction spectra of adjacent channel widths of modified channelizer with interleaved complementary bandwidth channels. 296

Figure 10.16 Synthesizing interleaved narrow and wide channel channelizer from dual analysis channelizers. 297

Figure 10.17 Spectra of synthesized broadband channel with interleaved alternating wide and narrow bandwidth channel filters. 297

Figure 10.18 Block diagram of channelizer-based variable bandwidth filter extracting positive frequencies and reducing sample rate with half-length synthesis filter. 298

Figure 10.19 Frequency response of super channels synthesized from an odd number (1, 3, 7, and 15) of baseband channels: even stacking of BW intervals. 298

Figure 10.20 Block diagram of channelizer-based variable bandwidth filter: BW adjusted by binary mask between analysis and synthesis filter banks. 299

Figure 10.21 Frequency response of super channels synthesized from an even number (2, 4, 8, and 16) of baseband channels: Odd stacking of BW intervals. 300

Figure 10.22 Block diagram of channelizer-based variable bandwidth filter: BW adjusted by even or odd indexed binary mask between analysis and synthesis filter banks. 301

Figure 10.23 Spectrum, zoom to in-band ripple of super channels synthesized from three alternate even stacked and four alternate odd stacked subchannels. 301

Figure 10.24 Analysis filter bank with multiple arbitrary bandwidth output signals down-converted to baseband from arbitrary input center frequencies. 303

Figure 10.25 Synthesis filter bank with wide bandwidth output signal containing multiple up-converted input signals with arbitrary bandwidths and center frequencies. 304

Figure 10.26 Spectra of four QPSK input signals presented to 30-channel synthesis channelizer with symbol rates of 10, 20, 40, and 80 MHz at sample rates of 40, 40, 120, and 160 MHz, respectively. 305

Figure 10.27 Top subplot: input spectrum of 40 MHz symbol rate input signal with spectral overlays of six-path channelizer filter responses. Six lower subplots show spectra of baseband channel output time series, each sampled at 40 MHz. 305

Figure 10.28 (a) Spectra of band centered input signal to eight-path analysis channelizer and spectra of baseband output channels sampled at 40 MHz. 307

Figure 10.28 (b) Spectra of frequency offset band input signal to eight-path analysis channelizer and spectra of baseband output channels sampled at 40 MHz. . . . 307

Figure 10.29 Top subplot: spectrum of synthesized super channel formed from multiple input signals with different bandwidths. Thirty lower subplots: spectrum of baseband signal presented to 30-path synthesizer channelizer by pre-processing analysis channelizers. 308

Figure 10.30 Spectra of two channelizers with different center frequencies but with same channel shape and same frequency spacing. Upper subplot centers match DFT center frequencies and are centered on the even integers on this figure. Lower subplot centers are offset by half their spacing and are centered on the odd integers in this figure. 310

Figure 10.31 Aligning spectra of input signal with spectral responses of filter bank by complex heterodyne of input signal in upper subplot or by complex heterodyne of filter coefficient weights in lower subplot. 311

Figure 10.32 Two unit circles with roots of $(Z^{15} - 1)$, the frequencies corresponding to a 15-point DFT. The left subplot indicates the location of DC or zero frequency of an unaltered input sequence presented to the DFT. The right subplot indicates the location of DC or zero frequency heterodyned to the half sample rate by an alternating sign heterodyne of the input sequence. 312

Figure 10.33 Polyphase filter input sample indices and sign of input heterodyne for two successive 15-point data samples in 15-path polyphase filter. Note the sign reversals of the two new input vectors. 312

Figure 10.34 Polyphase filter input sample indices and sign of input heterodyne for two successive 10-point data sample sequences in 15-path polyphase filter. Note that there are no sign reversals of the two new input vectors. 313

Figure 10.35 Aligning spectra of input signal with spectral responses of odd length, non-maximally decimated filter bank by alternating sign heterodyne of input signal in upper subplot, or by alternating sign heterodyne of filter coefficient weights in lower subplot. 314

Figure 10.36 Spectra of input signal and channel centers of 15-path polyphase channelizer performing 10-to-1 down-sampling with alignment of channelizer spectra with half-channel bandwidth offset performed by embedding alternating sign heterodyne in filter weights. Lower 15 subplots show spectra obtained at each baseband channel output port. 315

Figure 10.37 Two unit circles with roots of $(Z^{18} - 1)$, the frequencies corresponding to an 18-point DFT. The left subplot indicates the location of DC or zero frequency of an unaltered input sequence presented to the DFT. The right subplot indicates the location of DC or zero frequency heterodyned to the quarter sample rate by $\exp(j\,n\,\pi/2)$ heterodyne of the input sequence. 316

Figure 11.1 M-path polyphase all-pass filter structure. 325

Figure 11.2 M-path polyphase filter with delays Z^{-K} allocated
to the Kth path.   325

Figure 11.3   Pole diagram and phase as function of frequency for
single delay Z^{-1}.   327

Figure 11.4   Pole-zero diagram of first-order all-pass networks
$[(1 + \alpha Z)/(Z + \alpha)]$: showing phase angles that form
output phase angle.   327

Figure 11.5   Phase response of first order in Z all-pass network
as function of pole position α ($\alpha = 0.9, 0.8, \ldots, 0,$
$\ldots, -0.8, -0.9$).   328

Figure 11.6   Pole-zero diagram of first order in Z^2 all-pass
networks $[(1 + \alpha Z^2)/(Z^2 + \alpha)]$.   329

Figure 11.7   Phase response of first-order (in Z^2) all-pass
network as function of pole position α ($\alpha = 0.9, 0.8,$
$\ldots, 0, \ldots, -0.8, -0.9$).   330

Figure 11.8   Visualization of components in Equation (11.13). .   332

Figure 11.9   Canonic implementation of first-order all-pass filter.   333

Figure 11.10   Reordered all-pass filter structure.   334

Figure 11.11   Single coefficient all-pass filter structure.   334

Figure 11.12   Single coefficient all-pass filter dual structure. . . .   335

Figure 11.13   Alternate single coefficient all-pass structure. . . .   336

Figure 11.14   Spectral variance for two forms of all-pass filter
for 16-bit and 20-bit coefficients as function of
coefficient α.   336

Figure 11.15   Two-path polyphase filter.   337

Figure 11.16   Redrawn two-path polyphase filter.   338

Figure 11.17   Two-path all-pass filter.   339

Figure 11.18   Direct implementation of cascade first-order filters
in Z^2.   340

Figure 11.19   Folded implementation of cascade first-order filters
in Z^2.   340

Figure 11.20   Pole-zero diagram of two path, five-pole, two-
multiplier filter.   343

Figure 11.21   Two-path phase slopes, frequency, impulse, and
phase responses of two-path, five-pole, two-
multiplier filter.   344

Figure 11.22   Pole-zero diagram of two-path, nine-pole, four-
multiplier filter.   345

Figure 11.23 Two-path phase slopes, frequency, impulse, and phase responses of two-path, nine-pole, four-multiplier filter. 346

Figure 11.24 Phase response of both paths of two-path, nine-pole, four-multiplier filter. 347

Figure 11.25 Pole-zero diagram of 2-path, 33-pole, 8-multiplier filter. 349

Figure 11.26 Two-path phase slopes, frequency, impulse, and phase responses of 2-path, 33-pole, 8-multiplier filter. 349

Figure 11.27 Variation of out-of-band attenuation versus transition bandwidth for non-uniform phase two-path filters containing 2−8 coefficients. 351

Figure 11.28 Variation of out-of-band attenuation versus transition bandwidth for uniform phase two-path filters containing 2−12 coefficients. 351

Figure 11.29 Pole-zero plot of 65-tap FIR filter. 352

Figure 11.30 Impulse response and frequency response of 65-tap FIR filter. 353

Figure 11.31 Comparison of impulse responses of linear phase IIR and FIR filters. 354

Figure 11.32 Comparison of frequency responses of linear phase IIR and FIR filters. 354

Figure 11.33 Detailed comparison of in-band magnitude ripple of linear phase IIR and FIR and of group delay of equal ripple linear phase IIR filter. 355

Figure 11.34 Magnitude response of low-pass and high-pass half-band filter. 356

Figure 11.35 Block diagram of general second-order all-pass filter. 359

Figure 11.36 Pole-zero diagram of generalized second-order all-pass filter. 359

Figure 11.37 Effect on architecture of frequency transformation applied to two-path half-band all-pass filter. 360

Figure 11.38 Frequency response obtained by frequency transforming half-band filter to normalized frequency 0.1. . 360

Figure 11.39 Pole-zero diagram obtained by frequency transforming half-band filter to frequency 0.1. 361

Figure 11.40 Frequency response obtained by frequency transforming half-band filter to frequency 0.02. 362

Figure 11.41 Pole-zero diagram obtained by frequency transforming half-band filter to frequency 0.02. 363

Figure 11.42 Block diagram of general fourth-order all-pass filter. 364

Figure 11.43 Block diagram of low-pass to band-pass transformation applied to low-pass to low-pass transformed delay element. 365

Figure 11.44 Effect on architecture of low-pass to band-pass frequency transformation applied to two-path arbitrary-bandwidth all-pass filter. 366

Figure 11.45 Frequency response of two-path all-pass filter subjected to low-pass to low-pass and then low-pass to band-pass transformations. 366

Figure 11.46 Pole-zero plot of two-path all-pass filter subjected to low-pass to low- pass and then low-pass to band-pass transformation. 367

Figure 11.47 Half-band, all-pass polyphase filter with 2-to-1 down sampler moved from output to input and replaced with commutator. 368

Figure 11.48 Half-band, all-pass polyphase filter 1-to-2 up sampler moved from input to output and replaced with commutator. 369

Figure 11.49 Time and frequency response of eighth-order recursive filter. 370

Figure 11.50 Cascade of four second-order canonic filters to form eighth-order elliptic filter. 371

Figure 11.51 Structure of ninth-order two-path recursive all-pass filter. 372

Figure 11.52 Time and frequency response of ninth-order two-path recursive filter. 372

Figure 11.53 Structure of low bandwidth filter as a cascade of down-sampling half-band filters, the low-pass filter, and a cascade of up-sampling half-band filters. . . . 373

Figure 11.54 Time and frequency response of 8-to-1 down-sampled and 1-to-8 up-sampled ninth-order two-path recursive filter. 374

Figure 11.55 Frequency response for first of three segments of successive filtering: zero-packed bandwidth limiting filter and interpolating filter. 375

Figure 11.56 Low-pass half-band filter to positive frequency half-band filter. 377

Figure 11.57 Path phases of two-path recursive all-pass, non-uniform phase, and uniform phase Hilbert-filters. . 379

Figure 11.58 Pole-zero diagrams for non-uniform and uniform phase Hilbert-transform filters. 379

Figure 11.59 Frequency and phase response of non-uniform and uniform phase two-path recursive all-pass Hilbert-transform filters. 380

Figure 11.60 Phase profiles of eight paths of the eight-path non-uniform and uniform phase polyphase filter. 382

Figure 11.61 Pole-zero plots of composite eight-path recursive non-uniform and uniform phase filters. 383

Figure 11.62 Details of pole-zero plots of composite eight-path filters. 383

Figure 11.63 Frequency, impulse, and phase responses of non-uniform phase eight-path filter. 384

Figure 11.64 Frequency, impulse, and phase responses of uniform phase eight-path filter. 384

Figure 11.65 Iterated filter, up-sampled 1-to-4 by zero-packing, and then filtered to suppress spectral replicates. . . 385

Figure 11.66 Spectra of 1-to-4 zero-packed first stage, 1-to-2 zero-packed second stage, and non-zero-packed third stage of iterated filter. 386

Figure 11.67 Spectral response of cascade filters in iterated filter chain. 387

Figure 12.1 Boxcar filter impulse response and frequency response. 394

Figure 12.2 FIR filter implementation of boxcar filter. 395

Figure 12.3 Cascade integrator comb filter. 396

Figure 12.4 Frequency response of 10-tap comb filter. 396

Figure 12.5 Structures and associated impulse responses of boxcar, cascade comb, and integrator, and cascade integrator and comb. 398

Figure 12.6 Structures and associated step responses of boxcar, cascade comb, and integrator, and cascade integrator and comb. 400

Figure 12.7 Overflow behavior of 2's-complement binary counter. 402

Figure 12.8   Cascade of four length-M boxcar filters.   404

Figure 12.9   Scaled impulse responses of four-stage, 10-tap boxcar filter.   405

Figure 12.10   Scaled frequency responses of four-stage, 10-tap boxcar filter.   406

Figure 12.11   Zoom to zero of frequency responses of four-stage, length-10 boxcar.   406

Figure 12.12   Suppression spectral replicas by two-, three-, and four-stage CIC filter.   407

Figure 12.13   Order and reordering of resampling switch and subfilters of CIC filter for up sampling and for down sampling.   408

Figure 12.14   Down sampled three-stage CIC filter, rearranged, and converted to a Hogenauer filter.   409

Figure 12.15   Two-stage CIC filter, 20-tap comb, 16-Bit accumulator, and 7-bit input.   410

Figure 12.16   Two-stage CIC filter, 20-tap comb, 15-bit accumulator, and 7-bit input.   411

Figure 12.17   Four-stage up sampling CIC filter with integrator outputs identified.   411

Figure 12.18   Impulse responses at integrators of four-stage CIC filter with 20-tap comb.   412

Figure 12.19   Input sequences to maximize output from integrators in four-stage CIC.   413

Figure 12.20   Integrator responses for input sequence selected to maximize output level.   414

Figure 12.21   Impulse responses at comb filters of four-stage CIC filter with 20-tap comb.   414

Figure 12.22   Input sequences to maximize output from comb filters in four-stage CIC.   415

Figure 12.23   Graphs showing required bit field at input and output of each process in CIC.   416

Figure 12.24   Four-stage down sampling CIC filter with integrator inputs identified.   417

Figure 12.25   Impulse responses from integrators of four-stage CIC filter with 20-tap comb.   418

Figure 12.26   Graph showing required bit field at input and output of each process in CIC.   420

Figure 12.27   Cascade of 1-to-5 shaping filter and 1-to-16 CIC interpolating filter. 420

Figure 12.28   Time and frequency response of time series from 1-to-5 shaping filter. 421

Figure 12.29   Shaping pulse response of three-stage 1-to-16 Hogenauer filter. 421

Figure 12.30   Spectrum of composite impulse response with CIC spectral overlay. 422

Figure 12.31   (a) Coherent bit growth of integrators of second- and third-order CIC filters as function of rate change M. 423

Figure 12.31   (b) Coherent bit growth of integrators of fourth- and fifth-order CIC filters as function of rate change M.   424

Figure 12.32   (a) Incoherent bit growth of integrators of second- and third-order CIC filters as function of rate change M. 424

Figure 12.32   (b) Incoherent bit growth of integrators of fourth- and fifth-order CIC filters as function of rate change M. 425

Figure 12.33   (a) Coherent bit growth of integrators of second-, third-, fourth-, and fifth-order CIC filters as function of rate change M. 425

Figure 12.33   (b) Incoherent bit growth of integrators of second-, third-, fourth-, and fifth-order CIC filters as function of rate change M. 426

Figure 12.34   Spectrum of two stage 10-tap CIC filter's triangle shaped impulse response. 427

Figure 12.35   Spectrum of two-stage CIC filter and two-stage CIC with bifurcated spectral zeros. 428

Figure 12.36   Pole zero diagram of CIC filter with repeated and with bifurcated zeros. 428

Figure 12.37   Block diagram of modified CIC filter. Reduce center weight by delaying input $M - 1$ samples. Scale and subtract delayed sample. Insert one delay in each branch of filter. Merge delays in lower path. 429

Figure 12.38 Upper subplot: partition boxcar transfer function ratio of polynomials to cascade integrators and comb filters. Lower subplot: interchange order of M-to-1 down sample and polynomials in M with the noble identity. 430

Figure 12.39 Upper subplot: three stage bifurcated boxcar filter. Center subplot: partition of three-stage boxcar transfer function ratio of polynomials to cascade integrators and comb filters. Bottom subplot: interchange order of M-to-1 down sample and polynomials in M with the noble identity. 430

Figure 12.40 Spectrum of three-stage CIC filter and three-stage CIC with bifurcated spectral zeros. 431

Figure 12.41 Upper subplot: main lobe of sinc spectrum with desired pass band distorted by main-lobe droop. Lower subplot: same spectra with compensating filter spectrum, an offset, and scaled opposing polarity cosine. 432

Figure 12.42 Spectra of one-, two-, three-, and four-stage CIC filters, their droop responses, their compensation filters, and their compensated responses. 433

Figure 12.43 Finer detail spectra of one-, two-, three-, and four-stage CIC filters, their droop responses, their compensation filters and their compensated responses. 434

Figure 13.1 Resampling filtering option when ratio of sample rate to transition bandwidth is large. 438

Figure 13.2 Interpolated FIR filter. Up-sampled prototype and interpolating clean-up filter. 439

Figure 13.3 Spectral description of zero-packed, interpolating, and composite filters. 439

Figure 13.4 Impulse response and frequency response of prototype 150-tap filter. 440

Figure 13.5 Impulse response of two cascade filters and composite interpolated response along with frequency responses of each impulse response. 441

Figure 13.6 Impulse response of three cascade resampling filters and composite impulse response along with frequency responses of each impulse response. . . . 442

Figure 13.7   Block diagram of half-band spectral masking filter.   444

Figure 13.8   Spectra of top path, $H(Z^M)$ and masking filter $G_1(Z)$, spectra of bottom path, $H_C(Z^M)$, and masking filter $G_2(Z)$, and spectra of sum of two paths.   444

Figure 13.9   Impulse response and frequency response of prototype 151-tap filter.   446

Figure 13.10   Spectra of up-sampled 1-to-5 half-band low-pass filter and first masking filter, of up-sampled 1-to-5 half-band high-pass filter and second masking filter, and of composite sum filter.   447

Figure 13.11   Initial spectral representation of 1-to-4 up-sampled baseband filter.   449

Figure 13.12   Spectra of up-sampled 1-to-4 low-pass filter and first masking filter, of up-sampled 1-to-4 complementary filter and second masking filter, and of composite sum filter.   450

Figure 13.13   Spectral band, octave $K + 1$ (octave-2) sampled at f_s shifts to location of octave K (octave-1) when 2-to-1 down-sampled to $f_s/2$.   451

Figure 13.14   Multioctave processing iteratively filter and down-sample to access successive octaves.   451

Figure 13.15   Filter partition for 15-band proportional bandwidth analysis filter bank.   455

Figure 13.16   Frequency response of first 10 bands of 15-stage proportional bandwidth filter bank and four filter types distributed through half-band filter chain. . .   455

Figure 13.17   Frequency response of 12-fractional octave filter bank in top octave and in next lower octave and spectra of four successive half-band filter response.   457

Figure 13.18   Graphical representation of center frequencies in audiometric filter banks with 6, 9, 10, and 15 channel filters.   459

Figure 13.19   Graphical representation of center frequencies and crossover frequencies at the band edges in 10-channel audiometric filter bank.   460

Figure 13.20   Cascade of filter h1 high-pass and low-pass filter responses with filter h2 high-pass and low-pass filter responses with alternate stage 2-to-1 down-sampling to form impulse responses of channelizer filters.   462

Figure 13.21   Spectral description of interaction between low-pass filter H1a and high-pass filter H2b to form band-pass filter with steeper transition bandwidth on left side than on right side.   463

Figure 13.22   Spectral description of interaction between low-pass filter H1a and high-pass filter H2b to form band-pass filter with steeper transition bandwidth on left side than on right side.   464

Figure 13.23   Impulse response of cascade filters h1a and h2b. Original 219-point sequence with zoom to lower level amplitudes. Edited to 129-point sequence with zoom to lower level amplitudes. Spectra of original sequence and edited sequence.   464

Figure 13.24   Block diagram of audiometric 10-channel channelizer. Impulse responses g1 through g10 of filters formed in Figure 13.20 and edited to reduced lengths as done in Figure 13.23 are inserted in this block diagram. The half-band filters in the three 2-to-1 down-sampling arms at the input port and the various up-sampling filters and inserted time delay at the output port have spectral requirements that are functions of the transition bandwidths of the filter bank spectra.   465

Figure 13.25   Sum of four lowest channel time responses and reconstructed spectrum response. Note extremely small level of ripple in reconstructed spectrum. . .   466

Figure P13.8   Spectra of four proportional bandwidth filters at successive sample rates of 24, 12, and 6 kHz. . . .   470

Figure 14.1   Standard radio receiver architecture.   472

Figure 14.2   Second-generation radio receiver architecture. . . .   473

Figure 14.3   Standard digital down converter with CIC and half-band filters.   474

Figure 14.4   Frequency response of five-stage CIC filter, at input sample rate, at output sample rate illustrating main lobe folding due to 10-to-1 resampling, and main lobe response with overlaid compensating 4-to-1 down-sample filter.   475

Figure 14.5   Spectra of CIC main lobe, of 21-tap compensating half-band filter, and the composite response.   475

Figure 14.6   Aliasing spectra at 2.25 f_s in second Nyquist zone to 0.25 f_s in zeroth Nyquist zone by IF sampling at f_s.   479

Figure 14.7   Periodic spectra on scaled circle showing how 450 kHz from the second Nyquist zone folds through 250 kHz in the first Nyquist zone to 50 kHz in zeroth Nyquist zone.   479

Figure 14.8   Four-path polyphase filter: simultaneously translates frequency band from quarter sample rate to baseband, down-samples 4-to-1, and converts real input to complex output.   480

Figure 14.9   Indexing of successive zero-packed data samples to five-path polyphase filter.   481

Figure 14.10   Signal processing structure of down converter using aliasing to and from quarter sample rate by subsampling and polyphase filtering.   482

Figure 14.11   Spectrum of 25 tap prototype low-pass filter with frequency bands that alias to baseband when resampled 5-to-1 (upper figure) and spectrum of same filter translated to quarter sample rate with frequency bands that alias to quarter sample rate when resampled 5-to-1 (lower figure).   484

Figure 14.12   Signal processing structure of down converter using aliasing to quarter sample rate by subsampling and again by five-path polyphase filter.   484

Figure 14.13   Signal flow in modulator and demodulator for communicating through band-limited AWGN channel.   486

Figure 14.14   Signal flow in first-generation digital demodulator.   486

Figure 14.15 Three samples, early, punctual, and late samples, on correlation function for three operating conditions. (1) Positive Slope: peak is ahead; (2) zero slope: at peak; (3) negative slope: peak is behind. 487

Figure 14.16 Three samples, early, punctual, and late samples, on positive and on negative correlation. 487

Figure 14.17 Signal flow for timing recovery with polyphase interpolator processing and shifting asynchronous samples to desired time locations. 489

Figure 14.18 Signal flow for timing recovery with polyphase matched filter. timing recovery selects matched filter path aligned with input sample position. 489

Figure 14.19 Early, prompt, and late gate filters for timing recovery control signals. 490

Figure 14.20 Two polyphase filter banks forming filter and derivative filter outputs. 490

Figure 14.21 Two single-stage filters with selection of coefficient sets from polyphase filter bank for matched filter and derivative matched filter. 491

Figure 14.22 Polyphase address pointer for timing recovery loop with 40-path filter and two samples per symbol. In the upper figure, the pointer does not cross the address pointer boundary, and in the lower figure, the pointer crosses the address pointer boundary and converts Address 0 to Address 40 and switches from even-indexed to odd-indexed data sample. 492

Figure 14.23 Block diagram of quadrature up converter at transmitter and quadrature down converter at receiver. 494

Figure 14.24 Spectra of matched filter for square-root Nyquist filter and frequency derivative matched filter. 495

Figure 14.25 Spectra at input and output of band-edge filter for centered input signal and for input signal with spectral offset. 496

Figure 14.26 Band-edge filter and frequency error detector appended to carrier recovery process. 497

Figure 14.27 Time and frequency domain development of band edge filter. Top subplot: shaping filter; second subplot: frequency derivative matched filter; third subplot: positive frequency Hilbert filter, band edge filter, and Hilbert filtered frequency derivative filter. 497

Figure 14.28 Time and frequency domain development of band edge filter. Top subplot: shaping filter; second subplot: odd symmetric frequency derivative matched filter; third subplot: imaginary part of Hilbert filter, band edge filter, and even symmetric Hilbert filtered frequency derivative filter. 499

Figure 14.29 Asymmetric power spectra in output of band edge filter. 499

Figure 14.30 Receiver block diagram of non-data aided band edge filter synchronization of carrier frequency and modulation timing. 500

Figure 14.31 Spectra of conjugate product time series from band edge filters. A pair of symbol rate spectral lines aligned with frequency and phase of modulation waveform and DC spectral line proportional to frequency offset of modulation spectrum. 501

Figure 14.32 M-path filter implemented as recursive all-pass filters in Z^M. 503

Figure 14.33 Structure of each path in M-path recursive all-pass polyphase filter. 503

Figure 14.34 Phase shift for 10 paths of 10-path recursive filter. . 504

Figure 14.35 Phase slopes (group-delay) for 10 paths of 10-path recursive filter. 504

Figure 14.36 Phase shift for 10 paths of 10-path non-recursive filter. 505

Figure 14.37 Phase slopes (group-delay) for 10 paths of 10-path non-recursive filter. 505

Figure 14.38 Coefficient values for 10-path filter and fifth-order polynomial fitted to values. 506

Figure 14.39 Programmable time delay network with associated coefficient generator. 507

Figure 14.40 Original delays plus mid-value delays obtained by evaluating coefficient polynomials defined by original 10-path filter. 507

Figure 14.41 Continuously variable digital delay line in timing recovery loop. 508

Figure 14.42 Timing phase accumulator that controls timing delay of digital delay line and timing error, and the loop filter output input to phase accumulator. 508

Figure 14.43 Impulse response and spectra of shaping filter. . . . 511

Figure 14.44 Matched filter response and details showing ISI levels. 512

Figure 14.45 Block diagram of signal flow for shaping filter and 1-to-8 polyphase FIR filter interpolator. 512

Figure 14.46 Path phase responses of eight-path polyphase filter and frequency response of prototype filter. 513

Figure 14.47 Spectra of up-sampled shaping filter with spectra of interpolating filter and spectra of interpolated shaping filter response. 514

Figure 14.48 Block diagram of signal flow for shaping filter and cascade of three levels of half-band 1-to-2 up-sampling FIR interpolating filter. 515

Figure 14.49 Spectra of up-sampled shaping filter with spectra of first half-band FIR interpolating filter and spectra of 1-to-2 interpolated filter response. 515

Figure 14.50 Spectra of 1-to-4 up-sampled shaping filter with spectra of second half-band FIR interpolating filter and spectra of 1-to-4 interpolated filter response. . . 516

Figure 14.51 Spectra of 1-to-8 up-sampled shaping filter with spectra of third half-band FIR interpolating filter and spectra of 1-to-8 interpolated filter response. . . 516

Figure 14.52 Block diagram of signal flow for shaping filter and 1-to-8 polyphase, approximately linear, recursive filter interpolator. 517

Figure 14.53 Zoom to group delay ripple and magnitude ripple of 1-to-8 polyphase recursive, approximately linear, recursive filter interpolator. 517

Figure 14.54 Interpolated matched filter response and details showing ISI levels. 518

Figure 14.55 Path phase responses of eight-path polyphase IIR filter and frequency response of prototype filter. . . 519

Figure 14.56   Spectra of up-sampled shaping filter with spectra of IIR interpolating filter and spectra of interpolated shaping filter response.   519

Figure 14.57   Block diagram of signal flow for shaping filter and cascade of three levels of half-band 1-to-2 up-sampling IIR interpolating filter.   520

Figure 14.58   Spectra of up-sampled shaping filter with spectra of first half-band IIR interpolating filter and spectra of 1-to-2 interpolated filter response.   521

Figure 14.59   Spectra of 1-to-4 up-sampled shaping filter with spectra of second half-band IIR interpolating filter and spectra of 1-to-4 interpolated filter response. . .   521

Figure 14.60   Spectra of 1-to-8 up-sampled shaping filter with spectra of third half-band IIR interpolating filter and spectra of 1-to-8 interpolated filter response.   522

Figure 14.61   Block diagram of a noise feedback quantizer with predicting filter $P(Z)$.   524

Figure 14.62   Block diagram of a noise feedback quantizer with delay line as predicting filter.   524

Figure 14.63   Alternate block diagram of a noise feedback quantizer showing digital integrator in feedback loop.   525

Figure 14.64   Block diagram of a noise feedback quantizer with block diagram of digital integrator replaced by its transfer function.   525

Figure 14.65   Roots of NTF and power spectrum of composite output signal of a noise feedback quantizer.   526

Figure 14.66   Block diagram of two-loop, 1-bit sigma-delta modulator, and its companion resampling filter. . .   527

Figure 14.67   Time series of input and output of sigma-delta modulator and spectrum with zoomed detail of output spectrum from modulator.   528

Figure 14.68   Time Series and spectrum of CIC filtered output of sigma-delta modulator with overlaid filter response and zoomed detail of output spectrum.   528

Figure 14.69   Time series and spectrum of CIC filtered and down-sampled output of sigma-delta modulator and zoomed detail of output spectrum.   529

Figure 14.70 Time series and spectrum of second FIR filtered and down-sampled output from CIC filter with filter overlay and zoomed detail of output spectrum. . . . 530

Figure 14.71 Dual form filter showing accumulators in four-tap polyphase filter. 531

Figure 14.72 Polyphase filter implemented with short look-up tables addressed from binary input data delivered from sigma-delta modulator. 532

Figure 14.73 Time series and spectrum of polyphase filtered output of sigma-delta modulator with overlaid filter response and zoomed detail of output spectrum. . . 532

Figure 14.74 Time series and spectrum of polyphase filtered and down-sampled output of sigma-delta modulator and zoomed detail of output spectrum. 533

Figure 14.75 Time series and spectrum of polyphase filtered and down-sampled output of sigma-delta modulator and zoomed detail of output spectrum. 533

Figure 14.76 Input and output spectra of stereo FM receiver and block diagram of conventional FM receiver and stereo demodulator. 535

Figure 14.77 Block diagram of DSP-based FM channelizer. . . . 536

Figure 14.78 Spectral characteristics of prototype low-pass filter in polyphase receiver. 536

Figure 14.79 Standard heterodyne, filter, and down-sample architecture channel selector. 537

Figure 14.80 Heterodyned filter, down-sample, and down-convert architecture version of channel selector. 538

Figure 14.81 Signal flow diagram of modified polyphase filter to permit frequencies offset by quarter of output sample rate to alias to baseband. 540

Figure 14.82 Signal flow diagram of digital FM discriminator. . . 541

Figure 14.83 Filter specifications for pilot extraction filter. 542

Figure 14.84 Pilot extraction and doubling by polyphase down-sample, filter, frequency doubling, and polyphase up-sample. 542

Figure 14.85 Spectrum of input and output of polyphase down-sampler, at output of low-pass filter, at output of squaring circuit, and at output of polyphase up-sampling filter. Also time series of pilot component of input signal and double frequency pilot at output of process. 543

List of Tables

Table 3.1 Windows formed as weighted sum of cosines.   51

Table 3.2 Integrated side-lobe levels for equiripple and $1/f$ side-lobe FIR filters, unquantized, and 10-bit quantized versions.   68

Table 5.1 Filter specifications. 118

Table 7.1 Parameters required for interpolating filter. 170

Table 7.2 Indices for input, for commutator, and for output series. 184

Table 7.3 Polynomial filter for output at Δ. 211

Table 7.4 Coefficients to compute Taylor series expansion of input time series. 211

Table 8.1 Length of four successive half-band filters and number of arithmetic operations performed by each filter per input sample. 235

Table 8.2 Workload proportionality factor per output sample for cascade of half-band filters. 236

Table 8.3 Alternate presentation of workload proportionality factor per output sample for cascade of half-band filters. 236

Table 9.1 List of highly composite transform sizes considered for 50-output channels of M-path polyphase analysis channelizer. 254

Table 12.1 Decimal integers and their 3-bit binary representation. 401

Table 12.2 Decimal integers and their signed 4-bit binary representation. 402

Table 12.3 Comb filter operating on output of overflowing accumulator. 403

Table 12.4 Maximum integrator response levels for sequences that maximize selected integrator response. 413

Table 12.5 Maximum comb response levels for sequences that maximize selected comb response. 416

Table 12.6 Maximum comb and integrator response and required bit field width. 416

Table 12.7 Noise gain from each CIC stage to output. 418

Table 12.8 Bit locations below output LSB for register pruning. . 419

Table 13.1 Comparison of three filter structures: coefficient lengths, arithmetic workload, and length of composite filter impulse response. 443

Table 13.2 Possible ratios of 6 dB pass-band bandwidth-to-sample rate (f_c/f_s) for range of small integer (M) for 1-to-M zero-packing in masking filter. 445

Table 13.3 Computed and ISO standard center frequencies of 15-band proportional bandwidth filter bank, and also, lower and upper edge of same bands. 454

Table 13.4 Down-sampled sample rate and normalized band edge frequencies for 15-band proportional bandwidth filter bank. 454

Table 14.1 List of possible sample rates that alias 450 kHz to quarter sample rate. 478

Table 14.2 Two-state state machine input and inner product schedule for five-path 2.5-to-1 resampling filter. . . . 482

Table 14.3 Coefficients of all-pass filters of Paths $1-9$ of 10-path polyphase filter. 509

Table 14.4 Interpolator options and their comparative measures. . 510

List of Abbreviations

ADC	Analog to Digital Converter
CD	Compact Disc
CIC	Cascade Integrator Comb
DAC	Digital to Analog Converter
DSP	Digital Signal Processing
EVM	Error Vector Magnitude
FIR	Finite (Duration) Impulse Response
FIRPM	Finite Impulse Response, Parks McClellan
IIR	Infinite (Duration) Impulse Response
ISI	Inter-Symbol Interference
LTI	Linear Time Invariant
LTV	Linear Time Varying
MAC	Multiplier-Accumulator
MPEG	Motion Picture Expert Group
MP3	MPEG Level 3
PN	Pseudo Noise
P-M	Parks-McClellan
PTV	Periodically Time Varying
RMS	Root-Mean Square
S&H	Sample and Hold
TDR	Time Domain Reflectometers
ZOH	Zero-Order Hold

1

Why Multirate Filters?

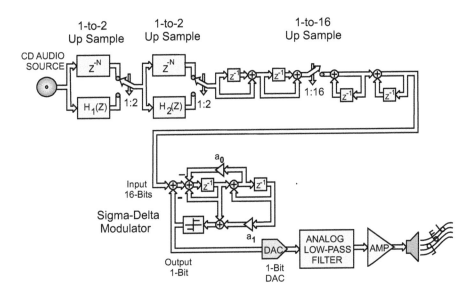

Why would we want to change the sample rate in a filter? There are two reasons. The first is performance. The second is cost. Multirate systems often perform a processing task with improved performance characteristics while simultaneously offering that performance at significantly lower cost than traditional approaches. *Multirate filters* are digital filters that operate with one more sample rate change embedded in the signal processing architecture. Occasionally, the use of a sample rate change in a filtering is the natural consequence of the signal processing chain. In other cases, the sample rate change is imposed to access the cost advantages related to multirate processing. We will develop examples of both scenarios throughout this book but will identify a few examples here.

1.1 Compact Disc 4-to-1 Oversample

A wonderful example of a multirate filtering application is the signal conditioning performed by a compact disc (CD) player. The CD player converts the digital representation of the music stored on the CD to analog audio for the listening pleasure of the CD user. Figure 1.1 presents the standard signal conditioning operations required when converting a digital signal to its equivalent analog representation. It entails a succession of three operators, a digital-to-analog converter (DAC), a sample and hold (S&H), and an analog-smoothing filter. The DAC converts the succession of digital sample values to a succession of corresponding analog amplitudes. The S&H suppresses glitch transients in the analog amplitudes due to bit race conditions in the multibit conversion, while the smoothing filter suppresses out-of-band spectral components.

Figure 1.2 presents the time domain representation of the signal at successive points in the signal-conditioning path. Figure 1.3 presents the spectra corresponding to the time signals of Figure 1.2. The two-sided bandwidth of the input signal is 40 kHz. The CD sample rate is 44.1 kHz, which results in the spectral replicates due to the sampling process being located at multiples of 44.1 kHz. The DAC replaces each sample value by a proportional DC term valid for the interval between sample values. The process of replacing a sample value with a data scaled rectangle is described as a *zero-order hold* (ZOH). The spectral response of the ZOH is a $\sin(x)/x$ or $\text{sinc}(\pi f/f_s)$ function with zero crossings located at multiples of the sample rate. As seen in the spectral plot, the zeros of the sinc suppress the center of the spectral replicates, the side lobes of the sinc attenuate the remaining spectral mass, and the sinc main lobe response distorts the desired baseband spectral region.

The analog-smoothing filter must satisfy a number of signal conditioning requirements. The first is to finish the incomplete filtering task started by the

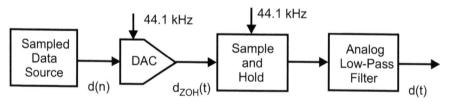

Figure 1.1   Signal conditioning tasks required to convert a digital signal to its analog representation.

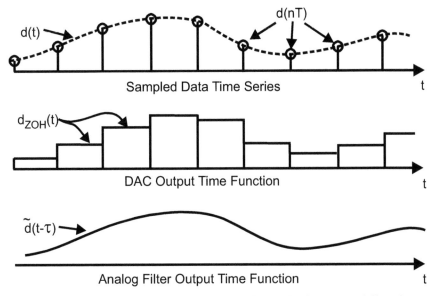

Figure 1.2   Three time signals: input and samples of input, DAC output, and filtered output.

DAC, the suppression of the residual spectral replicates. The filter required to perform this task is of high order ($N \cong 10$) and consequently relatively expensive. The high-order filter is required to obtain a narrow transition bandwidth, starting at 20 kHz and achieving the required 80 dB attenuation beyond 24 kHz for a composite DAC and filter attenuation of 96 dB. The second requirement is the correction of the in-band $\sin(x)/x$ distortion, which is accomplished by having a pass band response matching the inverse of the sinc response over the signal bandwidth. The third requirement is that the filter should not introduce severe group delay distortion in the vicinity of its band edge. A desired constraint is that the filter be one of a pair with matching gain and phase for the stereo audio signals. And, finally, the last requirement is that the pair of filters costs an absurdly low amount, say less than $0.50. If you are still chuckling over this list of requirements, you realize that they are not realistic specifications for an analog filter.

When faced with a problem we cannot solve, we invoke a trick that **Star Trek** enthusiasts will recognize as the *Kobyashi Maru Scenario* (**Wrath of Khan**). When faced with an unsolvable problem, change it into one you can solve, and solve that one instead. Figure 1.4 presents the modified signal conditioning tasks that employ this trick and inexpensively enable

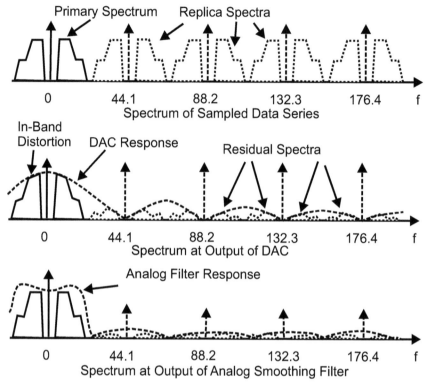

Figure 1.3   Three spectra: sampled input, DAC output, and filtered output.

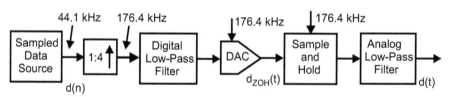

Figure 1.4   Modified signal conditioning tasks required to convert digital signal to analog representation.

the conversion of a sampled data representation of a signal to its analog representation.

Figure 1.5 presents the time domain representation of the signal at various points in the modified signal-conditioning path, while Figure 1.6 presents the spectral description of the corresponding signals shown in Figure 1.5. In the modified processing, we digitally raise the sample rate of the input

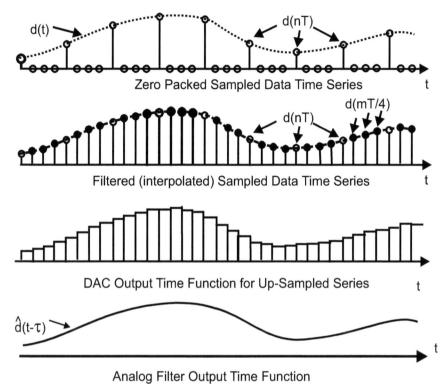

Figure 1.5 Four time signals: input samples zero-packed 1-to-4, 1-to-4 interpolated samples of input, DAC output of 1-to-4 interpolated samples, and analog filter output.

sequence by a factor of 4 from 44.1 to 176.4 kHz. This permits 4-spectral copies of the spectra to be presented to the digital filter with three of them to be suppressed in the sampled data domain. While we cannot build an analog filter with the desired specifications listed in the previous paragraph, we have no difficulty meeting these specifications in the digital filter. The output of the digital filter contains interpolated sample points at four times the original input sample rate. The spectra of the oversampled sequence are now separated by 176.4 kHz rather than the original 44.1 kHz. The spectral response of the DAC operating at the new output sample rate has a much easier task of suppressing the spectral replicates with fractional bandwidth one-eighth of the wider sinc main lobe width as opposed to the original one-half of the original sinc main lobe width. The transition bandwidth of the analog filter required to finish the incomplete spectral suppression is now 176.4−44.1 or 132.3 kHz, which is nearly four octaves, as opposed to the

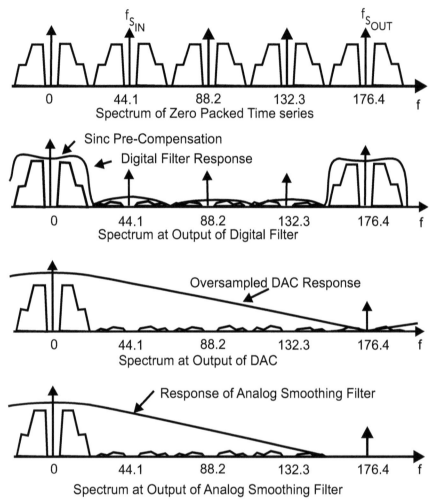

Figure 1.6   Four spectra: sampled input, 1-to-4 up-sampled output, DAC output, and filtered output.

original transition width of 4.1 kHz which is one-fifth of an octave. With a significantly larger transition bandwidth, the analog filter is of lower order and can be purchased at significantly reduced cost. The analog filter is no longer required to correct the DAC sinc distortion since the corrective pre-distortion is embedded in the digital up-sampling filter. In the process just described, the conversion from input digital to output analog representation is performed in two stages, a digital up-sampling by a low-cost digital filter and a low-cost analog-smoothing filter. The CD players that invoke this option are identified as being 4-to-1 oversampled. We will see shortly that 4-to-1 oversampling is a common option in most digital signal processing (DSP) based systems at the interface between the digital and analog versions of the signal. For instance, it is very common for digital modems to use four samples per symbol, a four-times oversampled version of the data, to reduce the distortion due to the analog filtering as well as the cost of the analog filter following the DAC.

1.2 Anti-Alias Filtering

This example is similar to the previous example in the sense we oversample to reduce the complexity and cost of an analog filter. This differs from the previous example by being its dual. The standard task we address here is the signal conditioning required to convert an analog signal to a digital signal. Figure 1.7 presents a block diagram of the primary components similar to that shown in Figure 1.1 for the digital-to-analog signal conditioning. Note the addition of an *automatic gain control* (AGC) block missing from Figure 1.1. Here we see the analog low pass filter that performs the anti-alias function, the AGC block that adjusts the input signal level to match the dynamic range of the S&H and ADC. The sample rate of the S&H and ADC must satisfy the Nyquist criterion of the input signal. The Nyquist rate of a high-quality data collection or recording system is the signal's two-sided bandwidth plus the transition bandwidth of the anti-alias filter. The two-sided signal bandwidth for high-end audio is 40 kHz, and standard sample rates are 44.1 kHz for the CD and 48 kHz for MP3 digital audio format. The transition bandwidths of the analog filters for CD and MP3 are 4.1 and 8 kHz, respectively.

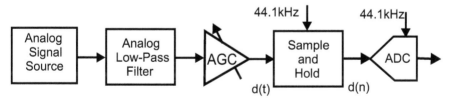

Figure 1.7 Signal conditioning tasks required to convert an analog signal to its digital representation.

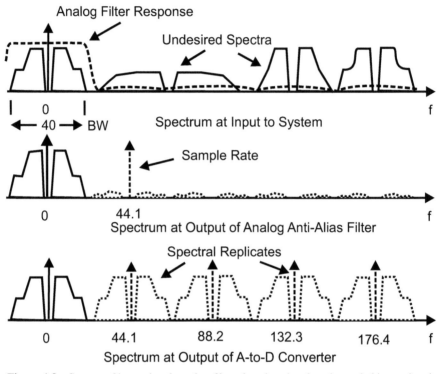

Figure 1.8   Spectra of input signal, analog filtered analog signal, and sampled input signal.

Figure 1.8 presents the frequency domain representation of the signal at successive points in the signal-conditioning path. The two-sided bandwidth of the input signal is 40 kHz. The CD sample rate is 44.1 kHz. The analog anti-alias filter has a pass band that extends to 20 kHz and a stop band that must attenuate adjacent spectra by 96 dB starting at 24.1 kHz. This filter must meet the performance requirements listed for the smoothing filter in the previous section. Here again, we find the requirements to be unrealistic. In particular, the narrow transition bandwidth of the filter requires it to be a high-order filter. High-order filters are characterized by severe group delay distortion near the band edge. High-order filters generally do not meet consumer price requirements and it is unlikely that the cost of the filter pair will meet the desired $0.50 target price.

Figure 1.9 presents the modified signal-conditioning task that uses multirate signal processing to enable a low-cost analog filter to perform the anti-aliasing task. Here we use a low-cost analog filter with wide transition

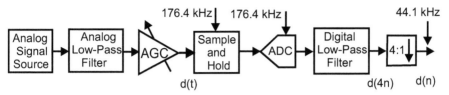

Figure 1.9 Modified signal conditioning to convert an analog signal to its digital representation.

bandwidth. The sample rate of the signal must now be increased to account for the wider transition bandwidth and for this example is 176.4 kHz which is four-times oversampled relative to the desired sample rate. A digital filter that meets the desired filter specifications filters the four-times oversampled sequence. After filtering, the filtered sequence is oversampled and the sample rate is reduced by a factor of 4 from 176.4 to 44.1 kHz. Reducing the sample rate by the factor of 4 reduces the distance between the spectral copies from the input sample rate separation of 176.4 kHz to the output sample rate separation of 44.1 kHz.

Figure 1.10 presents the frequency description of the signal at various positions in the modified signal-processing path. The analog anti-aliasing filter in the modified form has a 132 kHz transition bandwidth as opposed to the 4.1 kHz transition bandwidth of the non-oversampled version shown in Figure 1.5. Consequently, the degree of the analog filter can be reduced from a 10th order to 4th order. The lower degree filter will exhibit reduced group delay distortion. If the delay distortion has to be eliminated, the digital filter can be designed to absorb the phase response required to equalize the composite analog and digital filter to obtain linear phase.

Many DSP-based systems collect data with an oversampled ADC. This is often done to reduce the cost of the analog anti-aliasing filter while preserving the quality of the signal being processed by the filter. A high-quality digital anti-alias filter suppresses the spectral content in the excess spectral span of the extra-wide transition bandwidth. The sample rate of the processed data is then reduced to the desired output sample rate. In some systems, the excess data rate is maintained at the four-times oversampled rate to support additional processing related to interpolation from one sample rate to an arbitrary sample rate. As we will see in later sections, the processing task of an interpolator is reduced significantly when the input data is initially oversampled by a factor of 4.

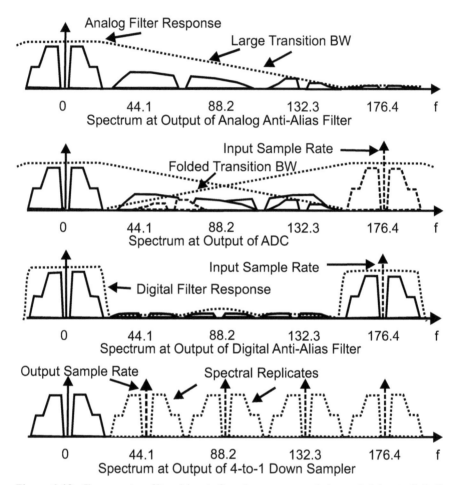

Figure 1.10 Four spectra: filtered input, four-times oversampled sampled input, digitally filtered output, and 4-to-1 down-sampled output.

References

Candy, James, and Gabor Temes. *Oversampled Delta-Sigma Data Converters*, Piscataway, NJ: IEEE Press, 1992

Nguyen, Khiem, Robert Adams, and Karl Sweetland. "A 113 dB SNR Oversampled Sigma-Delta DAC for CD/DVD Application", IEEE Transactions on Consumer Electronics, Volume: 44, Issue: 3 , Aug. 1998, Pages:1019-1023.

Norsworthy, Steven, Richard Schreier, and Gabor C. Temes. *Delta Sigma Data Converters: Theory, Design and Simulation,* Piscataway, NJ, IEEE Press, 1997.

Pohlman, Ken, *Principles of Digital Audio*, Indianapolis, Howard Sams & Co., 1985 Watkinson, John.,*The Art of Digital Audio*, London & Boston, Focal Press, 1989.

Problems

1.1 Determine the order of an analog Butterworth filter that can be used as the anti-alias filter for a CD quality signal. This filter must have its 3 dB pass band edge at 20 kHz and its 96 dB stop band edge at 24.1 kHz. If each component, capacitor, or inductor in this filter costs $0.05, estimate the cost for a pair of anti-alias filters. A digital filter, oversampled by 10 and satisfying the same performance requirements can be used as an estimate of the required order analog filter.

1.2 Determine the order of an analog elliptic filter that can be used as the anti-alias filter for a CD quality signal. This filter must have its 0.1 dB pass band edge at 20 kHz and its 96 dB stop band edge at 24.1 kHz. If each component, capacitor, or inductor in this filter costs $0.05, estimate the cost for a pair of anti-alias filters. A digital filter, oversampled by 10 and satisfying the same performance requirements, can be used as an estimate of the required order analog filter.

1.3 Determine the order of an analog elliptic filter that can be used as an anti-alias filter for a four-times oversampled CD quality signal. This filter must have its 0.1 dB pass band edge at 20 kHz and its 96 dB stop band edge at 156.4 kHz. If each component, capacitor, or inductor of this filter costs $0.05, estimate the cost for a pair of anti-alias filters. A digital filter, oversampled by 10 and satisfying the same performance requirements, can be used as an estimate of the required order analog filter.

1.4 Determine the order of an analog elliptic filter that can be used as an anti-alias filter for an eight-times oversampled CD quality signal. This filter must have its 0.1 dB pass band edge at 20 kHz and you have to determine the new sample rate and the spectral location of its 96 dB stop band edge. If each component of this filter, capacitor, or inductor costs $0.05, estimate the cost for a pair of anti-alias filters. A digital filter, oversampled by 10 and satisfying the same performance requirements, can be used as an estimate of the required order analog filter.

1.5 A signal uniformly occupying a full bandwidth of ± 20 kHz is sampled at 48 kHz. The signal samples are presented to a D-to-A converter, which performs the task of a ZOH with a frequency response equal to $\sin(\pi f/f_s)/(\pi f/f_s)$. To better appreciate the distortion and incomplete spectral suppression offered by the ZOH, generate an annotated and properly scaled figure that shows the in-band spectral droop due to the main lobe response and the residual spectra of the first spectral copies centered at 48 kHz.

1.6 A signal uniformly occupying a full bandwidth of ± 20 kHz originally sampled at 48 KHz is digitally up-sampled four-times to 196 kHz. The signal samples are presented to a D-to-A converter, which performs the task of a ZOH with a frequency response equal to $\sin(\pi f/f_s)/(\pi f/f_s)$. To better appreciate the improved levels of distortion and incomplete spectral suppression offered by the ZOH, generate an annotated and properly scaled figure that shows the in-band spectral droop due to the main lobe response and the residual spectra of the first spectral copies centered at 196 kHz.

2

The Resampling Process

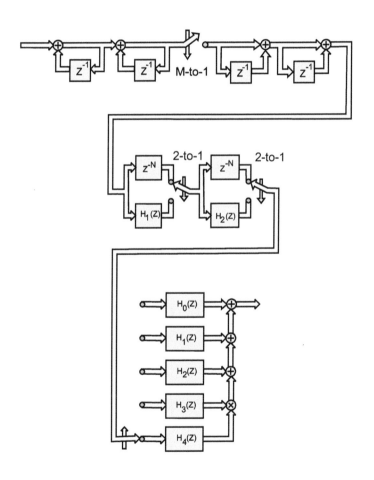

Central to multirate filters is the concept of sample rate change. In preparation for the study of filters that participate in this process, we first address the process of resampling a sampled signal as opposed to the process of sampling a continuous time signal. When a continuous time signal is sampled, there are no restrictions on the sample rate or the phase of the sample clock relative to the time base of the continuous time signal. On the other hand, when we resample an already sampled signal, the output sample locations are intimately related to the input sample positions. The resampling operation offers a convenient visualization of a precursor process to the desired process of changing the sample rate of a time series. A resampled time series contains samples of the original input time series separated by a set of zero-valued samples. The zero-valued time samples can be the result of setting a subset of input sample values to zero or as the result of inserting zeros between existing input sample values. Both options are shown in Figure 2.1. In the first example shown here, the input sequence is resampled 4-to-1, keeping every fourth input sample starting at sample index 0 while replacing the interim samples with zero-valued samples. In the second example, the input sequence is resampled 1-to-2, keeping every input sample but inserting a zero-valued sample between each input sample. These two processes are sometimes called down-sampling and up-sampling, respectively.

Two examples of 4-to-1 resampling of a time series are shown in Figure 2.2. The first resampling algorithm keeps every fourth sample staring at index 0 while setting the interim samples to zero. The second resampling

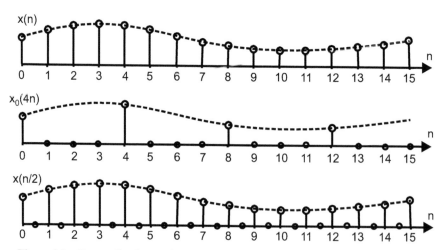

Figure 2.1   Resampling by zeroing sample values and by inserting zero-valued samples.

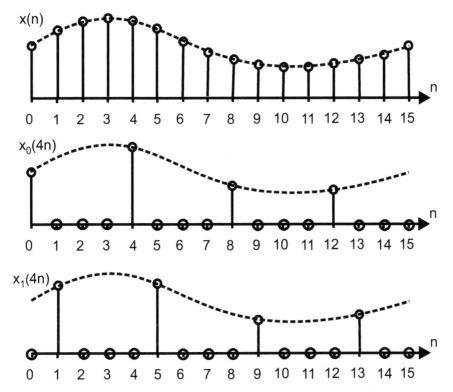

Figure 2.2 Two examples of 4-to-1 down-sampling of input series.

algorithm keeps every fourth sample starting at index 1 while setting the interim samples to zero. In general, there are Q initial starting locations for a Q-to-1 down-sampler. The spectra of the Q different down-sampled versions of the input signal have different phase profiles related to the time offset of the initial starting point relative to the time origin. The different phase profiles play a central role in multirate signal processing.

2.1 The Sampling Sequence

The process of performing sampling in the continuous time domain is often described with the aid of the generalized sampling function, a sequence of delayed impulses, as shown in the following equation:

$$S_T(t) = \sum_n \delta(t - nT). \tag{2.1}$$

In a similar manner, the resampling operation can be described with the aid of the discrete time sampling sequence formed by the inverse discrete Fourier transform (DFT) as shown in the following equation:

$$S_M(n) = \frac{1}{M} \sum_{n=0}^{M-1} \exp(j\frac{2\pi}{M}nm). \tag{2.2}$$

The sequence $S_M(n)$ is seen to be the sum of M complex sinusoidal sequences of amplitude $1/M$ with frequencies equally spaced around the unit circle at multiples of $(2\pi/M)$. The sum formed by the relationship shown in Equation (2.2) is the sequence shown in the following equation:

$$S_M(n) = \begin{cases} 1 & \text{for } n = \nu M, \nu \text{ an integer} \\ 0 & \text{Otherwise} \end{cases}. \tag{2.3}$$

The DFT relating the time and spectral description of the sampling sequence for $S_5(n)$ is shown in Figure 2.3.

The sampling sequence can be offset from index 0 to index r as shown in the following equation:

$$S_M(n-r) = \frac{1}{M} \sum_{m=0}^{M-1} \exp(j\frac{2\pi}{M}m(n-r)). \tag{2.4}$$

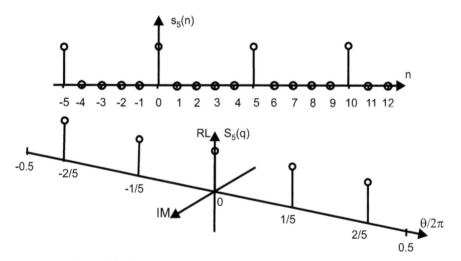

Figure 2.3   Sampling sequence $s_5(n)$ in time and frequency domain.

The sequence $S_M(n - r)$ is seen to be the sum of M complex sinusoidal sequences with amplitude $(1/M)$, with phase angles $\exp(-j \ r \ m2\pi/M)$ at frequencies equally spaced around the unit circle at multiples of $(2\pi/M)$. The sequence satisfies the relationship shown in the following equation:

$$S_M(n - r) = \begin{cases} 1 & \text{for } (n - r) = \nu M, \text{ or } n = r + \nu M, \nu \text{ an integer} \\ 0 & \text{Otherwise} \end{cases}.$$

(2.5)

The DFT relating the time and spectral description of the sampling sequence for $S_5(n - 1)$ is shown in Figure 2.4. Note that the time offset in the sampling sequence has resulted in a phase rotation of the spectral components forming the offset sequence.

The resampling operation can be visualized as the product of the input sequence $x(n)$ and the sampling sequence $s_M(n)$ to obtain the output sequence $x_0(Mn)$. The time domain product causes a frequency domain convolution. Hence, the spectrum of the resampled signal contains M offset replicates of the input spectrum. The spectrum of the resampled sequence is shown in the following equation:

$$X_0(\theta) = \frac{1}{M} \sum_{k=0}^{M-1} \exp[j(\theta - \frac{2\pi}{M}k)].$$

(2.6)

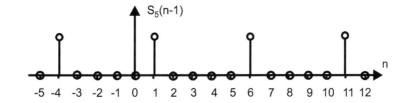

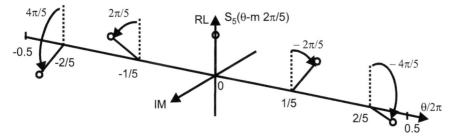

Figure 2.4 Sampling sequence $s_5(n - 1)$ in time and frequency domain.

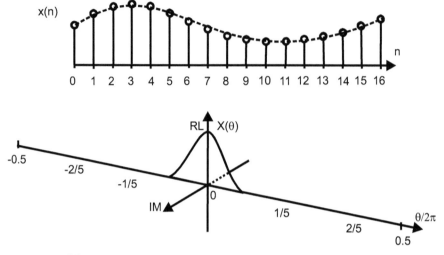

Figure 2.5   Time and frequency domain representation of sampled sequence $x(n)$.

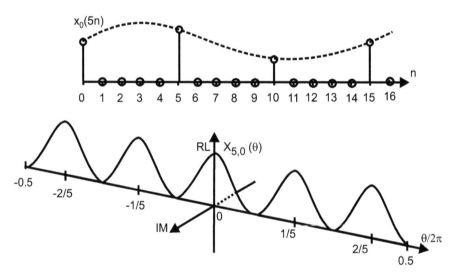

Figure 2.6   Time and frequency domain representation of sampled sequence $x_0(5n)$.

The time and frequency representation of an oversampled time series is shown in Figure 2.5. The time and frequency version of this series resampled by the resample sequence $s_5(n)$ is shown in Figure 2.6. The fivefold replication of the input spectrum is seen in the spectral representation of the resampled sequence. The time and frequency version of the input series

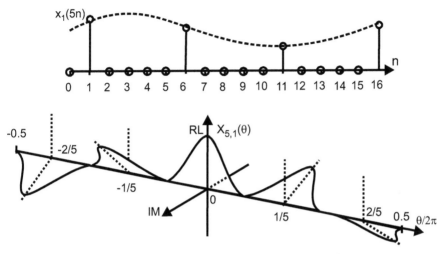

Figure 2.7 Time and frequency domain representation of sampled sequence $x_1(5n)$.

resampled by the resample sequence $s_5(n - 1)$ is shown in Figure 2.7. Here too, we see that the fivefold replication of the input spectrum is seen in the spectral representation of the resampled sequence. The difference in the spectral replicates of Figures 2.6 and 2.7 is the phase shifts of the spectral regions inherited from the spectral terms phase shifted by the one-sample offset of the resampled time series. Note that there has not been a sample rate change as we converted the input time series to either of the resampled time series. Sample rate changes are yet to come.

2.1.1 Modulation Description of Resampled Sequence

A different process can be used to describe the periodic replication of the baseband spectrum caused by the resampling process, illustrated in Figure 2.6. The time and frequency representation of a time series of the form illustrated in Figure 2.5 is listed as shown in the following equation:

$$h(n) \Leftrightarrow H(\theta). \tag{2.7}$$

The spectral replicates indicated in Figure 2.6 are related to heterodyned versions of the input time series as indicated in the following equation:

$$h(n) \cdot \exp(j\frac{2\pi}{M}nk) \Leftrightarrow H(\theta - \frac{2\pi}{M}k). \tag{2.8}$$

The replicated, translated, and phase shifted version of the resampled time series can be represented as a sum of heterodyned versions of the original baseband signal. This form is shown in the following equation:

$$h_r(nM) = \frac{1}{M} \sum_{m=0}^{M-1} h(n) \exp(j\frac{2\pi}{M}mr)$$

$$H_r(\theta) = \frac{1}{M} \sum_{m=0}^{M-1} H(\theta - \frac{2\pi}{M}m) \exp(j\frac{2\pi}{M}mr).$$

(2.9)

The importance of the relationship described in Equation (2.9) is that the resampling process appears to be equivalent to the spectral translation of the baseband spectrum. This connection suggests that resampling can be used to affect translation of spectral bands, up and down conversion, without the use of sample data heterodynes. In fact, we often embed the spectral translation of narrowband signals in resampling filters and describe the process as *aliasing*. Derivations that use this relationship will be presented in later sections. A final comment about the resampling process is that it can be applied to a time series or to the impulse response of a filter, which, of course, is simply another time series. When the resampling process is applied to a filter, the architecture of the filter changes considerably. The altered form of the filter is called a multirate filter.

2.2 What Is a Multirate Filter?

Multirate filters are digital filters that contain a mechanism to increase or decrease the sample rate while processing input sampled signals. The simplest of such filters performs integer up-sampling of 1-to-P or integer down-sampling of Q to-1. By extension, a multirate filter can employ both up-sampling and down-sampling in the same process to affect a rational ratio sample rate change of P-to-Q. More sophisticated techniques exist to perform arbitrary and perhaps slowly time varying sample rate changes. The integers P and Q may be selected to be the same so that there is no sample rate change between input and output but rather an arbitrary time shift or phase offset between input and output sample positions of the complex envelope. The sample rate change can occur at a single location in the processing chain or can be distributed over several subsections.

A number of symbols have been used to represent the sample rate change element in a block diagram. Three of the most common symbols are shown in

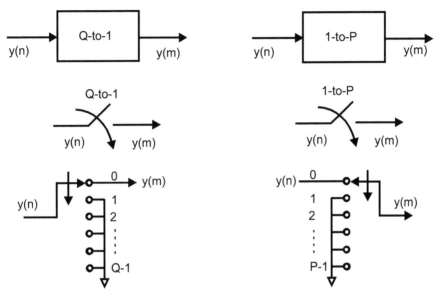

Figure 2.8   Symbols representing down-sampling and up-sampling elements.

Figure 2.8 for the down-sampling element and for the up-sampling element. The two elements are duals and the systems that employ them for sample rate changes will also be seen to be duals.

Conceptually, the process of down-sampling can be visualized as a two-step progression indicated in Figure 2.9. There are three distinct signals associated with this procedure. The process starts with an input series $x(n)$ that is processed by a filter $h(n)$ to obtain the output sequence $y(n)$ with reduced bandwidth. The sample rate of the output sequence is then reduced Q-to-1 to a rate commensurate with the reduced signal bandwidth. In reality, the processes of bandwidth reduction and sample rate reduction are merged in a single process called a multirate filter. The bandwidth reduction performed by the digital filter can be a low-pass process or a band-pass process.

The time and spectral descriptions of the three signal points in a 3-to-1 down-sampling version of Figure 2.9 are shown in Figure 2.10. Here, the filter limits the bandwidth to the band of interest and initially computes an output sample for each input sample. Reduction of output bandwidth to a third of the input bandwidth permits a corresponding reduction in the output sample rate. This is accomplished by having the resample switch output selected samples at the reduced rate and replacing discarded samples with zero-valued output samples. Since these zero-valued samples carry no information about the

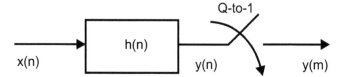

Figure 2.9   Down-sampling process filtering and sample rate reduction.

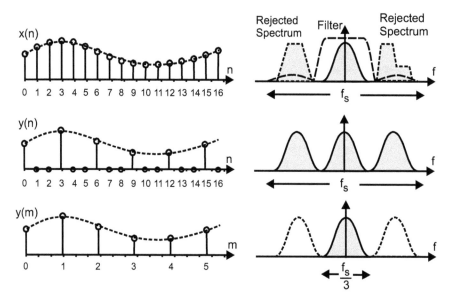

Figure 2.10   Time series and spectra for signal points in 3-to-1 down-sample process.

signal or its bandwidth, we are free to discard the zero-valued replacement samples. In reality, we do not insert the zero-valued samples since they are immediately discarded.

In a similar fashion, the process of up-sampling can be visualized as a two-step process indicated in Figure 2.11. Here too, there are three distinct time series. The process starts by increasing the sample rate of an input series $x(n)$ by resampling 1-to-P. The zero-packed time series with P-fold replication of the input spectrum is processed by a filter $h(n)$ to reject the spectral replicates and output sequence $y(m)$ with the same spectrum as the input sequence but sampled at the P-times higher sample rate. In reality, the processes of sample rate increase and selected bandwidth rejection are merged in a single process called a multirate filter. The bandwidth rejection performed by the digital filter can be a low-pass or a band-pass process.

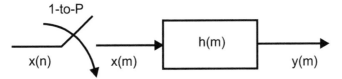

Figure 2.11 Up-sampling process filtering and sample rate increase.

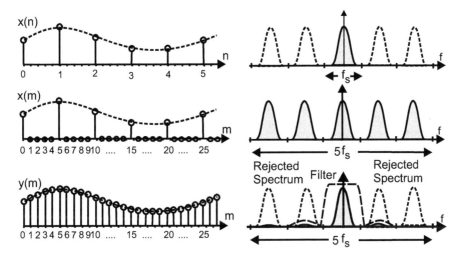

Figure 2.12 Time series and spectra for signal points in 1-to-5 up-sample process.

The time and spectral descriptions of the three signal points in a 1-to-5 up-sampling process of Figure 2.11. are shown in Figure 2.12. Here the resampler increases the sample rate by a factor of 5 by inserting four zero-valued samples between input samples. The filter limits the bandwidth to the band of interest and computes output samples at an increased rate (5/1) relative to input rate, replacing the zero-valued samples with interpolated values. Since these zero-valued samples do not contribute to the filter output, they are usually not inserted in the input data sequence. Their presence here is to give us perspective and addressing guidance when forming the multirate filter.

2.2.1 Properties of Resamplers

In the coming sections, we will be manipulating and rearranging the processes of resampling and filtering. A useful visualization tool that we have

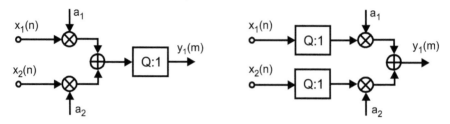

Figure 2.13   Down-sampled sum of scaled sequences equivalent to sum of scaled down-sampled sequences.

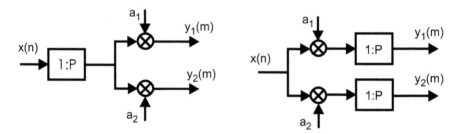

Figure 2.14   Scaled up-sampled sequences equivalent to up-sampled scaled sequences.

developed through exposure to linear systems is the block diagram and signal flow representations of digital filters. The block diagram is, of course, an equivalent (right brain) description of sets of difference equations connecting the variables in a set of difference equation defining the filter. The building blocks we use in a linear time invariant (LTI) filter structure are delay lines, multipliers, and adders. We now add to this list the resampler and identify relationships satisfied by the resamplers as they interact with traditional block diagram entities. We will present the relationships in dual pairs, the first for the up-sampler, and the second for the down-sampler. Since the addition of scaled sequences occurs without concern for their sample rate, the two operations of scaling and summing can commute. Figure 2.13 demonstrates the reordering of a scaled sum and a down-sampler. Similarly, since the application of a scale factor to a sequence occurs without regard to the sample rate, the two operations of up-sampling and scaling can also commute. Figure 2.14 demonstrates the reordering of up-sampling and scaling. Note that the relationships shown in Figures 2.13 and 2.14 are dual signal flow graphs. Dual graphs are formed by replacing nodes with summations, and summations with nodes, reversing the direction of the signal flow, and then interchanging input and output ports.

Figure 2.15 Q-units of delay and Q-to-1 down-sampler is equivalent to Q-to-1 down-sampler and 1-unit of delay.

Figure 2.16 1-to-P up-sampler followed by P-units of delay is equivalent to 1-unit of delay and P-to-1 up-sampler.

Figure 2.15 illustrates that Q-units of delay followed by a Q-to-1 down-sampler are the same as 1-unit of delay following a Q-to-1 down-sampler. This is true because Q clock cycles at the input to the resampler span the same time interval as one clock cycle at the output of the resampler. The resampler changes the number of samples in an interval but does not change the length of the interval. Figure 2.16 illustrates the dual relationship that P-units of delay following a 1-to-P resampler are the same as a 1-to-P resampler following a 1-unit delay. This is true since P clock cycles at the output of the up-sampler spans the same time interval as one clock cycle at the input to the up-sampler.

Figure 2.17 illustrates that a filter defined by polynomials in Z^Q followed by a Q-to-1 down-sampler is equivalent to a filter defined by polynomials in Z following a Q-to-1 down-sampler. We are essentially pulling the resampler through the filter and using the previous property to replace Q-delays at the input rate with 1-delay at the output rate. In a similar manner, Figure 2.18 illustrates that a filter defined by a polynomial in Z^P following a 1-to-P up-sampler is equivalent to a filter defined by a polynomial in Z preceding a 1-to-P up-sampler. In both cases, we can reverse the order of filtering and resampling so that the filtering is performed at the lower of the two rates. The equivalency of this interchange is known as the *noble identity*.

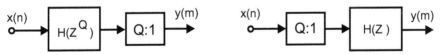

Figure 2.17 Exchanging order of filtering and down-sampling.

Figure 2.18   Exchanging order of up-sampling and filtering.

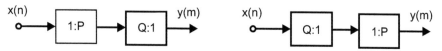

Figure 2.19   Reordering cascade up-samplers and down-samplers.

Figure 2.20   Interchange of relative prime input and output resamplers.

Figure 2.19 illustrates the interconnection of a 1-to-P up-sampler and a Q-to-1 down-sampler, cascaded to obtain a Q-to-P resampler. Reversing the order of the resamplers as shown is permitted if the integers P and Q are relatively prime and is not permitted if they share a common factor.

The interchange of order is useful in the following application where a multirate filter is to be designed that performs the resampling function of up P and down Q as shown in Figure 2.20. If P and Q are relatively prime, we can pull the 1-to-P up-sampler through the filter to its output port, then interchange the up-sampler and down-sampler and proceed to pull the down-sampler through the filter to its input port. The desired effect obtained by interchanging the input up-sampler with the output down-sampler is that the filter operates at the minimum processing rate. We will illustrate this exchange in a later example.

2.2.2 Examples of Resampling Filters

There is a very large class of multirate filters. Since this section has introduced what they are, we thought this would be an appropriate place to show important examples of multirate filters. Figure 2.21 presents three types of resampling filters used in down-sampling applications. The top subfigure shows an example of an M-stage polyphase down-sampling filter. The center subfigure is an example of a cascade of multiple half-band filters in which each stage performs a 1-to-2 down-sample operation. The bottom subfigure is a cascade of multiple digital integrators, a down-sampler, and multiple digital

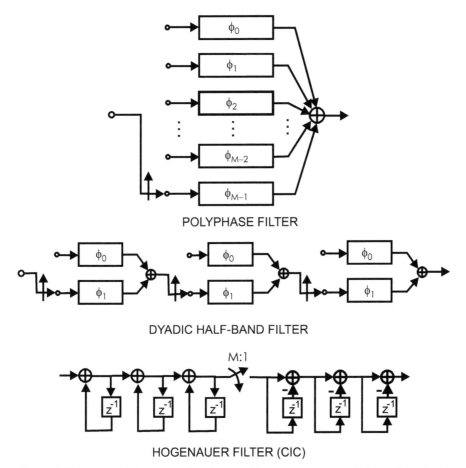

POLYPHASE FILTER

DYADIC HALF-BAND FILTER

HOGENAUER FILTER (CIC)

Figure 2.21 Standard down-sampling filter architectures: M-to-1 polyphase filter, 8-to-1 dyadic half-band filter chain, and M-to-1 Hogenauer (CIC with embedded resampler) filter.

differentiators configured in a structure known as the Hogenauer filter. When the down-sampler resides at the output to the cascade chain, the resulting filter is known as a Cascade Integrator Comb (CIC) filter.

Figure 2.22 presents the three types of resampling filters used in up-sampling applications. These structures are dual versions of their counterparts described in the down-sampling application. To form a dual filter, we replace summing junctions with nodes, replace nodes with summing junctions, and reverse the arrows in the signal flow as well as reverse the input and output terminals of the filter. In an LTI filter structure, a filter and its dual perform the

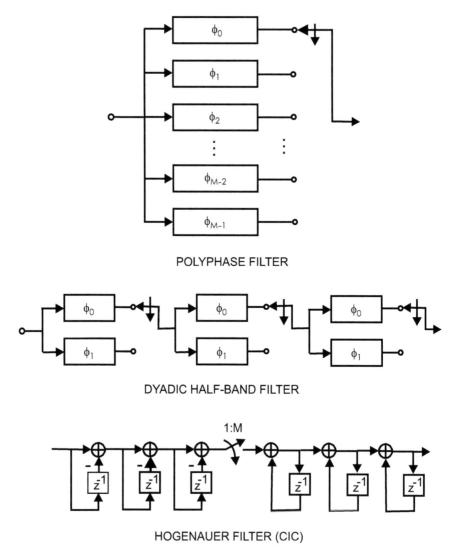

POLYPHASE FILTER

DYADIC HALF-BAND FILTER

HOGENAUER FILTER (CIC)

Figure 2.22   Standard up-sampling filter architectures: 1-to-M polyphase filter, 1-to-8 dyadic half-band filter chain, and 1-to-M Hogenauer (CIC with embedded resampler) filter.

same function and are indistinguishable at their input and output terminals. In the periodically time varying (PTV) filter structure, a filter and its dual perform opposite functions. If the filter performs down-sampling, its dual performs up-sampling.

The top subfigure of Figure 2.22 shows an example of an M-stage up-sampling polyphase filter. The center subfigure is an example of a cascade of half-band filters in which each stage performs a 2-to-1 down-sample operation. The bottom subfigure is a cascade of multiple digital differentiators, an up-sampler, and multiple digital integrators configured in the Hogenauer filter structure. Here too, when the up-sampler resides at the input to the cascade chain, the resulting filter is known as a CIC filter.

2.3 Useful Perspectives for Multirate Filters

We have seen that when the resampling switch modifies a time series, the new time series exhibits spectral replicates at equally spaced spectral intervals. This means that samples of a sinusoid located at a particular center frequency observed at the input to a resampler results in a new series with replicates of the sinusoid located at other frequency locations. Here we have output frequencies not equal to the input frequencies. This behavior would not be possible if the system is an LTI process. Thus, the multirate filter is not LTI but, in fact, is a linear time varying (LTV) process. If we examine the filter structure of the polyphase filter shown in Figure 2.22, we see that the impulse response of this system depends on which sub-filter is connected to the output port when the input impulse is presented to the filter. Since the output periodically revisits each commutator port, we say that the multirate filter is a PTV process.

The sample rate change that accompanies the multirate filter leads to an interesting quandary. We normally use the sample rate of a process as a reference interval when we discuss signal or process bandwidth. When we change the sample rate, we change the reference making it awkward to measure relative bandwidth with a flexible ruler. For a specific example, consider Figure 2.23 which presents a simple time series that of a low-pass filter, sampled at four times the bandwidth, and at five times the bandwidth. Glancing at the spectra of the two versions of the filter, it appears at first that raising the sample rate reduced the bandwidth of the filter. In fact, the bandwidth did not change, the sample rate was increased, and the image had to be rescaled to permit the larger spectral interval to fit into the same width display interval. We find it useful to avoid the use of the sample rate as a reference when discussing a multirate process but rather use the unchanged bandwidth as the reference. This advantage of this perspective will be obvious in applications presented in later sections.

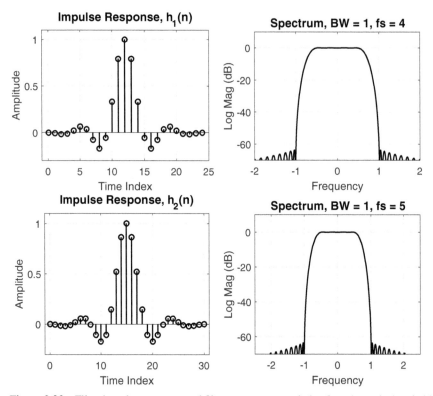

Figure 2.23 Filter impulse response and filter spectrum sampled at four times the bandwidth and at five times the signal bandwidth.

A similar relationship can be illustrated in the case of a single sine wave sampled at two different rates. Figure 2.24 presents four pairs of time and frequency descriptions of a sinusoid. The first pair presents the waveform and spectrum of a continuous sine wave of center frequency f_0. The second pair presents the sampled waveform and spectrum with the folding frequency 0.5 fs_1 indicated on the same frequency axis. A normalized frequency axis f/fs_1 is also shown for the sampled data spectrum. The third pair presents the sampled waveform and spectrum with the folding frequency 0.5 fs_2 indicated on the same frequency axis. Here too, a normalized frequency axis f/fs_2 is shown for the sampled data spectrum. The second and third figure pairs present their time series on the same axis but their spectra on different scaled axis with the analog frequency common to both axes. The fourth pair is a scaled version of the third with the scaled axis aligned to the same image width. This scaled

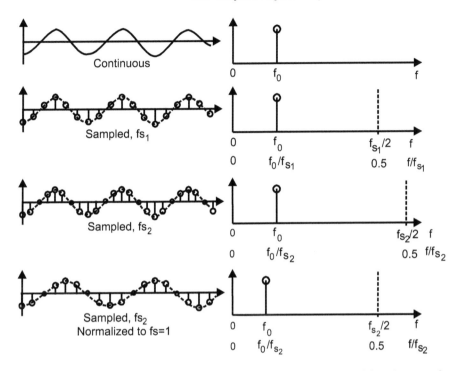

Figure 2.24 Time and spectral presentations of a continuous and sampled data sine wave for two different sample rates and two different scaling options.

spectrum with aligned axis gives the appearance that the sine wave center frequency has been reduced. In fact, the frequency in cycles per interval is the same in the two images, but the digital frequency in radians/sample has changed because the sample-to-sample interval has changed.

While on the topic of scaling, let us describe an experiment in which we hold the time interval fixed and then sample a complex sinusoid with different ratios of signal center frequency to sample frequency. In the first experiment, we accomplish this by holding the interval fixed, extending over 80 samples, at a fixed sample rate of unity and select different normalized center frequencies values of 0.1, 0.2, 0.4, and 0.8. In each case, the data set contains 80 samples of the sinusoid so that the spectrum of the time series contains the same energy. In the last case, the frequency of the sinusoid exceeds the half sample rate and aliases to the negative frequency. We see this scenario in Figure 2.25.

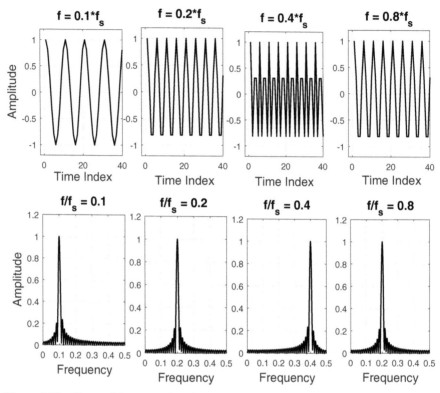

Figure 2.25   Time and frequency description of four, fixed time interval, fixed sample rate = 1, with successively higher center frequencies $f_0 = 0.1, 0.2, 0.4,$ and 0.8.

In the second experiment, we alter the data collection process by holding the interval fixed, extending over 80 samples at the maximum rate of unity and then selecting a sequence of sample rates of 1, 1/2, 1/4, and 1/8. As we collect data over the same interval at successively lower sample rates, we accumulate successively smaller number of samples so that the sequence contains proportionally less energy. As in the previous experiment, the sample rate of the last sinusoid violates the Nyquist criterion so that the signal aliases to the negative frequencies. This scenario is illustrated in Figure 2.26.

The point of this demonstration is that when we change the sample rate for data collected over a fixed interval, the energy or spectral amplitude of the sequence is changed and it may be necessary to apply a compensating scaling factor to the resampled data.

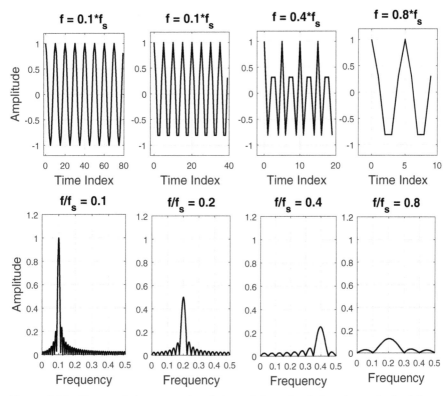

Figure 2.26 Time and frequency description of four, fixed time interval, fixed input frequency $f_0 = 0.1$, with successively lower sample rates $f_s = 1$, 1/2, 1/4, and 1/8.

2.4 Nyquist and the Sampling Process

Much of the signal processing discussed here deals with the process of changing the sample rate of a time sequence. When describing the sample process we have as a primary concern, or at least as a background concern, the task of reconstructing the continuous waveform from the sampled data sequence. We understand the conditions under which this task is possible and recognize the condition is a consequence of the sampling theorem. The sampling theorem states that a band-limited signal having no frequency components above f_{MAX} Hz can be determined uniquely by values sampled at uniform intervals of T_S, satisfying the relationship shown in the following equation:

$$T_S \leq \frac{1}{2f_{\text{MAX}}} \text{ sec}. \tag{2.10}$$

This relationship is known as the uniform sampling theorem, which is perhaps best known in terms of the sample frequency restriction shown in the following equation:

$$f_S \geq 2f_{MAX} \text{ Hz.} \tag{2.11}$$

The sampling rate restriction is known as the *Nyquist criterion* with the minimum sample rate from Equation (2.11) $f_S = 2f_{MAX}$, known as the *Nyquist rate*. The intuitive description of the sampling theorem is related to our understanding that when a signal is uniformly sampled, its spectrum is replicated at all multiples of the sample rate f_S. If the signal is strictly band limited with a two-sided spectral support or bandwidth of 2 BW, we can prevent overlap of the spectral copies by separating them by more than their width. This statement of the sampling criterion is shown in the following equation:

$$f_S \geq \text{ Two Sided Bandwidth.} \tag{2.12}$$

The engineer's response to Equation (2.12) is, "By how much should the sample rate exceed the two-sided bandwidth"? We have to ask this because the assumption embedded in any statement of the sampling theorem is that the spectrum is isolated and free standing. In fact, most data collection schemes require an anti-alias filter to isolate the spectrum residing in a selected spectral span from other spectra in adjacent spectral spans. We require a filter to perform the spectral separation, and the spectral response of this filter affects the sample rate. Figure 2.27 presents a spectrum comprising three adjacent spectral spans and the effect of the signal conditioning on the signal with this spectrum. We see that the filter rejects the adjacent channels down to a level matching the dynamic range of the data collection process. Typical values of dynamic range are 60, 72, and 96 dB for an ADC with spurious free dynamic range of 10, 12, and 16 bits, respectively. When the filtered signal is sampled, the spectral copies must be separated sufficiently so that the levels of the folded remnants that fall back into the signal bandwidth are below the dynamic range of the ADC. This is assured when the sample rate satisfies Equation (2.13). In words, the engineer's version of the sampling theorem is "The sample rate must equal the two-sided bandwidth plus the transition bandwidth of the anti-aliasing filter." This is the design criterion to which we will adhere as we develop our applications of multirate filters.

$$
\begin{aligned}
f_S &= \text{Two Sided Bandwidth} + \text{Filter Transition Bandwidth} \\
&= 2\text{BW} + \Delta f.
\end{aligned}
\tag{2.13}
$$

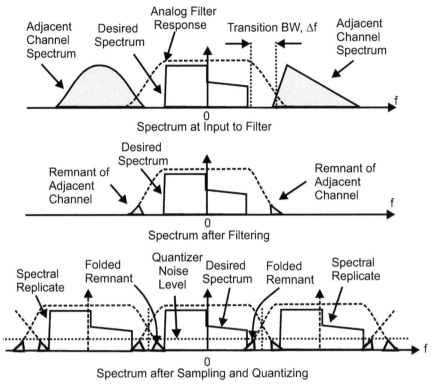

Figure 2.27 Spectrum of input signal and of filter, of filtered input signal, and of sampled input signal.

References

Crochiere, Ronald, and Lawrence Rabiner. *Multirate Signal Processing*, Englewood Cliff, NJ: Prentice-Hall Inc., 1983.

Fliege, Norbert, *Multirate Digital Signal Processing: Multirate Systems, Filter Banks, Wavelets*, West Sussex:, John Wiley & Sons, Ltd,, 1994.

Jovanovic-Dolecek, Gordana, *Multirate Systems: Design and Applications*, London, Idea Group, 2002. Mitra, Sanjit, *Digital Signal Processing: A Computer-Based Approach*, 2nd ed, NY, NY: McGraw-Hill, 2001.

Mitra, Sanjit, and James Kaiser, *Handbook for Digital Signal Processing*, NY, NY: John Wiley & Sons, 1993.

Vaidyanathan, P. P., *Multirate Systems and Filter Banks*, Englewood Cliff, NJ: Prentice-Hall Inc., 1993.

Problems

2.1 Determine the Z-transform and discrete time Fourier series expansion of

 a) $\delta(n)$
 b) $\delta(n - 10)$
 c) $\delta(n - 20)$
 d) $s_0(n) = \sum\limits_{k=0}^{99} \delta(n - 10k)$

2.2 Determine the Z-transform and discrete time Fourier series expansion of

 a) $\delta(n - 1)$
 b) $\delta(n - 11)$
 c) $\delta(n - 21)$
 d) $s_1(n) = \sum\limits_{k=0}^{99} \delta(n - 10k - 1)$

2.3 Determine the Z-transform and discrete time Fourier series expansion of

 a) $\delta(n - r)$
 b) $\delta(n - 10 - r)$
 c) $\delta(n - 20 - r)$
 d) $s_r(n) = \sum\limits_{k=0}^{99} \delta(n - 10k - r)$

2.4 Determine the Z-transform and sampled data time sequence correspo-nding to

 a) $H(\theta) = 1$
 b) $H(\theta) = \exp(j\,20\,\theta)$
 c) $H(\theta) = \exp(j\,40\,\theta)$
 d) $H(\theta) = \sum\limits_{k=0}^{49} \exp(j\,10\,k\,\theta)$

2.5 A complex time series $h(t) = \exp(j\,2\,\pi\,500\,t)$ is sampled at a 4.0 kHz rate and 200 samples collected for processing. Determine

 a) $h(n)$, (sample time values)
 b) $H(k)$, (DFT frequency samples)
 c) The number of cycles of the input time series contained in the span of collected data

    d) Index k of the DFT containing the non-zero component of the DFT

    e) The amplitude of the non-zero DFT sample

2.6 A complex time series $h(t) = \exp(j\ 2\ \pi\ 1000\ t)$ is sampled at a 4.0 kHz rate and 200 samples collected for processing. Determine

    a) $h(n)$, (sample time values)

    b) $H(k)$, (DFT frequency samples)

    c) How many cycles of the input time series are contained in the span of collected data?

    d) Index k of the DFT containing the non-zero component of the DFT

    e) The amplitude of the non-zero DFT sample

2.7 A complex time series $h(t) = \exp(j\ 2\ \pi\ 2000\ t)$ is sampled at a 4.0 kHz rate and 200 samples collected for processing. Determine

    a) $h(n)$, (sample time values)

    b) $H(k)$, (DFT frequency samples)

    c) The number of cycles of the input time series contained in the span of collected data

    d) Index k of the DFT containing the non-zero component of the DFT

    e) The amplitude of the non-zero DFT sample

2.8 A complex time series $h(t) = \exp(j\ 2\ \pi\ 4000\ t)$ is sampled at a 4.0 kHz rate and 200 samples collected for processing. Determine

    a) $h(n)$, (sample time values)

    b) $H(k)$, (DFT frequency samples)

    c) The number of cycles of the input time series contained in the span of collected data

    d) Index k of the DFT containing the non-zero component of the DFT

    e) The amplitude of the non-zero DFT sample

2.9 A complex time series $h(t) = \exp(j\ 2\ \pi\ 500\ t)$ is sampled at a 2.0 kHz rate and 200 samples collected for processing. Determine

    a) $h(n)$, (sample time values)

    b) $H(k)$, (DFT frequency samples)

    c) The number of cycles of the input time series contained in the span of collected data

    d) Index k of the DFT containing the non-zero component of the DFT

    e) The amplitude of the non-zero DFT sample

2.10 A complex time series $h(t) = \exp(j\,2\,\pi\,500\,t)$ is sampled at a 1.0 kHz rate and 200 samples collected for processing. Determine

    a) $h(n)$, (sample time values)
    b) $H(k)$, (DFT frequency samples)
    c) The number of cycles of the input time series contained in the span of collected data
    d) Index k of the DFT containing the non-zero component of the DFT
    e) The amplitude of the non-zero DFT sample

2.11 A complex time series $h(t) = \exp(j\,2\,\pi\,500\,t)$ is sampled at a 0.50 kHz rate and 200 samples collected for processing. Determine

    a) $h(n)$, (sample time values)
    b) $H(k)$, (DFT frequency samples)
    c) The number of input time series is contained in the span of collected data
    d) Index k of the DFT containing the non-zero component of the DFT
    e) Determine the amplitude of the non-zero DFT sample

2.12 Use a windowed $\sin(x)/x$ to generate the impulse response of a sampled low-pass filter with the following specifications: pass-band bandwidth, 0-to-40 Hz, stop-band bandwidth, 60-to-200 Hz, sample rate, 400 Hz, and stop-band attenuation greater than 60 dB. The following MATLAB script will accomplish this:

```
h1 = sinc(−10.0:0.25:10.0.*kaiser(81,6)';
h1 = h1/sum(h1);
```

Zero-pack the impulse response h1 to form a new sequence h2. The following MATLAB script will accomplish this:

```
h2 = reshape([h1;zeros(1,81)],1,162);
```

Plot in two figures, the time series, and the log magnitude spectrum of h1 and of h2. The following MATLAB script will accomplish this for h1. The time and frequency axes must be changed for h2 to reflect the double sample rate:

```
subplot(2,  1,  1)
plot(0:0.25:20+0.25, (h1))
grid on;
axis([-2 22 –0.1 0.30])
```

```
subplot(2,1,2)
freq = -(0.5:1/1024:0.5-1/1024)*400
plot(freq, fftshift(20*log10(fft(h1,1024))))
grid on
axis([-200 200 –80 10])
```

Examine the two spectra and comment on the number and the locations of the filter bandwidths in the two figures.

2.13 Repeat Problem 2.12 except zero-pack 1-to-3 instead of 1-to-2. The following MATLAB script accomplishes the zero-packing:

```
h3 = reshape([h1;zeros(2,82)],1,243);
```

2.14 Use a windowed sin(*x*)/*x* to generate the impulse response of a sampled low-pass filter with the following specifications: pass-band bandwidth, 0-to-100 Hz, stop-band bandwidth, 150-to-1000 Hz, sample rate, 2 KHz, and stop-band attenuation greater than 60 dB. The following MATLAB script will accomplish this:

```
h1 = sinc(-8:01:8)*kaiser(161,8)';
h1 = h1/sum(h1);
```

Also form h2 and h3, down-sampled time series by the following MATLAB script:

```
h2 = zeros(1,161);
h2(1:5:161) = h1(1:5:161);
h3 = zeros(1,161);
h3(2:5:161) = h1(2:5:161);
```

Plot in three subplots the spectrum log magnitude of the sequence h1, h2, and h3 and comment on the bandwidths and locations of the observed pass band(s).

2.15 Use a windowed sin(*x*)/*x* to generate the impulse response of a sampled low-pass filter with the following specifications: pass-band bandwidth, 0-to-100 Hz, stop-band bandwidth, 150-to-1000 Hz, sample rate, 2 KHz, and stop-band attenuation greater than 60 dB. The following MATLAB script will accomplish this:

```
h1 = sinc(-8:01:8)*kaiser(161,8)';
h1 = h1/sum(h1);
```

Also form h2 and h3, down-sampled time series by the following MATLAB

script: h2 = h1(1:5:161);
h3 = h1(2:5:161);

Plot in three subplots the log magnitude spectrum of the sequence h1, h2, and h3 and comment on the bandwidths and locations of the observed pass band(s).

3

Digital Filters

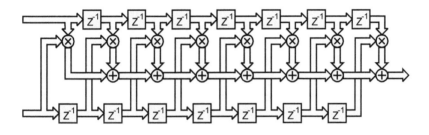

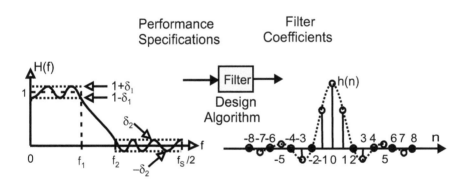

The intent of this chapter is to review the properties and performance constraints of digital filters so that we are better able to embed them in a multirate system. There is no intent here to present a comprehensive overview and detailed description of the many ways to design and implement the various digital filters. What we will do is identify the important system considerations that a designer should consider in the design process and suggest various options as well as present guidelines to help select structures and performance tradeoffs required to finalize a digital filter design.

Digital filters can be classified in many ways, including allusion to their general characteristics such as low-pass, band-pass, band-stop, and others and secondary characteristics such as uniform and nonuniform group delay. An important classification is the filter's architectural structure with a primary consideration being that of finite (duration) impulse response (FIR) and infinite (duration) impulse response (IIR). Except for special cases, involving pole-zero cancelation, FIR and IIR filters are implemented by nonrecursive and recursive structures. Further subclassifications, such as canonic forms, cascade forms, lattice forms, and the like are primarily driven by consideration of sensitivity to finite arithmetic, memory requirements, ability to pipeline arithmetic, and hardware constraints.

The choice to perform a given filtering task with a recursive or a nonrecursive filter is driven by a number of system considerations, including processing resources, clock speed, and various filter specifications. Performance specifications, which include operating sample rate, pass-band and stop-band edges, pass-band ripple, and out-of-band attenuation, all interact to determine the complexity required of the digital filter. We first examine how the performance parameters of the FIR filter interact to determine the filter length and hence the FIR filter's complexity. We then examine the similar relationship between the parameters of IIR filters that we eventually cast into multirate architectures.

3.1 Filter Specifications

The relationships between filter length and filter specifications are valid for any FIR filter. Since the most common filtering task is that of a low-pass filter, we examine the prototype low-pass filter to understand the coupling between the filter parameters. These interactions remain valid for other filter types. The frequency response of a prototype low-pass filter is shown in Figure 3.1. The pass band is seen to be an ideal rectangle that has unity gain between frequencies $\pm f_1$ Hz with zero gain elsewhere. The filter is designed to operate at a sample rate of f_S Hz. For the convenience of dealing with a specific example, we chose the single-sided band edge to be 10 kHz and the sample rate to be 100 kHz. The attraction of the ideal low-pass filter $H(f)$ as a prototype is that we have an exact expression for its impulse response $h(t)$ from its closed form inverse Fourier transform, the ubiquitous $\sin(x)/x$ as shown in the following equation:

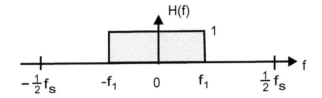

Figure 3.1 Frequency response prototype low-pass filter.

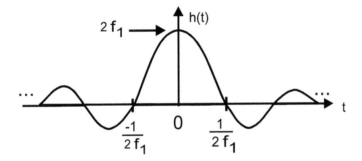

Figure 3.2 Impulse response of prototype low-pass filter.

$$h(t) = 2f_1 \frac{\sin(2\pi \frac{2f_1}{2}t)}{(2\pi \frac{2f_1}{2}t)}. \tag{3.1}$$

In words, the argument of the sin(x)/x function is always the product of 2π, half the spectral support $(2f_1)/2$, and the independent variable t. The numerator is periodic and becomes zero when the argument is a multiple of π, in general at $\pm k\pi$. The location of the first zero occurs as shown in the following equation at:

$$\text{First Zero}: \quad 2\pi \frac{2f_1}{2}t = \pi, \quad t_{\text{Zero}} = \frac{1}{2f_1}. \tag{3.2}$$

The nonrealizable impulse response of the prototype filter is shown in Figure 3.2. For the two-sided bandwidth of 20 kHz, the first zero occurs at 50 μs.

The sin(x)/x filter shown in Figure 3.2 is a continuous function which we have to sample to obtain the prototype sampled data impulse response. To preserve the filter gain during the sampling process, we scale the sampled function by the sample rate as shown in Equation (3.3). The sampled impulse response of the prototype low pass is shown in Equation (3.4), while Figure 3.3 presents a visualization of the sampling process.

$$h(n) = \frac{1}{fs} h(t)|_{t=nT} = \frac{1}{fs} h(t)|_{t=n\frac{1}{fs}} \tag{3.3}$$

$$h(n) = \frac{2f_1}{fs} \frac{\sin(2\pi \frac{f_1}{fs} n)}{(2\pi \frac{f_1}{fs} n)}$$

$$= \frac{2f_1}{fs} \frac{\sin(n\theta_1)}{(n\theta_1)}, \quad \text{where } \theta_1 = 2\pi \frac{f_1}{fs}. \tag{3.4}$$

An important observation is that when used in a fixed-point arithmetic processor, the composite scale factor $2f_1/f_s$ shown in Equation (3.4) is removed from the impulse response weights and is reinserted as a scaling factor when clearing the accumulator after the accumulation process. Without the scaling factor, the filter exhibits processing gain proportional to the ratio of sample rate to bandwidth. Accumulators are designed with extra bit width to accommodate this expected bit growth due to processing gain. While the scaling factor is required to cancel the processing gain of the filter weights, it should not be applied to the coefficient set since it reduces the precision with which the coefficients are represented, which leads to an increase in arithmetic noise of the filter process. This is a common source of error in filter design routines, one that is easily corrected by scaling the filter coefficients by the maximum weight so that the maximum weight is unity rather than $2f_1/f_s$. For the specific example we are using, this scale factor is 20/100 or 0.2, which represents a loss in coefficient precision of more than 2 bits. This loss in coefficient precision is significant when the filter bandwidth is a small fraction of the sample rate. This scaling factor will be seen as an important concern when the filter is used in a resampling configuration.

An insightful observation is to be had by examining Figure 3.3 and asking, "how many samples are there between the peak and first zero crossing of the prototype impulse response?" The number of samples is seen to be $f_S/(2f_1)$, the ratio of sample rate to two-sided bandwidth, which for our specific example is 5. We thus have an interesting measure of the filter bandwidth: if we examine a filter impulse response and note that there are 50 samples from peak to first zero crossing, we can conclude that the two-sided bandwidth is 1/50th of the sample rate. If we can count, we can estimate the fractional bandwidth of a FIR filter!

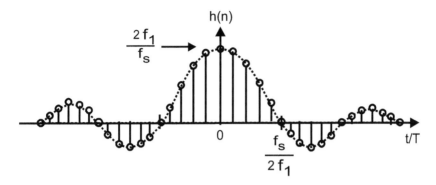

Figure 3.3 Sampled data impulse response of prototype filter.

The sampling process, of course, causes the spectra to be periodically extended with spectral replicates at all multiples of the sample rate. The expression for the spectrum of the sampled data impulse response is shown in Equation (3.5) where the sampled data frequency variable ωT_s is denoted by θ with units of radians/sample. In this coordinate system, the spectrum is periodic in 2π.

$$H(\theta) = \sum_{n=-\infty}^{+\infty} h(n)\, e^{-jn\theta} \qquad (3.5)$$

3.2 Windowing

The problem with the sample set of the prototype filter is that the number of samples is unbounded and the filter is noncausal. If we had a finite number of samples, we could delay the response to make it causal. Our first task then is to form a finite list of filter coefficients from the unbounded set. The process of pruning an infinite sequence to a finite sequence is called *windowing*. In this process, a new sequence is formed as the product of the finite sequence $w(n)$ and the infinite sequence as shown in Equation (3.6) where the .* operator is the standard MATLAB point-by-point multiply.

$$h_W(n) = w(n).*h(n). \qquad (3.6)$$

The expression for the spectrum of the windowed impulse response is shown in (3.7) as the transform of the product $h(n)$ and $w(n)$ and once again in Equation (3.8) as the circular convolution of their spectra $H(\theta)$ and $W(\theta)$.

We will examine the effect of the convolution shortly.

$$H_W(\theta) = \sum_{n=-\infty}^{+\infty} w(n) \cdot h(n) \, e^{-jn\theta}$$

$$= \sum_{n=-N/2}^{+N/2} w(n) \cdot h(n) \, e^{-jn\theta} \tag{3.7}$$

$$H_W(\theta) = \frac{1}{2\pi} \int_{-\pi}^{+\pi} H(\phi) \, W(\phi - \theta) d\phi. \tag{3.8}$$

Our first contender for a window is the symmetric rectangle, sometimes called the default window. This weighting function abruptly turns off the coefficient set at its boundaries. The sampled rectangle weighting function has a spectrum described by the Dirichlet kernel as shown in Equation (3.9). The Dirichlet kernel is seen to be the periodic extension of the $\sin(2\pi f T_{\text{support}}/2)/(2\pi f T_{\text{support}}/2)$ spectrum, the transform of a continuous time rectangle function.

$$W_{\text{Rect}}(\theta) = \sum_{n=-N/2}^{+N/2} 1 \, e^{-jn\theta} = \frac{\sin(N\frac{\theta}{2})}{\sin(\frac{\theta}{2})}. \tag{3.9}$$

Figure 3.4 presents the spectrum of a 100-tap rectangle window as well as a zoom to the neighborhood of its main lobe. The frequency axis here is normalized frequency f/f_S so that the first spectral zero occurs at frequency $1/N = 1/100 = 0.01$. Note that the first side lobe has an amplitude of approximately –22 which, relative to the peak of amplitude 100, represents a power ratio of –13.2 dB, a handy relationship to remember.

The convolution between the spectra of the prototype filter with the Dirichlet kernel forms the spectrum of the rectangle windowed filter coefficient set. This convolution is shown in Figure 3.5. The convolution shows the main lobe of the Dirichlet kernel in three distinct spectral regions: out-of-band, straddling the band edge, and in-band. The contribution to the corresponding output spectrum is seen to be stop-band ripple, transition bandwidth, and pass-band ripple. The pass-band and stop-band ripples are due to the side lobes of the Dirichlet kernel moving through the pass band of the prototype filter while the transition bandwidth is due to the main lobe of the kernel moving from the stop band to the pass band of the prototype. Note

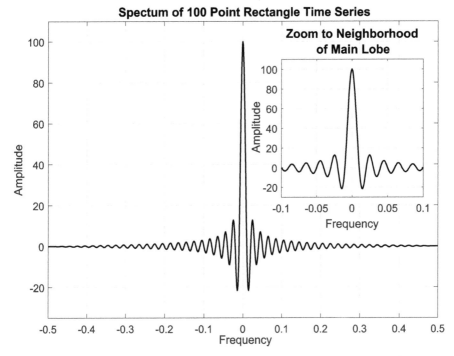

Figure 3.4 Spectrum of 100-point rectangle window with zoom to main lobe.

that the transition bandwidth is the same as the main-lobe width of the kernel, approximately $1/N$th of the sample rate for a filter of length N.

A property of the Fourier series is that a truncated version of the series forms a new series exhibiting the minimum mean-square approximation to the original function. We thus note that the coefficient set obtained by a rectangle window exhibits the minimum mean square (MMS) approximation to the prototype frequency response. The problem with MMS approximations in numerical analysis is that there is no mechanism to control the location or value of the error maxima. The local maximum errors are attributed to the Gibbs phenomena, the failure of the series to converge in the neighborhood of a discontinuity. These errors can be objectionably large. Figure 3.6 presents a log-magnitude display of the spectrum formed by the rectangle windowed coefficient set. We see here that the stop-band side lobes only present 22 dB attenuation near the filter band edge.

A process must now be invoked to control the objectionably high side-lobe levels. We have two ways to approach the problem. First, we can

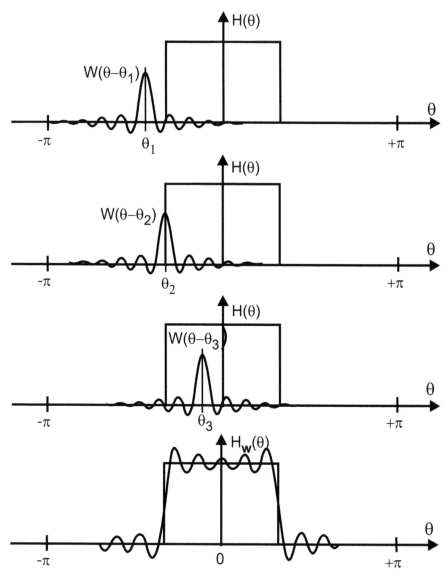

Figure 3.5 Spectrum of rectangle windowed prototype filter obtained as convolution between spectrum of prototype filter and spectrum of rectangle window.

redefine the frequency response of the prototype filter so that the amplitude discontinuities are replaced with specified transition bandwidth tapers. In this process, we exchange transition bandwidth for side-lobe control. How

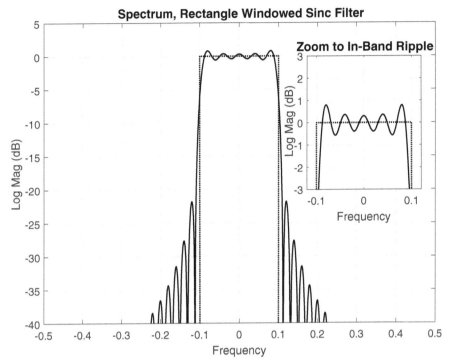

Figure 3.6 Log display of spectrum to emphasize high levels of out-of-band side-lobe response of rectangle windowed prototype filter.

to make that change effectively is the next question. Equivalently, knowing the objectionable stop-band side lobes are caused by the side lobes in the spectrum of the window, we can replace the rectangle window with other even symmetric functions with reduced amplitude side-lobe levels. Here the question to be addressed is, how do we select weighting functions with low spectral side lobes? As we will now show, the two techniques, side-lobe control and transition-bandwidth control, are tightly coupled. The easiest way to visualize control of the side lobes is by destructive cancelation between the spectral side lobes of the Dirichlet kernel associated with the rectangle and the spectral side lobes of translated and scaled versions of the same kernel. Figure 3.7 presents the window formed by the addition of a single cycle of a cosine to the rectangle window as well as the transforms of the time domain components. Note that the single cycle of cosine is the lowest frequency sinusoid orthogonal to the rectangle. This orthogonality is observed in the frequency domain as the placement of the cosine's spectral components at

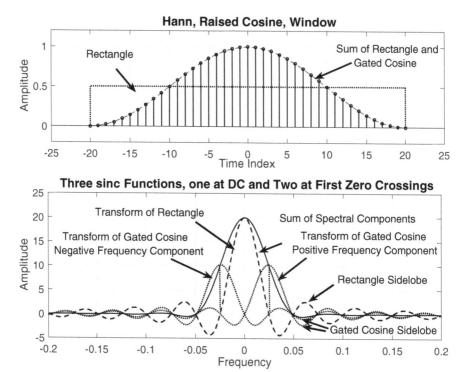

Figure 3.7   Raised cosine window and transform illustrating sum of translated and scaled Dirichlet kernels.

the first spectral zeros of the rectangle's spectrum. We note that the side lobes of the DC-centered kernel and the side lobes of the translated kernels have opposite polarities in the spectral region outside the main-lobe spectral support. By judicious choice of weighting terms, the side-lobe amplitudes are significantly reduced.

In Figure 3.7, we easily see that adding the translated kernels to the original spectrum has doubled the distance from peak to first spectral null. Thus, the cost we incur to obtain reduced side-lobe levels is an increase in main-lobe bandwidth. Table 3.1 presents a list of window functions formed as the sum of translated kernels with their peak side-lobe levels along with their main-lobe widths.

Scanning Table 3.1, we can estimate the rate at which we can trade side-lobe levels for main-lobe width. This rate is approximately -22 dB/spectral bin so that in order to obtain -60 dB side lobes, we have to increase the main-lobe bandwidth to $2.7 f_S/N$. Remembering that the window's two-sided

Table 3.1 Windows formed as weighted sum of cosines.

Window name	Weights	Maximum side lobe	Main-lobe width peak to first zero
Rectangle	$a_0 = 1.0$	-13.5 dB	1
Hann	$a_0 = 0.5$	-32 dB	2
	$a_1 = -0.5$		
Hamming	$a_0 = 0.54$	-43 dB	2
	$a_1 = -0.46$		
Blackman (Approximation)	$a_0 = 0.42$	-58 dB	3
	$a_1 = -0.50$		
	$a_2 = 0.08$		
Blackman (Exact)	$a_0 = 0.426\ 59$	-68 dB	3
	$a_1 = -0.496\ 56$		
	$a_2 = 0.076\ 85$		
Blackman–harris (3-Term)	$a_0 = 0.423\ 23$	-72 dB	3
	$a_1 = -0.497\ 55$		
	$a_2 = 0.079\ 22$		
Blackman–harris (4-Term)	$a_0 = 0.358\ 75$	-92 dB	4
	$a_1 = -0.488\ 29$		
	$a_2 = 0.141\ 28$		
	$a_3 = -0.011\ 68$		

main-lobe width is an upper bound to the filter's transition bandwidth, we can estimate the transition bandwidth of a filter required to obtain a specified side-lobe level. Equation (3.10) presents an empirically derived approximate relationship valid for window-based design, while (3.11) rearranges (3.10) to obtain an estimate of the filter length required to meet a set of filter specifications.

$$\Delta f_{\text{Minimum}} = \frac{f_S}{N}$$

$$\Delta f = \frac{f_S}{N} K(\text{Atten}) \cong \frac{f_S}{N} \frac{\text{Atten(dB)} - 8}{14}$$

(3.10)

$$N \cong \frac{f_S}{\Delta f} \frac{\text{Atten(dB)} - 8}{14}.$$

(3.11)

The primary reason we examined windows and their spectral description as weighted Dirichlet kernels was to develop a sense of how we trade window main-lobe width for window side-lobe levels and in turn filter transition bandwidth and side-lobe levels. Some windows perform this exchange of bandwidth for side-lobe level very efficiently while others do not. The Kaiser–Bessel window is very effective while the triangle (or Fejer) window is not.

The Kaiser–Bessel window is in fact a family of windows parameterized over β, the time-bandwidth product of the window. The main-lobe width increases with β, while the peak side-lobe level decreases with β. The Kaiser–Bessel window is a standard option in filter design packages such as MATLAB. For completeness, we describe it here in Equation (3.12), where I_0 is the zero-order modified Bessel function of the first kind. The series shown converges quite rapidly due to the $k!$ term in the denominator of the expansion. Typical range of the parameter β is 3 to 10 to obtain filter side-lobe levels in the range $40-100$ dB. We note that the window defaults to a rectangle for $\beta = 0$.

$$w(n) = \frac{I_0 \left[\pi \beta \sqrt{1.0 - \left(\frac{n}{N/2} \right)^2} \right]}{I_0(\pi \beta)} \tag{3.12}$$

$$\text{where} \quad I_0(x) = \sum_{k=0}^{\infty} \left[\frac{(x/2)^k}{k!} \right]^2 .$$

The transform of the Kaiser–Bessel window is approximately that shown in the following equation:

$$W(\theta) = \frac{N}{I_0(\pi \beta)} \frac{\sinh \left(\sqrt{(\pi \beta)^2 - (N\theta/2)^2} \right)}{\sqrt{(\pi \beta)^2 - (N\theta/2)^2}} . \tag{3.13}$$

We still have to relate the window side-lobe levels, which are integrated in the convolution process, to the form the filter side-lobe levels. As an example, the first side lobe of the Dirichlet kernel is -13.5 dB relative to the spectral peak, while the first side lobe in the resulting low-pass filter is -22 dB relative to pass-band gain. We will only concern ourselves with this relationship for the Kaiser–Bessel class of windows. Figure 3.8 presents a curve showing the spectral side-lobe levels realized by windowing a prototype impulse response with the Kaiser–Bessel window of specified parameter β along with a second curve showing the spectral side-lobe levels of the corresponding window. Note that at $\beta = 0$, the two levels correspond to the rectangle window.

Example 3.1. Window Design of Low-Pass FIR Filter
Design a FIR filter with the following specifications:

Sample Rate	100 kHz
Pass-Band Band Edge	±10 kHz
Stop-Band Band Edge	±15 kHz
Minimum Attenuation	60 dB

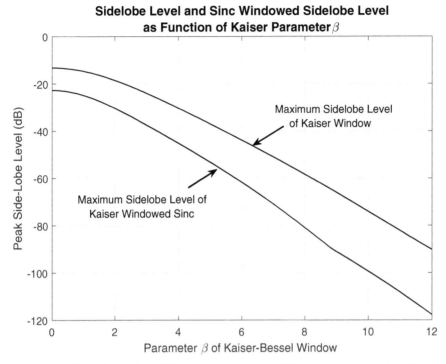

Figure 3.8 Side-lobe levels of Kaiser–Bessel windowed FIR filter and of the Kaiser–Bessel window as a function of window parameter β.

From the filter specifications and from Equation (3.11), we estimate the

filter length N: $N \cong (f_s/\Delta f) * (\text{Atten(dB)} - 8)/14$
$= (100/5) * (60 - 8)/14 = 75$ taps

Using the sinc function in MATLAB to form samples of the prototype impulse response and the Kaiser–Bessel window to control the spectral side-lobe levels, we estimate from Figure 3.8, or as a result of a few trials, that 60 dB side-lobe levels are obtained with parameter $\beta = 5.7$. Two versions of the filter were designed. The first was designed for the pass-band parameter, which resulted in the band edges being centered at about the 6 dB frequency of the filter, this being a consequence of the Fourier transform converging to the midpoint of a discontinuity. The second design shifted the pass-band parameter to the midpoint of the transition band. Both designs are described compactly in the following two sets of MATLAB script. The time and frequency responses of the two designs are shown in Figure 3.9.

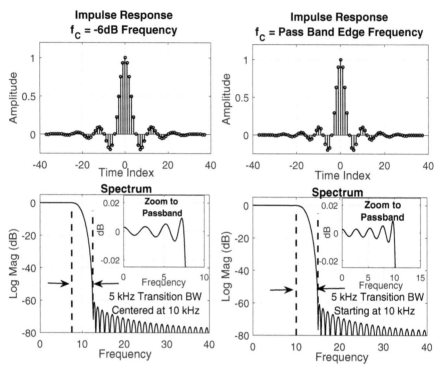

Figure 3.9 Time and frequency response of FIR filter windowed with Kaiser–Bessel: first designed for 6 dB band edge and second for pass-band and stop-band edges.

```
    hh1=sinc((2*f₁)/fs)*(-0.5*(N-1):0.5*(N-1));
%   hh1=hh1.*Kaiser(N,beta)';
    hh1=sinc(0.2*(-37:1:37)).*kaiser(75,5.7)';
    hh2=sinc(((2f₁+Δf/2)/fs)*(-0.5*(N-1):0.5*(N-1));
%   hh2=hh2.*Kaiser(N, beta)';
    hh2 = sinc(0.25*(-37:1:37)).*kaiser(75,5.7)';
```

Note in both filter designs that the in-band ripple has the same structure as the out-of-band ripple with peak values on the order of 1-part-1000. For the first design, the ripple pass band extends from 0 to 7.5 kHz, while in the second design, the ripple pass band extends from 0 to 10 kHz.

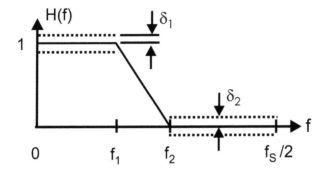

Figure 3.10 Parameters required to specify sampled data low-pass filter.

3.3 The Remez-Firpm Algorithm

In the previous section, we learned that FIR filters always exhibit ripple in the pass band and in the stop band as well as bandwidth to transition between the pass band and the stop band. Thus, filters must be specified in accordance with the parameters indicated in Figure 3.10 and identified in the following parameter list.

Filter Specification Parameters f_s: Sample Rate
f_1: Frequency at End of Pass Band f_2: Frequency at Start of Stop Band
δ_1: Maximum Pass-Band Ripple
δ_2: Maximum Stop-Band Ripple

$$N = \text{function}(f_S, f_1, f_2, \delta_1, \delta_2). \tag{3.14}$$

As indicated in Equation (3.14), N, the number of coefficient taps required of the FIR filter to meet the specifications, is a function of five parameters. How we select most of the parameters is self-evident. The sample rate must satisfy the Nyquist criterion, the pass-band and stop-band frequencies must satisfy the filtering requirements, and the stop-band ripple must satisfy the out-of-band attenuation requirement. The pass-band ripple requirement is related to a signal distortion criterion modeled as signal echoes caused by the pass-band ripple. We discuss the pass-band ripple criterion in terms of system performance in the next section. Maximum pass-band ripple values for many system designs are on the order of 1-part in 100 to 5-parts in 100 (i.e. 1%−5%). These levels are significantly larger than the stop-band ripple values that are on the order of 1-part in 1000 to 1-part in 10,000 (i.e. 60−80 dB). FIR filters designed by the windowing technique exhibit equal pass-band and stop-band ripple levels. We now seek a design process that permits

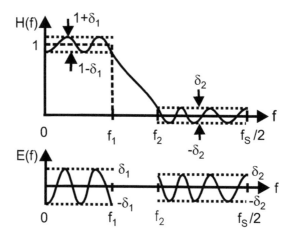

Figure 3.11   Frequency response of equiripple filter and error frequency profile.

different levels of pass-band and stop-band ripple. Filters with relaxed pass-band ripple requirements will require fewer coefficients and, hence, require fewer resources to implement.

The window design of a FIR filter occurs in the time domain as the point-by-point product of a prototype impulse response with the smooth window sequence. The quality of the resultant design is verified by examining the transform of the windowed impulse response. By contrast, the equiripple design is performed entirely in the frequency domain by an iterative adjustment of the location of sampled spectral values to obtain a Tchebyschev approximation to a desired spectrum. The desired, or target, spectrum has accompanying tolerance bands that define acceptable deviations from the target spectrum in distinct spectral regions. The alternation theorem assures us that there exists a Tchebyschev approximation to the target spectrum and that this solution exhibits equiripple errors with local extrema of alternating signs meeting the tolerance boundaries. This approximation and error function are shown for a desired set of tolerance bands in Figure 3.11.

When the FIR filter has $2M + 1$ even symmetric coefficients, its spectrum can be expanded as a trigonometric polynomial in θ as shown in the following equation:

$$H(\theta) = \sum_{n=0}^{N-1} a(n) \cos(n\theta). \tag{3.15}$$

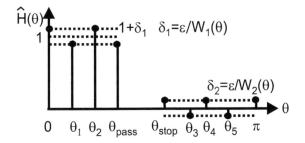

Figure 3.12 Distribution of initial estimates of extrema for multiple exchange algorithm.

Defining a positive valued weighting function $W(\theta)$ and the target function $T(\theta)$, we can define the weighted error function $E(\theta)$ as shown in the following equation:

$$E(\theta) = W(\theta) \cdot [H(\theta) - T(\theta)]. \tag{3.16}$$

The Mth order polynomial $H(\theta)$ is defined by $M + 1$ coefficients $a(n)$ or equivalently by the $M - 1$ local extrema $H(\theta_k)$ and the two boundary values at $H(\theta_{\text{pass}})$ and $H(\theta_{\text{stop}})$. The problem is that we do not know the locations of the local extrema. The Remez multiple exchange algorithm rapidly locates these extremal positions by iterating from an initial guess of their positions to their actual positions. A ubiquitous design algorithm written by McClellan, Parks, and Rabiner expanded on the original Parks and McClellan design and has become the standard implementation of the Remez algorithm. It is embedded in most FIR filter design routines. The process proceeds as follows. An initial estimate of the extremal frequencies θ_k is assigned to the target function $T(\theta_k)$ with alternating sign offsets of the form shown in Equation (3.17) and in Figure 3.12. A polynomial is generated that passes through these initial points using the Lagrange interpolator as shown in (3.18).

$$\hat{H}(\theta_k) = T(\theta_k) + (-1)^k e / W(\theta_k) \tag{3.17}$$

$$H(\theta) = \sum_{k=0}^{M} \hat{H}(\theta_k) P_k(\theta)$$

$$\text{where } P_k(\theta) = \frac{\displaystyle\prod_{m=0, m \neq k}^{M} (\theta - \theta_m)}{\displaystyle\prod_{m=0, m \neq k}^{M} (\theta_k - \theta_m)}. \tag{3.18}$$

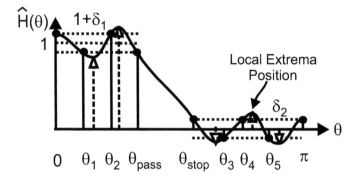

Figure 3.13   Polynomial passed through initial estimate sample points.

The polynomial $\hat{H}(\theta)$ passes through the initial sample points, but generally these points do not correspond to the local extrema points. The polynomial is sampled at a dense grid, on the order of 16 times the number of extremal points, and the samples are searched to locate the true extremal points as indicated in Figure 3.13. The locations of the previous estimate are replaced with the locations of the extremal positions of the polynomial formed from the previous estimate. This exchange is illustrated in Figure 3.14. The process of exchanging previous estimates with improved estimates continues until the amplitudes of the local extrema are within the specified error exit criteria. The algorithm converges quite rapidly, typically on the order of 4-to-6 iterations. After convergence, the polynomial is sampled at $M + 1$ equally spaced positions and inverse transformed to determine its impulse response.

The example we cited to describe the manner the Remez algorithm iteratively converges to the Tchebyschev solution is an even symmetric filter with an odd number of coefficients. A slight variation of the design process is required when the filter is even symmetric with an even number of taps, or when the filter is odd symmetric with an even or an odd number of coefficients. The variation is transparent to the user of standard design tools; so we will not discuss the details here. The interested reader should read the material presented in *Handbook for Digital Signal Processing* edited by Mitra and Kaiser.

The Remez algorithm is also known by other names. These include the Parks–McClellen or P-M, the McClellen, Parks, and Rabiner or MPR, the Equiripple, and the Multiple Exchange. MATLAB has renamed its Remez algorithm *firpm* in recognition of the original work performed by Parks and

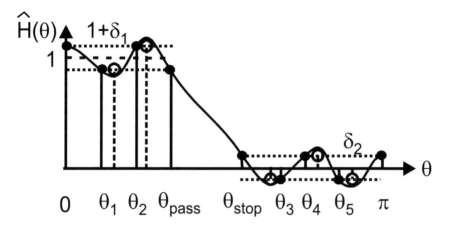

Figure 3.14 Exchanging sample points for iteration $k + 1$ with sample points located at extremal values obtained from estimate at iteration k.

McClellan. As mentioned earlier, the P-M version of the algorithm permeates the community. It is very versatile and capable of designing FIR filters with various frequency responses including multiple pass bands and stop bands and with independent control of ripple levels in the multiple bands. We limit our discussion to low-pass filters.

The first question we address is, what is the functional description of Equation (3.14) which relates the filter length and the filter parameters? A number of empirically derived approximations have been published that provide an estimate of filter lengths designed by the Remez algorithm from the filter specifications. A simple estimate based on the relationship between transition bandwidth stop-band side-lobe levels presented in Table 3.1 published by harris is shown in Equation (3.19) and one published by Herrmann is shown in Equation (3.20). The Herrmann approximation is used in the MATLAB function *firpmord* with the standard caveat that the estimate often underestimates the filter order and the user should verify performance and increment the filter length if necessary and repeat the design and verify process.

$$N = \frac{f_S}{\Delta f} K(\delta_2) \cong \frac{f_S}{\Delta f} \frac{\text{Atten(dB)}}{22} \qquad (3.19)$$

$$N = \frac{f_S}{\Delta f} K(\delta_1, \delta_2, \Delta f, f_S)$$

$$K(\delta_1, \delta_2, \Delta f, f_S) = c_1(\delta_1) \cdot \log(\delta_2) + c_2(\delta_1) + c_3(\delta_1, \delta_2) \left(\frac{\Delta f}{f_S}\right)^2$$

$$c_1(\delta_1) = (0.0729 \cdot \log(\delta_1))^2 + 0.07114 \cdot \log(\delta_1) - 0.4761$$

$$c_2(\delta_1) = (0.0518 \cdot \log(\delta_1))^2 + 0.59410 \cdot \log(\delta_1) - 0.4278$$

$$c_3(\delta_1, \delta_2) = 11.01217 + 0.541244 \cdot (\log(\delta_1) - \log(\delta_2)).$$

$$(3.20)$$

Figure 3.15 is a parameterized set of plots showing how the Herrmann estimate (3.20) of the multiplier factor $K(\delta_1, \delta_2)$ varies with pass-band ripple δ_1, stop-band ripple δ_2, and transition bandwidth $\Delta f / f_s$. The curves correspond to values of in-band ripple equal to 10%, 1%, 0.1%, and 0.01% and further for three values of transition bandwidth equal to 1%, 5%, and 10% of sample rate. We note that the filter length increases when the filter requires reduced levels of either in-band or out-band ripple as well as reductions in transition bandwidth.

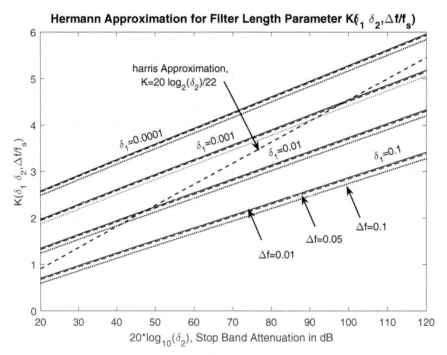

Figure 3.15   Multiplier parameter $K(\delta_1, \delta_2, \Delta f)$ as function of pass-band ripple δ_1, stop-band ripple δ_2, and transition bandwidth $\Delta f / f_s$.

The lesson here is that we should not oversatisfy filter specifications because doing so results in additional processing load for the filter. Plotted on the same figure is the harris estimate (3.19), which supplies estimates in the center of the solution space from which interaction with design routines can be used to refine the estimate.

The weighting function $W(\theta)$ shown in Equation (3.16) is embedded in the P-M version of the Remez algorithm as the penalty function $P(\theta)$ and the weight vector W in the MATLAB implementation of the same algorithm. The MATLAB call to the *firpm* algorithm uses a frequency vector and gain vector to form a connect-the-dot target function and a weight vector specifying a weight value per interval. MATLAB has two curious conventions of which the user should be aware. The first is that the N in the call to the *firpm* algorithm is the polynomial order rather than the number of filter coefficients. An Nth-order polynomial is defined by $N + 1$ coefficients; thus, if we want a 57-tap filter, we use 56 in the function call. MATLAB also uses a frequency axis normalized to the half sample rate rather than normalized to the sample rate; i.e. $f_{norm} = f/(f_s/2)$. A MATLAB call to design a low-pass filter would have the following form:

```
hh=firpm(N-1, [0 f1 f2 fS/2]/(fS/2), [1 1 0 0], [w1 w2]);
hh=firpm(N-1, [0 f1 f2 f3]/(f3), [1 1 0 0], [w1 w2]);
```

The *firpm (remez)* function call builds the initial arrays $T(f)$ and $W(f)$ introduced in Equation (3.17) and shown in Figure 3.16. The algorithm then builds the function $H(f)$ with ripple levels δ_1 and δ_2 that satisfy the following equation:

$$\delta_1 W_1 = \delta_2 W_2$$

$$\delta_1 = \delta_2 \frac{W_2}{W_1}.$$

(3.21)

To obtain a design with out-of-band ripple δ_2 equal to 1/10th of in-band ripple δ_1, we set W_2 to be 10 W_1, or the weight vector $W = [110]$.

Example 3.2. Firpm Design of Low-Pass FIR Filter
Design a low-pass FIR filter with the firpm algorithm that meets the same specifications as the window design Example 3.1.

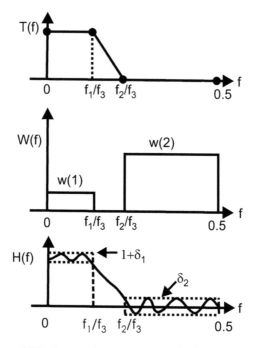

Figure 3.16   Input and output arrays in the Remez algorithm.

The expanded filter specifications are:

Sample Rate	100 kHz
Pass Band	±10 kHz
Stop Band	±15 kHz
Min. Atten.	60 dB
Pass-Band Ripple	0.1 dB (1.2%)

From the filter specifications, we estimate the filter length N from Figure 3.15 to be

$$N \cong (f_s/\Delta f) * K(\delta_1,\delta_2)$$
$$= (100/5) * 2.5 = 50 \text{ taps}$$

We can also obtain a comparable estimate from the Herrmann estimate in MATLAB by using the call to Firpmord as shown:

$$NN = \text{firpmord}([f_1\, f_2],[a_1\, a_2],[\delta_1\, \delta_2],f_s)$$
$$NN = \text{firpmord}([10\ 15], [1\ 0], [0.01\ 0.001], 100)$$

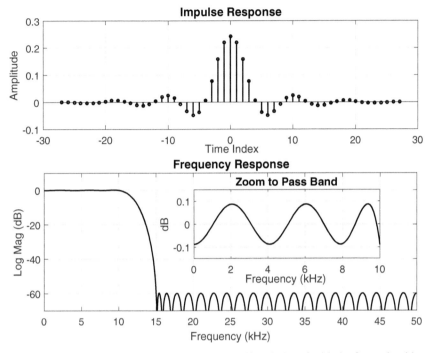

Figure 3.17 Time and frequency response of FIR filter designed with the firpm algorithm.

The response to this call is: NN = 51.

Note this filter requires a smaller number of coefficients, 51 as opposed to 75, for the same filter designed by the Kaiser–Bessel window design. The filter length is smaller because the in-band ripple for the Remez design is larger than the in-band ripple of the window design. We now use the MATLAB Firpm function to design the filter. The estimated filter length of 51 taps did not meet the 60 dB specifications. The filter length had to be increased to 55 to satisfy the attenuation requirement. The call is of the form shown next and the time and frequency response of the filter is shown in Figure 3.17. Note the equal ripple spectral response in both pass band and stop band. By design, the pass-band ripple is 1-part-100 (0.1 dB), while the stop-band ripple is 1-part-1000 (60 dB).

```
h3=firpm(NN-1,[0 f1 f2 fs/2]/(fs/2),[1 1 0 0],[w1 w2]);
h3=firpm(54, [0 10 15 50]/50, [1 1 0 0], [1 10]);
```

3.3.1 Equiripple vs. 1/f Ripple Designs

We note that the filter designed by the equiripple design routine exhibits equal ripple in both pass band and stop band. We would be surprised had it not since we designed it for that property, the property being optimum in the weighted Tchebyschev sense. The stop-band spectrum has a rate of attenuation of 0 dB per octave. We recall that a spectrum has a rate of decay related to the order of the discontinuity of its time domain signal. For instance, a time signal with discontinuous amplitude (such as rectangle) has a spectrum that falls off as 1/f or –6 dB per octave. Similarly, a time signal that has a discontinuous first derivative (such as a triangle) has a spectrum that falls like 1/f^2 or –12 dB per octave. The rate of decay for the envelope of a spectrum is shown in Equation (3.22), where k is the order of the time derivative in which a discontinuity appears. Thus, if the discontinuity resides in the zeroth derivative, the signal itself is discontinuous, and the spectrum decays as 1/f.

$$\text{Asymtotic Decay Rate } = \frac{1}{f^{(K-1)}}. \tag{3.22}$$

This leads to an interesting observation! If the rate of decay is zero, then the discontinuity resides in the –1 derivative, in fact, in the first integral of the signal. But if the integral is discontinuous, then the function must contain an impulse. Stated more directly, an FIR filter that exhibits constant level side lobes has impulses in its time series. We may recall that the Dolph–Tchebyschev window, the window with minimum main-lobe width for a given side-lobe level, is characterized by constant level side lobes. It exhibits a pair of end-point impulses that prevented its use as a shading function in analog beam-forming applications. The Taylor window was devised to suppress the boundary value impulses and is a common shading function in the radar community.

If we pay particular attention to the end points of the filter designed by the Remez algorithm, we often find what appear to be end-point outliers but are in fact the impulses responsible for the constant level spectral side lobes. The size of the impulse is on the order of the size of the spectral side lobes and might be overlooked on the scale of the filter coefficient set. A bit of humor: when we first noted the outlier impulses at the end of long FIR filters designed by the remez algorithm, we thought it was due to machine finite arithmetic. We redesigned the filters with quadruple precision 128 bit words. The impulses were still there! We thought, "who invited you to this party?" Our immediate answer was, "you did when you asked for constant level side lobes!"

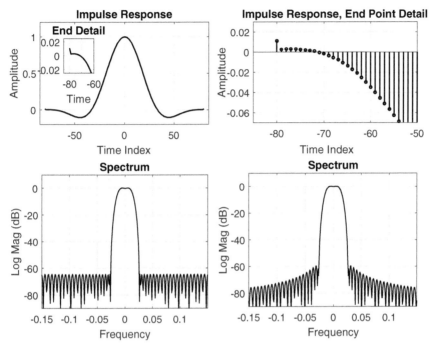

Figure 3.18 Remez impulse response showing details of end point, close-up of end point, and spectra of original filter and of filter with clipped end point.

Figure 3.18 shows the impulse response and frequency response of a filter designed with the Remez algorithm. A close-up detail of the end segment of the impulse response clearly shows the outlier. When this sample is clipped to match the amplitude of its neighbor, the filter loses its constant side-lobe characteristic and exhibits a $1/f$ rate of spectral decay. The $1/f$ asymptotic spectral decay is usually accompanied by a 6 dB increase of near-in side lobes. If we use the clipping to obtain the $1/f$ side lobe we can compensate for the spectral rise by designing the Remez filter with 6 dB additional attenuation for which the spectrum after clipping the end point can rise by the allotted margin. While most filters exhibit this outlier, occasionally, it is not apparent and for those cases, rather than clipping the end point, we can attenuate the boundary samples to modify the side-lobe behavior.

Why would we want to have the spectrum have a $1/f$ decay rate rather than exhibiting equiripple? There are two reasons, both related to system performance. The first is integrated side-lobe levels. We often build systems, as shown in Figure 3.19, comprising a digital filter and a resampling switch.

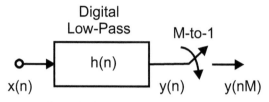

Figure 3.19　Resampling low-pass filter.

Here the digital filter reduces the bandwidth and is followed by a resampling switch that reduces the output sample commensurate with the reduced output bandwidth. When the filter output is resampled, the low-level energy residing in the out-of-band spectral region aliases back into the filter pass band. When the reduction in sample rate is large, there are multiple spectral regions that alias or fold into the pass band. For instance, in a 16-to-1 reduction in sample rate, there are 15 spectral regions that fold into the pass band. The energy in these bands is additive and if the spectral density in each band is equal, as they are in an equiripple design, the folded energy level is increased by a factor of sqrt(15). To prevent the piling-up of the aliased energy, we redesign the filter so that it exhibits $1/f$ side-lobe attenuation.

For a specific example, the filter presented in Figure 3.20 designed for 60 dB side-lobe levels is used in a 32-to-1 down sampling application. If the side lobes are equiripple at 60 dB, the integrated side-lobe level is –36.1 dB, which, when distributed over the remaining bandwidth of 1/32 (–15.1 dB) of input sample rate, results in an effective alias side-lobe suppression of –51.2 dB, equivalent to a 9 dB loss. The filter was redesigned for –67.5 dB equiripple and the numbers obtained for this design are integrated side-lobe level of –43.7 dB and an effective alias side-lobe level of –58.8 dB, which matched the expected 7.5 dB improvement. After clipping the endpoint of the redesigned filter, the close-in side lobe rose to –62 dB as the side lobes acquired the $1/f$ attenuation rate. The numbers for this variant are impressive, the filter exhibiting –51.1 dB integrated side lobes and an effective alias side-lobe level of –66.2 dB. This represents a 14 dB improvement in aliased spectral levels relative to the uniform side-lobe filter operating in the same resampling mode.

The second reason we may prefer FIR filters with $1/f$ side-lobe attenuation as opposed to uniform side lobes is finite arithmetic. A filter is defined by its coefficient set and an approximation to this filter is realized by a set of quantized coefficients. Given two filter sets $h(n)$ and $g(n)$, the first with equiripple side lobes and the second with $1/f$ side lobes, we form two

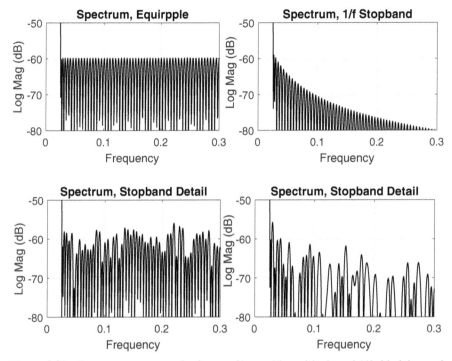

Figure 3.20 Frequency response of reference filters with equiripple and $1/f$ side lobes and of 10-bit quantized version of same filters.

new sets, $h_Q(n)$ and $g_Q(n)$, by quantizing their coefficients. The quantization process is performed in two steps. First, we rescale the filters by dividing by the peak coefficient. Second, we represent the coefficients with a fixed number of bits to obtain the quantized approximations. These operations are shown in the following equation:

$$h_{\text{scaled}} = h/\text{sum}(h), \quad \text{(for low pass filter)}$$
$$\text{h}_{\text{Quantized}} = \text{floor}(h_{\text{scaled}} \cdot 2^{(\text{bits}-1)})/2^{(\text{bits}-1)}. \tag{3.23}$$

The zeros of an FIR filter residing on the unit circle perform the task of holding down the frequency response in the stop band. The interval between the zeros contains the spectral side lobes. When the interval between adjacent zeros is reduced, the amplitude of the side lobe between them is reduced and when the interval between adjacent zeros is increased, the amplitude of the side lobe between them is increased. The zeros of the filters are the roots of the polynomials $H(Z)$ and $G(Z)$. The roots of the polynomials formed by the

Table 3.2 Integrated side-lobe levels for equiripple and $1/f$ side-lobe FIR filters, unquantized, and 10-bit quantized versions.

	Equiripple side lobes	$1/f$ Side lobes
Unquantized	−36.1 dB	−49.3 dB
Quantized (10-bits)	−35.4 dB	−45.1 dB

quantized set of coefficients $H_{quant}(Z)$ and $G_{quant}(Z)$ differ from the roots of the unquantized polynomials.

For small changes in coefficient size, the roots exhibit small displacements along the unit circle from their nominal position. The amplitude of some of the side lobes must increase due to this root shift. In the equiripple design, the initial side lobes exactly meet the design side-lobe level with no margin for side-lobe increases due to root shift caused by coefficient quantization. On the other hand, the filter with $1/f$ side-lobe levels has plenty of margin for side-lobe increases due to root shift caused by coefficient quantization. Figure 3.20 presents the frequency response of two reference filters, one with equiripple side lobes and the other with $1/f$ side lobes. We see their spectra for unquantized and for their 10-bit quantized versions. As expected, the side lobes of the quantized version of the equiripple filter exceed the 60 dB attenuation level, while the side lobes of the quantized $1/f$ side lobes continue to meet the required 60 dB attenuation level. For reference, the integrated side lobes for the four cases are listed in Table 3.2.

The final question we address is, how do we design FIR filters with $1/f$ side lobes? We need a design technique to replace the trick we illustrated earlier that obtained the desired $1/f$ side lobes by clipping the end-point outliers because the process also affects the in-band ripple and may not work for a particular filter design. The tool we apply to side-lobe control is the weighting function described in Equation (3.17) and in Figure 3.16. If we have access to the weight function array, we can modify the weight function in the stop band so that it increases linearly with frequency as shown in Figure 3.21. When the weight function increases with frequency, the resultant side-lobe levels vary inversely with frequency. The frequency-dependent weighting function is used in a number of filter design routines. If the weight function array is not directly accessible, we can use a step-wise weighting function in small adjacent frequency intervals to form a staircase approximation to the frequency-dependent ramp. A MATLAB call to the *firpm* algorithm that uses the staircase weight function is shown here and its spectrum is shown in the lower right subplot of Figure 3.22.

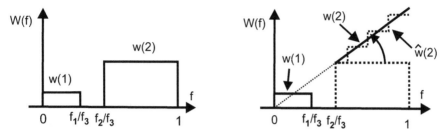

Figure 3.21 Modified weight functions for Remez algorithm to obtain increasing penalty function in stop band to obtain $1/f$ side lobes.

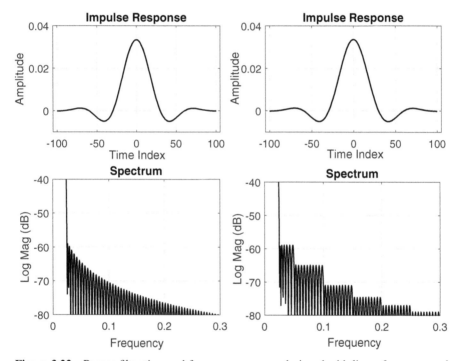

Figure 3.22 Remez filter time and frequency response designed with linear frequency and with staircase frequency weight function.

```
ff=[0 0.6 3.4 5.0 5.1 7.5 7.6 10.0 10.1 15.0. 60.1 64.0]
ff=ff/64.0;
aa=[ 1  1  0  0  0  0  0  0  0  0  0  0  ];
ww=[   1    1.5    3    4.5    6    21   ];
hh=firpm(154,ff,aa,ww);
```

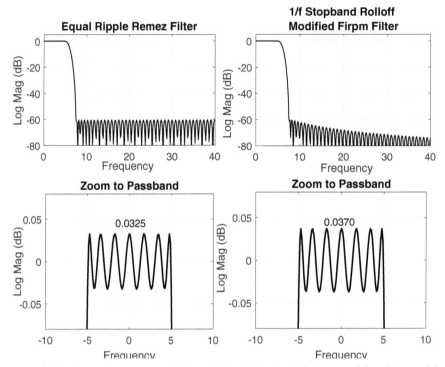

Figure 3.23   Spectral response of filters designed by the default remezfrf and by myfrf routines in Remez call.

The MATLAB script file that sets the penalty function for the *firpm* algorithm is remezfrf.m (here frf means frequency response function). A modified version, called myfrf.m, forms a 1/*f* stop-band ripple. The MATLAB call for a standard filter and for a modified filter design that uses the script file is shown here.

```
h1=firpm(154,[0 0.6 3.4 64]/64,[1 1 0 0],[1 10]);
h2=firpm(154,[0 0.6 3.4 64]/64,{'myfrf',[1 1 0 0]},[1 10]);
```

Figure 3.23 shows frequency response of FIR filters designed with the default *remezfrf.m* and the *myfrf.m* frequency-dependent penalty functions. Also shown is a zoom to the pass-band ripple to illustrate that there is only minor effect in the pass band when tilting the stop-band ripple.

3.3.2 Acceptable In-Band Ripple Levels

This section addresses the matter of selecting the in-band ripple specifications for an FIR filter. The classic problem is that we know how to specify the in-band ripple in a filter but have little guidance of how to determine that desired level. To assist in the task of selecting an acceptable level of in-band ripple, it is useful to understand the effect of in-band ripple on signals moving through the filter. We start with the requirements for a distortion-less channel, one for which the output signal is at most a delayed and scaled version of the input signal. Figure 3.24 identifies the input and output parameters of a linear filter.

In order for a wave shape to pass through a filter without distortion, often referred to as distortion-less transmission, we require the relationship of Equation (3.24) to be valid for all $x(t)$ with bandwidth less than the filter's bandwidth.

$$y(t) = A x(t - \tau). \tag{3.24}$$

Equation (3.25) is the Fourier transform of the two sides of Equation (3.24).

$$Y(\omega) = A X(\omega) e^{-j\omega\tau} = X(\omega) A e^{-j\omega\tau}. \tag{3.25}$$

Since the output transform $Y(\omega)$ is the product of the input transform $X(\omega)$ and the filter transform $H(\omega)$, we conclude that $H(\omega)$, the distortion less filter, satisfies the following equation

$$H(\omega) = A e^{-j\omega\tau} \tag{3.26}$$

We recognize that the distortion-less filter must exhibit a constant, but otherwise arbitrary, nonzero amplitude gain A, and a phase shift proportional to frequency, often identified as linear phase shift, with the proportionality factor being the time delay. We usually emphasize the requirement for linear phase because linear phase is not an attribute of recursive analog filters and is purchased by the use of additional filters known as phase equalizer filters.

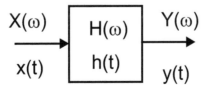

Figure 3.24   Input and output of a linear filter.

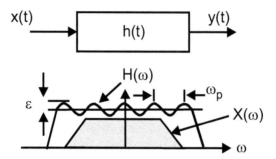

Figure 3.25   Frequency response of equiripple FIR filter and spectrum of input signal.

We know that linear phase shift, a property equivalent to pure time delay, can never be achieved exactly with lumped linear circuit components but can be achieved with distributed components which form transmission lines that respond with solutions to the wave equation. Analog phase equalizers in the analog domain are used to obtain equiripple approximations to linear phase slope.

The attraction, and an often-cited advantage, of nonrecursive filters is the ease with which they can achieve linear phase shift. To achieve linear phase in an FIR filter, its impulse response must exhibit symmetry with respect to its center point. We thus find that linear phase shift, a difficult attribute to achieve in the analog domain, is essentially free in the sampled data domain. For reasons that escape us, additional discussion of distortion effects seems to stop here as if access to linear phase has solved the problem. This is a bit premature since we still have to address the effect of equiripple deviation from constant amplitude gain as well as the effect of the equiripple deviation from uniform phase shift of the phase equalized recursive filter.

Figure 3.25 presents the frequency response of an equiripple FIR filter and the spectrum of an input signal with bandwidth completely contained within the filter bandwidth. Here the amplitude response is modeled as a nominal gain of unity with a cosine ripple of amplitude ε and of period $\omega_P = 2\pi/T_P$. The filter also has a uniform group delay of T_D seconds to reflect the causality of its impulse response. The spectral response of the filter is described in Equation (3.27). Note that the output spectrum is composed of two components, one due to the nominal unity gain with its linear phase shift and one due to the cosine ripple in addition to its linear phase shift. We partition the cosine modulation into a pair of complex

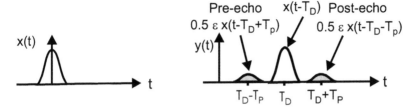

Figure 3.26 Time signals at input and output of filter with equiripple spectral response.

exponential terms and combine these terms with the linear group delay phase term to obtain a total of three spectral components observed at the filter output.

$$
\begin{aligned}
Y(\omega) &= X(\omega) \cdot H(\omega) \\
&= X(\omega) \cdot [1 + \varepsilon \cos(\omega T_P)] \, e^{-j\omega T_D} \\
&= X(\omega) e^{-j\omega T_D} + \varepsilon X(\omega) \cos(\omega T_P) e^{-j\omega T_D} \\
&= X(\omega) e^{-j\omega T_D} + \frac{\varepsilon}{2} X(\omega) \cos(\omega T_P) e^{-j\omega(T_D - T_P)} \\
&\quad + \frac{\varepsilon}{2} X(\omega) \cos(\omega T_P) e^{-j\omega(T_D + T_P)}
\end{aligned}
\tag{3.27}
$$

$$
y(t) = x(t - T_D) + \frac{\varepsilon}{2} x(t - (T_D - T_P)) + \frac{\varepsilon}{2} x(t - (T_D + T_P)). \tag{3.28}
$$

When we interpret the time domain response from the spectral description of the output, we find three distinct time response contributions. The major component of the output signal is the time-delayed version of the input signal denoted by $x(t - T_D)$. The remaining two components are a pair of scaled and translated versions of the input signal. These components are called *paired echoes*. A pre-echo and a post-echo, each of amplitude $\varepsilon/2$, form the paired echoes residing on each side of the primary response time and translated by the reciprocal period of the filter ripple frequency. The structure of the paired echoes is shown in Equation (3.28) and in Figure 3.26. Higher frequency (i.e. shorter period) spectral ripple causes larger amounts of echo time offset. Similar paired echo responses can be derived for phase ripple, except that phase ripple exhibits odd symmetry from which we establish that the echoes are also odd symmetric about the main response.

Note the dual relationship: when we multiply a time function by a time domain cosine wave, its transform, a spectrum, is scaled and translated in the

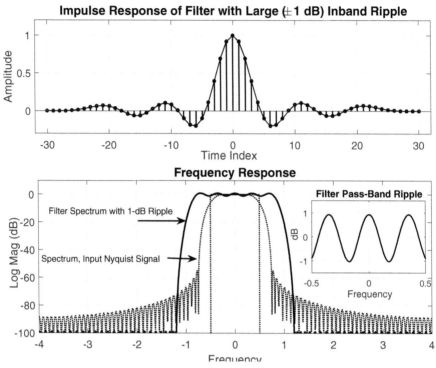

Figure 3.27   Impulse and frequency response of a FIR filter. Details of pass band show ± 1 dB in-band ripple. Spectrum of input signal is fully contained in filter pass band.

frequency domain by the reciprocal period of the temporal cosine. Similarly, when we multiply a frequency function by a frequency domain cosine, its transform, a time signal, is scaled and translated in the time domain by the reciprocal period of the spectral cosine.

While on the topic of dual relationships, we recognize that time domain echoes cause a periodic ripple in the frequency response of a channel often modeled as multipath related frequency selective fading. Hence, in retrospect, it is not surprising that filters exhibiting periodic frequency domain ripple are characterized by time domain echo structures. We may recall from our first course in transmission lines that the time domain interaction between the incident and reflected sinusoid due to a reflection causes periodic frequency-dependent constructive and destructive cancelation. We also note that time domain reflectometers (TDRs) use this coupling between echo time-response and spectral ripple to extract time

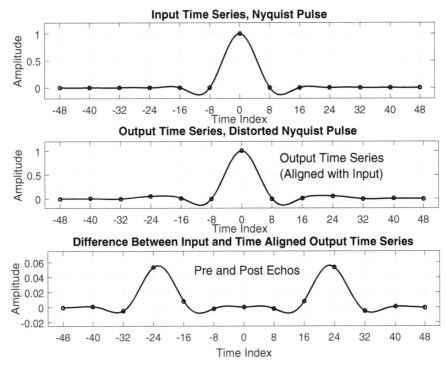

Figure 3.28 Time-aligned input and output signal waveforms and difference of two waveforms to show paired echoes.

position and amplitude information from frequency domain measurement of a transmission line.

We now know why the amplitude of the filter pass-band ripple is of concern to us. When used in a communication system, the ripple is a source of distortion called inter-symbol interference (ISI). We use the next figure to demonstrate how a filter in a receiver signal path generates ISI. Figure 3.27 shows the impulse response and frequency response of an FIR filter that processes a selected input signal. The spectral response of the input signal is overlaid on the filter response, and as we see, the bandwidth of the input signal is fully contained in the filter bandwidth.

We also see a detail of the filter pass-band region where we see that the filter exhibits 1 dB ripple. One-decibel ripple is approximately 12% that presents amplitude of 0.12 ripple to the input spectrum. We also see that the filter contains nearly three cycles of ripple in the bandwidth of the input signal. The importance of the number of ripple cycles in the bandwidth of the input signal is important, and we will explain why in a moment. Figure 3.28 shows the input signal from which the input spectrum of the previous figure was formed, and the time-aligned output obtained from the filter output. Also shown is the difference between the input and the time-aligned output. This difference clearly shows the pre- and post-echoes due to the filter in-band ripple. These echoes are seen to be located at three sample intervals on either side of the main response with three cycles per signal bandwidth being the frequency of the filter ripple.

Incidentally, pre- and post-echoes are formed by ripple in the phase response as well as by ripple in the magnitude response. The subtle difference is that magnitude response ripple results in an even symmetric echo pair, while phase response ripple results in an odd symmetric echo pair. This is, of course, related to the even and odd symmetry of the magnitude and phase characteristics of a filter. Recursive filters, with their nonuniform phase response, are the contributors of the odd symmetric echoes. To illustrate this, we examine Figure 3.29, which in the left-hand column presents the impulse response, frequency response, and the in-band group delay of an eighth-order elliptic filter. The right-hand column presents the same curves for the elliptic filter in cascade with an eighth-order phase equalizer. Note the symmetric impulse response and the equal-ripple group delay response of the equalized filter. Figure 3.30 presents the input and time-aligned output time series from the elliptic filter. Also seen is the difference between the two series in which we see the odd symmetric echoes caused by the filter's group delay.

Figure 3.31 presents the input and time-aligned series from the phase equalized elliptic filter. The final subplot of this figure shows the difference between the input and time-aligned output and the odd symmetric echo pair is clearly seen. We note that the IIR echo components are not exactly odd symmetric and we attribute this to even symmetric echo components due to the amplitude ripple response of the elliptic filter which also contributes echoes to the composite response.

We can now compare the filter responses we examined to illustrate the pre- and post-echo distortion. As mentioned earlier, the input signal was a Nyquist pulse, oversampled by a factor of 8, so that the pulse has exactly eight samples per symbol. Every eighth sample of this input pulse coincides with an expected zero crossing. The distance between the pulse peak and the zero crossing is the symbol time for this pulse. Symbols or waveforms separated by exactly this interval are orthogonal and do not interact. When signaling with the Nyquist pulse, the receiver can collect and measure each

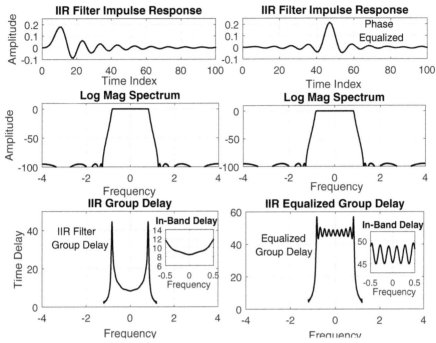

Figure 3.29 Impulse, magnitude response, and group delay response of an IIR filter and of a phase equalized version of same filter. Details of equalized phase show ± 2.2 degree in-band ripple.

waveform independently and is thus able to communicate through band-limited channels without ISI. When we illustrated the echoes with the FIR filter, the input and output were time aligned to extract the added echoes. The filter ripple parameters of 0.12 amplitude ripple with frequency of three cycles per input bandwidth told us to expect a pair of echoes of amplitude 0.06 separated by three symbol durations from the primary time response. To verify this, the bottom graph in Figure 3.29 was formed as the difference between the time-aligned input and output waveforms. Here we can verify that the amplitude and location of the paired echoes closely match the values predicted from the parameters in the ripple. The offsets tell us that the period of the frequency domain ripple was slightly less than three cycles per signal bandwidth. Similarly, the ripple parameters of the IIR filter, the $2.2 * 2\pi/360$ peak phase ripple with frequency of five cycles per input bandwidth, told us that we should expect a pair of echoes of amplitude 0.019 separated by five

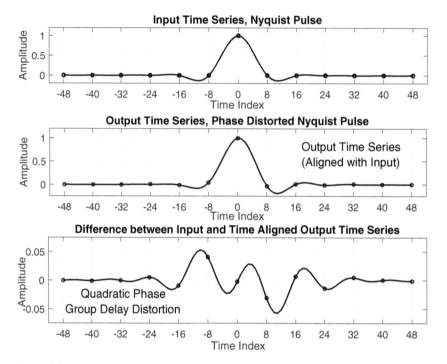

Figure 3.30   Time-aligned input and output waveforms from nonuniform phase IIR filter with the difference of two waveforms to show quadratic phase distortion.

symbol durations from the primary time response. This echo pair is seen in the third subplot of Figure 3.31.

A receiver attributes the ISI caused by the filters in the signal flow path to channel distortion, and if the receiver contains an equalizer, it will attempt to remove the distortion by forming the opposing pass-band gain and phase ripple. There are receivers that may not include an equalizer for which the filter-induced ISI would likely cause performance degradation. Examples include digital signal processing based processing of video signals as part of source coding, signal reconstruction, and simple signal processing of National Television Standards Committee (NTSC) or Phase Alternating Line (PAL) composite video signals. Radar signal processing for imaging radars suffers the same degradation due to inadequate attention to paired echoes from equiripple digital FIR filters as well as equiripple analog Tchebyschev filters in the signal-processing path.

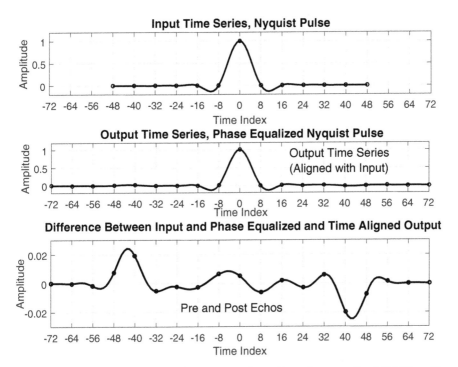

Figure 3.31 Time-aligned input and output waveforms from phase equalized IIR filter with the difference of two waveforms to show paired echoes.

A properly designed system would have part of its implementation loss budget assigned to the pass-band ripple and then the specifications would have to convert budget into acceptable ripple level. We commented earlier that 1 dB is 12% ripple from which it follows that 0.1 dB is 1.2% and if a system has budgeted 0.1 dB to filter losses, the composite filter chain must exhibit less than 1.2% ripple. We often design systems with 0.1 dB in-band ripple, which, except for the most stringent specifications, satisfies most system requirements. Reasonable specifications for an FIR filter might have in-band ripple of 1-part-in-100 and out-of-band ripple of 1-part-in-1000 or 1-part-in-10,000. This is the justification for the use of the Remez algorithm and its variants that modify the out-of-band ripple slope.

Figure 3.32 is a block diagram of an end-to-end modulator and demodulator that will be used to illustrate the effect of in-band ripple. Here the output of the shaping filter is fed directly to the input of the matched filter without the intermediate channel that would add noise and channel

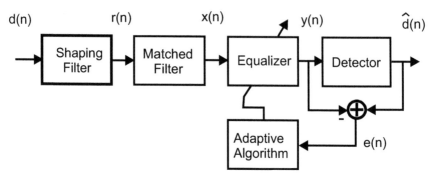

Figure 3.32   Block diagram of 16-QAM modulator, demodulator chain used to illustrate ISI due to in-band ripple of shaping filter and matched filter.

distortion. The output of the matched filter is passed to the equalizer that would normally remove the distortion caused by the channel. The equalizer delivers its processed signal to a detector that forms an estimate of the data. This estimate is compared with the input to the detector and differences between the detector input and output are attributed to noise and channel distortion. The error, along with the data, directs the adaptive algorithm to adjust the equalizer weights in the direction that results in a reduction in the average detector error.

The upper left subplot of Figure 3.33 presents the 16-QAM constellation set observed at the matched filter output. The constellation points correspond to the range of amplitudes, four for each of the in-phase and quadrature-phase components of the received and processed waveform. In the absence of noise and distortion, these amplitudes are $\pm 1/3$ and ± 1.0. We see a small variance cloud centered around the constellation points. This cloud is the observed effect of the ISI caused by in-band ripple of the shaping filter and matched filter. The upper right subplot presents the log-magnitude of the energy in the error sequence as the adaptive equalizer whitens the error sequence by acquiring the inverse of the distortion process. The subplot in the lower left shows the constellation diagram obtained at the equalizer output after convergence. Note the reduction in the variance cloud due to the ISI cancelation by the equalizer. Finally, the lower right subplot presents a close-up of the spectral ripple exhibited by the cascade of the shaping filter and matching filter along with the frequency response of the adaptive equalizer. Note that the spectral gain of the equalizer is the inverse of the spectral gain of the filter pair.

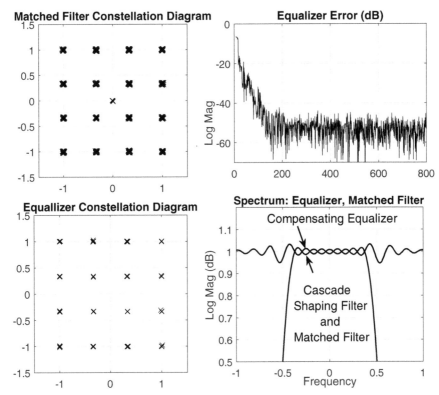

Figure 3.33 Constellation at output of matched filter, error profile of converging equalizer, constellation at output of equalizer, and spectra of shaping filter and compensating equalizing.

References

harris, Fred. "DATA WINDOWS: Finite Aperture Effects and Applications in Signal Processing," Encyclopedia of Electrical Engineering, John Wiley & Sons, Editor, John Webster 1999

harris, Fred. "On the Use of Windows for Harmonic Analysis with the Discrete Fourier Transform," *Proceedings of the IEEE*, Vol. 88, No. 1, January 1978, pp 51-83.

Herrmann, O., L.R. Rabiner, and D.S.K. Chan. "Practical Design Rules for Optimum Finite Impulse Response Low-pass Digital Filters," Bell Syst. Tech. J., 52, 769-799, 1973.

Kaiser, Jim. "Nonrecursive Digital Filter Design Using I_0-sinh Window Function," *Proceedings of the IEEE International Symposium on Circuits and Systems (ISCAS'74)*, pp. 20-23, April 1974

McClellan, J. H., T.W Parks, and L.R. Rabiner. "FIR Linear Phase Filter Design Program", Programs for Digital Signal Processing, Chapter 5.1 pp 5.1-1-5.1-13, IEEE Press 1979.

Mitra, Sanjit, and James Kaiser, *Handbook for Digital Signal Processing*, New York, John Wiley & Sons, 1993.

Mitra, Sanjit, *Digital Signal Processing: A Computer-Based Approach*, 2nd ed, New York, McGraw- Hill,, 2001.

Problems

3.1 First examine the MATLAB sinc function by typing help in the command window. Now examine the call sinc(−4:0.1:4) and describe how the sinc call responds to the arguments. Plot the subplots for the stem of the sequence and the log-magnitude spectrum. The following MATLAB script accomplishes this:

```
subplot(2,1,1);
dat=sinc(-4:0.1:4);
stem(-4:0.1:4,dat); grid subplot(2,1,2);
ff=(-0.5:1/1024:0.5-1/1024)*10;
f_dat=fft(dat/sum(dat),1024);
plot(ff, fftshift(20*log10(abs(f_dat)))) grid
axis([-5 5 –60 10])
```

Examine the time series and the spectrum and then describe what will change if the *dat* line is changed to sinc(-8:0.1:8), and then again if it changed to sinc(-4:0.02:4)? How should the scale factor on the *ff* line change for the two new sinc options? Why in the *f_dat* line is *dat* divided by *sum(dat)*? Replace *f_dat* with *f_dat=fft(dat,1024)* and examine its spectral plot.

3.2 The discrete time Fourier (TFF) series expansion of an N point sampled rectangle is shown next.

$$H(\theta) = \exp\left(-j\frac{N-1}{2}\theta\right)\frac{\sin\left(\frac{N}{2}\theta\right)}{\sin\left(\frac{1}{2}\theta\right)}$$

Determine the locations of this function's zero crossings.

Determine the amplitude of this function at $\theta = 0$.

Determine the value of this function at $\theta = \pi$.

Determine the location and amplitude of this function's first side lobe.

Is the ratio of the peak value to the first side-lobe level of this function dependent on N? Show this!

3.3 Form and plot the magnitude of the 200-point fast Fourier transform (FFT) for a 20-point rectangle sequence (ones (1,20)), and on the same plot, plot the magnitude of the 200-point FFT of the 20-point sequence exp(j*2*pi*(0:19)/20). What can you say about the location of each transform and the zeros of the two transforms?

3.4 Form and plot the magnitude of the 200-point FFT for a 20-point sequence exp(j*2*pi*(0:19)/20) and on the same plot, plot the magnitude of the 200-point FFT of the 20-point sequence exp(-j*2*pi*(0:19)/20). What can you say about the location of each transform and the zeros of the two transforms?

3.5 Form and plot the magnitude of the 200-point FFT for a 20-point rectangle sequence (ones (1,20)), and on the same plot, plot the magnitude of the 200-point FFT of the 20-point sequence exp(j*2*pi*(0:19)/20). What can you say about the location of each transform and the zeros of the two transforms?

3.6 Using the log-magnitude spectrum of the 512-point FFT of 51-point sequences, determine the main-lobe widths and maximum side-lobe levels of the following traditional window functions. Rectangle, Hann, Hamming, Blackman–harris 3-term, Blackman–harris 4-term, and Kaiser with parameter 10.

3.7 Form a plot showing the progression of main-lobe width and side-lobe levels (in dB) for the Kaiser window over a range of the parameter β equal to the integer values $1-10$.

3.8 The expression defining the Fourier transform of the Kaiser window is listed in Equation (3.13). Determine and form a plot of the first 10 zeros of this spectrum for a range of parameter β equal to the integers 0–10.

3.9 Form plots of log-magnitude spectrum of 512-point FFT of 51-point rectangle windowed by Kaiser window for range of parameter values $0-10$.

3.10 Use the Remez algorithm to design a low-pass FIR filter satisfying the following specifications:

F_s: 20 kHz
Pass Band: 0–3 kHz,   In-Band Ripple: 0.1 dB
Stop Band: 5–10 kHz;   Stop Band: 60 dB

Estimate the length of the filter using Figure 3.15, then verify estimate with MATLAB *remezord*.

Verify the results of the Remez design by examining in-band ripple and out-of-band attenuation levels. Change filter length and weighting vector as appropriate to achieve filter design specifications.

3.11 Design a low-pass filter with the Remez algorithm to meet the following three specifications:

	Filter 1	Filter 2	Filter 3
F_s	20 kHz	20 kHz	20 kHz
Pass Band	0–3 kHz	0–4 kHz	0–5 kHz
Stop Band	5–10 kHz	7–10 kHz	8–10 kHz
Pass-Band Ripple	0.1 dB	0.1 dB	0.1 dB
Stop Band	60 dB	60 dB	60 dB

The three filters have the same specifications, including transition bandwidth, but different pass-band edges: using six subplots, plot the impulse response and frequency response of the three filters.

Comment on how transition bandwidth affects filter length and on how pass-band bandwidth affects the length and shape of filter impulse response.

3.12 Repeat problem 3.11 but change stop-band attenuation from 60 to 80 dB.

Comment on how increased out-of band attenuation affects length and shape of filter impulse response.

3.13 Repeat problem 3.11 but change in-band ripple from 0.1 to 0.01.

Comment on how decreased in-band ripple affects length and shape of filter impulse response.

3.14 A low-pass FIR filter is to be designed with the following specifications:

F_s 100 kHz
Pass Band   0–20 kHz,   In-band Ripple: 0.1 dB
Stop Band   25–50 kHz;   Stop Band: 60 dB

a) Use the standard MATLAB remez algorithm with weights appropriate to obtain the 0.1 dB in-band ripple and the 60 dB out-of-band attenuation.

b) Use the MATLAB remez algorithm with the modified *myfrf* script file to obtain sloping side lobes.

c) Plot and compare the pass-band ripple, and the in-band ripple of the two designs.

3.15 A low-pass FIR filter is to be designed with the following specifications:

F_s: 100 kHz
Pass Band   0-to-25 kHz,   In-Band Ripple: 0.1 dB
Stop Band   35-to-50 kHz   Stop Band: 60 dB

a) Use the MATLAB remez algorithm with the modified *myfrf* script file to obtain a design with sloping side lobes.

b) Use a Kaiser window for a windowed design to meet the same filter specifications.

c) Plot and compare the impulse response and the log-magnitude spectrum of the pass-band ripple and the in-band ripple of the two designs.

3.16 Use the Remez algorithm to design an equiripple low-pass filter with the following specifications:

F_s 100 kHz
Pass Band   0–10 kHz,   In-Band Ripple: 0.1 dB
Stop Band   15–50 kHz;   Stop Band: 60 dB

Use the following MATLAB script to simulate quantizing coefficients to *b*-bits:

    hh=remez(N,[0 10 15 50]/50,[1 1 0 0],[1 10]);
    hh_q=floor(2^(b-1)*hh)/2^(b-1);

Plot a sequence of spectra corresponding to a range of quantization bit levels of 8, 10, 12, and 14 bits: how many bits are required to maintain 60 dB side lobes?

3.17 Use the Remez algorithm and the *myfrf* script file to design a –6 dB/octave stop-band low-pass filter with the following specifications:

F_s 100 kHz
Pass Band   0–10 kHz,   In-Band Ripple: 0.1 dB
Stop Band   15–50 kHz;   Stop Band: 60 dB

Use the following MATLAB script to simulate quantizing coefficients to *b*-bits:

    hh=remez(N,[0 10 15 50]/50,{'myfrf',[1 1 0 0]},[1 10])
    hh_q=floor(2^(b-1)*hh)/2^(b-1)

Plot a sequence of spectra corresponding to a range of quantization bit levels of 8, 10, 12, and 14 bits: how many bits are required to maintain 60 dB side lobes?

3.18 Repeat problem 3.16 but scale the impulse response by its maximum value prior to quantization.

The following MATLAB script can accomplish this:

    hh=remez(N,[0 10 15 50]/50,[1 1 0 0],[1 10]);
    hh_scl=hh/max(hh);
    hh_scl_q=floor(2^(b-1)*hh_scl)/2^(b-1);

3.19 Repeat problem 3.17 but scale the impulse response by its maximum value prior to quantization.

The following MATLAB script can accomplish this:
hh=remez(N,[0 10 15 50]/50,{'myfrf',[1 1 0 0]},[1 10]);
hh_scl=hh/max(hh); hh_scl_q=floor(2^(b-1)*hh_scl)/2^(b-1);

3.20 Use the Remez algorithm to implement the impulse response of the following two filters:

    hh1=remez(55,[0 10 11.5 40]/40,[1 1 0 0], [1 100]);
    hh2=remez(15,[0 1 10 40]/40,[1 1 0 0]);

Use subplots to show the impulse response and the frequency response of the two filters. Note and comment on the in-band ripple of the hh1 filter.

Now pass the time series hh2 through the filter hh1 and plot the response. Relate the position, amplitude, and sign of the pre- and post-echoes to the frequency, amplitude, and sign of the in-band ripple of filter hh1.

3.21 Repeat problem 3.20 except change the two filters to match the following: hh1=remez(65,[0 10 11.5 40]/40,[1 1 0 0], [1 100]); hh2=remez(15,[0 1 10 40]/40,[1 1 0 0]);

4

Useful Classes of Filters

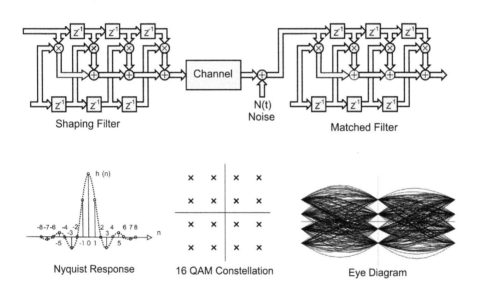

Shaping Filter

Channel

N(t) Noise

Matched Filter

Nyquist Response

16 QAM Constellation

Eye Diagram

Filters come in all flavors and sizes. We generally describe them with broad stroke coverage such as low pass, high pass, band pass, and the like. We then apply secondary qualifiers such as recursive and nonrecursive or nonlinear phase and linear phase. Certain subclasses of digital filters appear so often in systems that we hardly apply the qualifiers because we know the filter structure by where it resides in the system. One example of such a filter is the ubiquitous filter that shapes the spectrum in modems designed to operate over band-limited channels without inter symbol interference (ISI). This shaping filter is the cosine-tapered square root Nyquist filter compactly described by the term *SQRT Nyquist filter* or simply the *SQRT* filter. Every book in communication theory describes the properties of the continuous versions of this family of filters while a number of digital signal processing

(DSP) books describe the coefficients of the sampled data version of this same filter. Every designer who has worked on a cable modem or satellite modem has brushed against the same SQRT Nyquist filter. In this chapter, we look very carefully at the digital versions of the SQRT Nyquist filter from the viewpoint of a filter and then as a system component. Another filter that we see very often in systems is the half-band filter. The half band appears in many variants, in particular as the quadrature mirror filter and the Hilbert transform filter. Both of these filters are common building blocks in multirate systems and warrant our careful attention and understanding. Another common workhorse, the Cascade Integrator Comb (CIC), appears in its own chapter.

4.1 Nyquist Filter and Square-Root Nyquist Filter

The Nyquist pulse is the wave shape required to communicate over band-limited channels with no ISI. In many communication systems, the waveform delivered to the receiver's detector is the sum of scaled and offset waveforms as shown in the following equation:

$$s(t) = \sum_n d(n)h(t - nT). \tag{4.1}$$

The amplitude terms $d(n)$ are selected from a small finite alphabet such as $\{-1, +1\}$ or $\{-1, -1/3, +1/3, +1\}$ in accordance with a specified mapping scheme between input bits and output levels. The signal $s(t)$ is sampled at equally spaced time increments identified by a timing recovery process in the receiver to obtain output samples as shown in the following equation:

$$s(mT) = \sum_n d(n) \cdot h(mT - nT). \tag{4.2}$$

We can partition this sum as shown in Equation (4.3) to emphasize the desired and the undesired components of the measurement. Here the desired component is $d(m)$ and the undesired component is the remainder of the sum which, if nonzero, is the ISI.

$$S(MT) = d(m)h(0) + \sum_{n \neq m} d(n)h(m - n). \tag{4.3}$$

In order to have zero ISI, the wave shape $h(t)$ must satisfy the specifications identifying sample values at the equally spaced sample

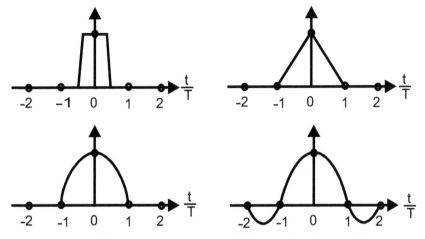

Figure 4.1 Various wave shapes satisfying condition for zero ISI.

increments $h(nT)$ shown in Equation (4.4). This relationship is known as the Nyquist pulse criterion for zero ISI.

$$h(nT) = \begin{cases} 0: & n \neq 0 \\ 1: & n = 0 \end{cases} \qquad (4.4)$$

There are an infinite number of functions that satisfy this set of restrictions. Examples of wave shapes that exhibit equally spaced zeros are shown in Figure 4.1. Note here that one wave shape has duration of exactly one symbol, two of the wave shapes have durations of exactly two symbols, and one wave shape has duration of four symbols. We restrict the range of possible wave shapes by considering spectral characteristics as well as time domain characteristics. We start by identifying the wave shape with minimum bandwidth. This wave shape is the ubiquitous sin(x)/x in the following equation:

$$h(t) = \frac{\sin\left(2\pi \cdot t \frac{1}{2T}\right)}{\left(2\pi \cdot t \frac{1}{2T}\right)}. \qquad (4.5)$$

The sin(x)/x wave shape is zero at every sample time, $t = nT$ except for $t = 0$, and is nonzero elsewhere. This pulse is variously known as the cardinal pulse when used for band limited interpolation and the Nyquist pulse when used in pulse shaping. The transform of this wave shape is the unit area

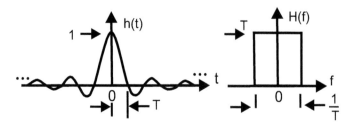

Figure 4.2   Sin(*x*)/*x* waveform and uniform pass-band spectrum.

rectangle with spectral support 1/*T* Hz. A segment of the sin(*x*)/*x* waveform and its Fourier transform is shown in Figure 4.2.

The problem with the sin(*x*)/*x* waveform is that it is noncausal and further resides on an infinite support. If the pulse resided on a finite support, we could delay the response sufficiently for the response to be causal. We have to form finite support approximations to the Nyquist pulse. Our first approximation to this pulse is obtained by convolving the rectangular spectrum $H(f)$ with an even symmetric, continuous spectrum $W(f)$ with finite support α/T. The convolution between $H(f)$ and $W(f)$ in the frequency domain is equivalent to a product in the time domain between the $h(t)$ and $w(t)$, where $w(t)$ is the inverse transform of $W(f)$. The spectral convolution and time product is shown in Figure 4.3 where we see that the effect of the spectral convolution is to increase the two-sided bandwidth from 1/*T* to $(1 + \alpha)/T$. The excess bandwidth α/T is the cost we incur to form filters on finite support. The term α is called the roll-off factor and is typically on the order of 0.1–0.5 with many systems using values of $\alpha = 0.2$. The transition bandwidth caused by the convolution is seen to exhibit odd symmetry about the half amplitude point of the original rectangular spectrum. This is a desired consequence of requiring even symmetry for the convolving spectral mass function. When the windowed signal is sampled at the symbol rate 1/*T* Hz, the spectral component residing beyond the 1/*T* bandwidth folds about the frequency $\pm 1/2T$ into the original bandwidth. This folded spectral component supplies the additional amplitude required to bring the spectrum to the constant amplitude of $H(f)$.

We also note that the significant amplitude of the windowed wave shape is confined to an interval of approximate width $4T/\alpha$ so that a filter with $\alpha = 0.2$ spans approximately 20*T* or 20 symbol durations. We can elect to simply truncate the windowed impulse response to obtain a finite support filter, and often choose the truncation points at $\pm 2T/\alpha$. A second window, a rectangle,

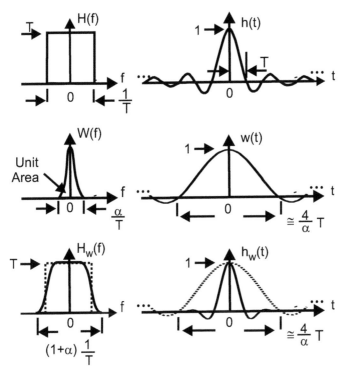

Figure 4.3 Spectral convolution of prototype Nyquist spectrum with even symmetric spectral mass, and equivalent time domain product of Nyquist pulse with window.

performs this truncation. The result of this second windowing operation is a second spectral convolution with its transform. This second convolution induces pass-band ripple and out-of-band side lobes in the spectrum of the finite support Nyquist filter. Before performing this truncation, we first address one additional aspect in the design of the Nyquist pulse, which is how the pulse is actually used in a transmitter–receiver pair.

4.2 The Communication Path

A communication system can be modeled most simply by the signal flow shown in Figure 4.4. Here, $d(n)$ represents the sequence of symbol amplitudes presented at symbol rate to the shaping filter $h_1(t)$.

The superposition of scaled and translated versions of $h_1(t)$ formed by the shaping filter combine to form the transmitter signal $s(t)$ described in

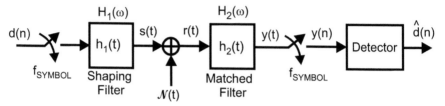

Figure 4.4   Simple model of signal flow in communication.

Equation (4.6). The channel adds noise to the transmitted signal to form the received signal $r(t)$ as shown in Equation (4.7). The received signal $r(t)$ is processed in the receiver filter $h_2(t)$ to reduce the contribution of the channel noise at $y(t)$, the output of the filter. The receiver filter output is shown in Equation (4.8) which we see is a sum of scaled and translated versions of $g(t)$, the combined impulse response of the transmitter filter and receiver filter plus filtered noise $N_2(t)$. The filter output $y(t)$ is sampled at the symbol rate and time offset to obtain $y(n)$ as shown in Equation (4.9). The sampled output contains three terms, the first proportional to the desired input sample $d(m)$, the second being a sample of filtered noise, and the third containing a weighted sum of earlier and later input samples $d(m - n)$. This last term is the combined ISI due to the memory of the shaping filter $h_1(t)$ and the receiver filter $h_2(t)$.

$$s(t) = \sum_m d(m) \cdot h_1(t - mT) \tag{4.6}$$

$$r(t) = s(t) + \mathcal{N}(t)$$
$$= \sum_m d(m) \cdot h_1(t - mT) + \mathcal{N}(t) \tag{4.7}$$

$$y(t) = r(t) * h_2(t)$$
$$= \int r(t - \tau) \cdot h_2(\tau) d\tau$$
$$= \int \sum_m d(m) h_1(t - mT - \tau) h_2(\tau) d\tau$$
$$+ \int \mathcal{N}(t - \tau) h_2(\tau) d\tau \tag{4.8}$$
$$= \sum_m d(m) g[(n - m)T] + N(nT)$$

where $g(t) = \int h_1(t - \tau)h_2(\tau)d\tau$

and $\quad N(t) = \int \mathcal{N}(t - \tau)h_2(\tau)d\tau$

$$y(nT) = \sum_{,m} d(m)g[(n - m)T] + N(nT)$$

$$= d(n)g(0) + N(nT) + \sum_{m \neq n} d(m)g[(n - m)T].$$

(4.9)

We can obtain from the terms in Equation (4.9) an unbiased estimate of $d(n)$ if $g(0)$ is 1, and zero ISI if $g([n - m]T)$ is 0 for all m not equal to n. This is the same requirement presented in Equation (4.4) as the requirement for a zero ISI filter. Thus, the convolution of the shaping filter at the transmitter and the noise control filter at the receiver filter must form the Nyquist filter as shown in the following equation:

$$h_1(t) * h_2(t) = h_{\text{NYQ}}(t - T_D)$$
$$H_1(\omega) \cdot H_2(\omega) = H_{\text{NYQ}}(\omega) \cdot e^{-j\omega T_D}.$$

(4.10)

To maximize the signal-to-noise ratio (SNR) in Equation (4.10), the receiver filter must be matched to the transmitter-shaping filter. The matched filter is a time-reversed and delayed version of the shaping filter, which is described in the frequency domain as shown in the following equation:

$$H_2(\omega) = H_1^*(\omega) \cdot e^{-j\omega T_D}.$$

(4.11)

Combining the requirements in Equations (4.10) and (4.11), we obtain the result in Equation (4.12) from which we determine the relationship between the shaping filter and the desired Nyquist filter response shown in Equation (4.13). The shaping filter is called a SQRT Nyquist filter.

$$H_1(\omega) \cdot H_2^*(\omega) \cdot e^{-j\omega T_D} = H_{\text{NYQ}}(\omega) \cdot e^{-j\omega T_D}$$

(4.12)

$$|H_1(\omega)|^2 = H_{\text{NYQ}}(\omega)$$

$$H_1(\omega) = \sqrt{H_{\text{NYQ}}(\omega)} = \text{SQRT}[H_{\text{NYQ}}(\omega)].$$

(4.13)

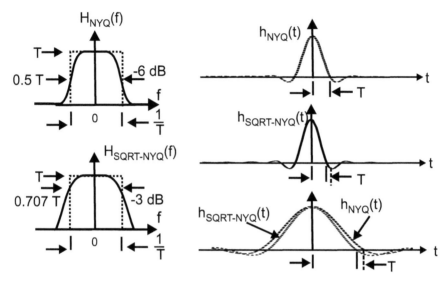

Figure 4.5   Spectrum and time response of the Nyquist filter and SQRT Nyquist filter.

Reviewing the result of this last derivation, we expect the shaping filter and the receiver filter to accomplish two tasks. The two filters in cascade form a Nyquist filter and further interact to maximize SNR. We say that the SQRT Nyquist filter performs half the spectral shaping at the transmitter and half the spectral shaping at the receiver.

We now examine the frequency response and the time domain response of the SQRT Nyquist filter. The square root operation applied to the magnitude spectrum of the Nyquist filter does not affect the zero-valued or the unit-valued segments of the filter response. Thus, the square root affects only the spectrum in the transition bandwidth. The Nyquist filter has a gain of 0.5 or –6.0 dB at the nominal band edge, while the square root filter has a gain of 0.707 or –3.0 dB at the same frequency. The shaping filter and the matched filter each applies 3.0 dB attenuation at the band edge to obtain the desired band-edge attenuation of 6.0 dB. Hence, the –6.0 dB bandwidth of the SQRT filter is wider than the –6.0 dB bandwidth of the Nyquist filter and the wider bandwidth square root Nyquist filter must have a narrower main lobe impulse response than the Nyquist filter. These relationships are illustrated in Figure 4.5.

We now address the taper in the excess bandwidth Nyquist pulse. As mentioned earlier, any even symmetric spectral mass can be used to perform the spectral convolution that forms the transition bandwidth of the tapered Nyquist spectrum. The most common spectral mass selected for communication systems is the half cosine of width $\alpha \cdot f_{\text{SYM}}$. The half cosine convolved with the spectral rectangle forms the spectrum known as the cosine-tapered Nyquist pulse with roll-off α. The description of this band-limited spectrum normalized to unity pass-band gain is presented in the following equation:

$$H_{\text{NYQ}}(\omega) = \begin{cases} 1 & : \quad \text{for } \dfrac{|\omega|}{\omega_{sym}} \leq (1-\alpha) \\ 0.5 \cdot \left\{ 1 + \text{COS} \left\{ \dfrac{\pi}{2\alpha} \left[\dfrac{\omega}{\omega_{sym}} - (1-\alpha) \right] \right\} \right\} & : \\ \quad \text{for } (1-\alpha) \leq \dfrac{|\omega|}{\omega_{sym}} \leq (1+\alpha) \\ 0 & : \quad \text{for } \dfrac{|\omega|}{\omega_{sym}} > (1+\alpha) \end{cases} . \quad (4.14)$$

The continuous time domain expression for the cosine-tapered Nyquist filter is shown in Equation (4.15). Here we see the windowing operation of the Nyquist pulse as a product with the window that is the transform of the half-cosine spectrum.

$$h_{\text{NYQ}}(t) = f_{\text{SYM}} \frac{\sin(\pi f_{\text{SYM}} t)}{(\pi f_{\text{SYM}} t)} \frac{\cos(\alpha \pi f_{\text{SYM}} t)}{[1 - (2\alpha f_{\text{SYM}} t)^2]}. \quad (4.15)$$

4.3 The Sampled Cosine Taper

Since the Nyquist filter is band limited, we can form the samples of a digital filter by sampling the impulse response of the continuous filter. Normally, this involves two operations. The first is a scaling factor applied to the impulse response by dividing by the sample rate, and the second is the sampling process in which we replace t with $n \cdot T_{\text{SMPL}}$ or n/f_{SMPL}. The sample rate must exceed the two-sided bandwidth of the filter that, due to the excess bandwidth, is wider than the symbol rate. It is standard to select the sample rate f_{SMPL} to be an integer multiple of the symbol rate f_{SYM} so that the filter operates at M samples per symbol. It is common to operate the filter at four or eight samples per symbol, but, for generality, we select $f_{\text{SMPL}} = M f_{\text{SYM}}$ so that $f_{\text{SYM}} t$ is replaced by $f_{\text{SYM}} n/(M \cdot f_{\text{SYM}})$ or n/M. After applying these operations to Equation (4.15), we obtain the results shown in the following equation:

$$h_{NYQ}(n) = f_{SYM} \frac{\sin\left(\dfrac{\pi}{M}n\right)}{\left(\dfrac{\pi}{M}n\right)} \frac{\cos\left(\alpha\dfrac{\pi}{M}n\right)}{\left[1 - \left(2\alpha\dfrac{1}{M}n\right)^2\right]}. \tag{4.16}$$

The filter described in Equation (4.16) has a two-sided bandwidth that is approximately 1/Mth of the sample rate. A digital filter exhibits a processing gain proportional to the ratio of input sample rate to output bandwidth; in this case, a factor of M. The 1/M scale factor in Equation (4.16) cancels this processing gain to obtain unity gain. When the filter is used for shaping and up-sampling, as it is at the transmitter, we remove the 1/M scale factor since we want the impulse response to have unity peak value rather than unity processing gain.

The square root of the cosine-tapered Nyquist filter results in a quarter cycle cosine-tapered filter. This description title is normally contracted to square root raised cosine or root raised cosine Nyquist filter. The description of this band-limited spectrum normalized to unity pass-band gain is shown in the following equation:

$$H_{SQRT\text{-}NYQ}(\omega) = \begin{cases} 1 & : \quad \text{for } \dfrac{|\omega|}{\omega_{sym}} < (1-\alpha) \\[2mm] 0.5 \cdot \left\{1 + \text{COS}\left\{\dfrac{\pi}{4\alpha}\left[\dfrac{\omega}{\omega_{sym}} - (1-\alpha)\right]\right\}\right\} : \\[2mm] \quad \text{for } (1-\alpha) \le \dfrac{|\omega|}{\omega_{sym}} < (1+\alpha) \\[2mm] 0 & : \quad \text{for } \dfrac{|\omega|}{\omega_{sym}} \ge (1+\alpha) \end{cases}. \tag{4.17}$$

The continuous time domain expression for the square root raised cosine Nyquist filter is shown in the following equation:

$$h_{SQRT\text{-}NYQ}(t)$$
$$= f_{SYM} \frac{(4\alpha f_{SYM}t)\cos[\pi(1+\alpha)f_{SYM}t] + \sin[\pi(1+\alpha)f_{SYM}t]}{[1 - (4\alpha f_{SYM}t)^2][\pi f_{SYM}t]}. \tag{4.18}$$

We perform the same scaling and sampling operation we performed in Equation (4.18) to obtain the sampled data version of the square root raised cosine Nyquist pulse shown in the following equation:

$$h_{\text{SQRT-NYQ}}(n) = \frac{1}{M} \frac{\left(4\alpha \frac{n}{M}\right) \cos\left[\frac{\pi}{M}(1+\alpha)n\right] + \sin\left[\frac{\pi}{M}(1+\alpha)n\right]}{\left[1 - \left(4\alpha \frac{1}{M}n\right)^2\right]\left[\pi \frac{n}{M}\right]}.$$

(4.19)

When the impulse response is used as an up-sampler and shaping filter at the modulator, Equation (4.19) must be rescaled. For this application, we want unity peak impulse response rather than unity processing gain. Multiplying by the term shown in Equation (4.20) scales the coefficients of Equation (4.19).

$$\text{Modulator Scale Factor} : \frac{M}{1 + \left(\frac{4}{\pi} - 1\right)\alpha}.$$

(4.20)

When the impulse response is used in a matched filter at the demodulator, Equation (4.19) must be rescaled again to account for the scaling applied at the transmitter. Multiplying by the term shown in Equation (4.21) scales the coefficients of Equation (4.19).

$$\text{Demodulator Scale Factor} : \quad 1 + \left(\frac{4}{\pi} - 1\right)\alpha.$$

(4.21)

4.3.1 Root Raised Cosine Side-Lobe Levels

We commented earlier that when we implement the SQRT Nyquist filter, we actually apply two windows; the first window is a smooth continuous function used to control the transition bandwidth and the second is a rectangle used to limit the impulse response to a finite duration. This second window forces side lobes in the spectrum of the SQRT Nyquist filter. These side lobes are quite high on the order of $24-46$ dB below the pass-band gain depending on roll-off factor and the length of the filter in a number of symbols.

The reason for the poor side-lobe response is the discontinuous first derivative at the boundary between the half-cosine transition edge and the start of the stop band. Consequently, the envelope of the time function falls off, as seen in Equation (4.18), as $1/t^2$ enabling a significant time

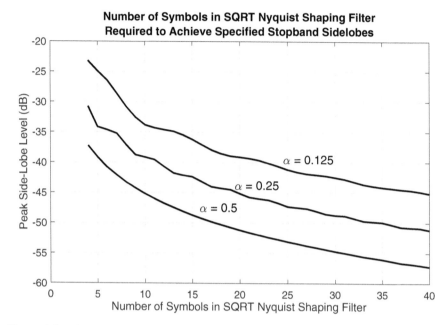

Figure 4.6   Highest out-of-band side-lobe levels for SQRT raised cosine Nyquist filter for a range of roll-off bandwidths and filter lengths.

discontinuity when the rectangle window is applied to the filter impulse response. In retrospect, the cosine-tapered Nyquist pulse was a poor choice for the shaping and matched filter in communication systems. Harry Nyquist off-handedly suggested that the half cycle cosine shape might make a good spectral taper to limit the filter's time duration. It was an unfortunate off-hand remark that has burdened our community with poor performing shaping filters. The half cycle cosine is not sufficiently smooth to enable good square root Nyquist filters. We will comment on the cosine taper's shortfalls and introduce two alternative options that do not exhibit the same shortfalls.

Figure 4.6 presents measured levels of side-lobe levels for a range of roll-off factors as a function of filter length in a number of symbols. We see that side-lobe levels fall very slowly with increased filter length and increases with reduced transition bandwidth. These levels of attenuation will not meet realistic spectral mask requirements for out-of-band attenuation that are typically on the order of 60–80 dB. Some mechanism must be invoked to control the filter out-of-band side-lobe levels related to the rectangle window. Whatever process is invoked should preserve or improve

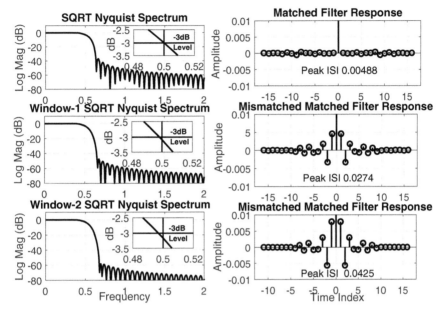

Figure 4.7 Spectrum of windowed SQRT raised cosine Nyquist filter, Kaiser(N,0) (-37 dB), Kaiser(N,2) (-50 dB), and Kaiser(N,3) (-60 dB), inserted details of spectrum near 3-dB bandwidth and details of ISI in matched filter outputs.

the ISI levels obtained by convolving the fixed length SQRT Nyquist filter with itself.

Attempting to control the spectral side lobes by applying a window other than the rectangle to the weights of the prototype SQRT Nyquist filter results in a significant increase in the ISI levels at the receiver output. This is illustrated in Figure 4.7, which illustrates the effect on spectral side lobes and ISI levels as a result of applying windows to the prototype impulse response. The increase in ISI is traced to the shift of the filter's 3-dB point away from the nominal band edge. The requirement for zero ISI at the output of the matched filter requires that the shaping and matched filters each exhibit 3-dB attenuation at the filter band edge, half the symbol rate. A design technique must control side-lobe levels while maintaining the 3-dB frequency at the symbol band edge.

4.3.2 Improving the Stop-Band Attenuation

The important spectral attributes of the SQRT Nyquist spectrum are the transition bandwidth or roll-off defined by α and the 3-dB attenuation at the band edge. Equally important considerations are the levels of spectral in-band ripple and out-of-band attenuation related to limiting the filter's time duration. There have been limited success efforts to improve these parameters while preserving the use of the cosine taper. The problem here is that the spectral cosine taper has a finite support which assures us that the time domain does not have a finite support and will require the second, rectangle window, to obtain finite durations. A more promising option is to abandon the cosine taper and pursue designs that start with a specified time domain length and apply techniques to form the desired width spectral transitions between pass band and stop band while controlling spectral side-lobe levels. A number of candidates' time domain widows immediately come to mind; these are the prolate spheroidal wave function, the closely related Kaiser Bessel window, and the minimum bandwidth Remez filter weights with selected spectral side-lobe decay rates. The Gaussian is not a candidate because it does not have a finite support. We implemented our designs with the MATLAB's Kaiser window.

The top subplot of Figure 4.8 presents the spectra of the cosine and harris tapers. Note that both tapers have the same bandwidth, and the harris taper rises gently and slowly toward its peak reflecting the fact that it has multiple (near) zero-valued derivatives at its boundaries. The center subplot shows the Nyquist filters formed by convolving the two spectral tapers with the unit width rectangle spectrum. Both filters exhibit odd symmetric transitions passing through the half amplitude level at the rectangles' discontinuity position. The bottom subplot shows the spectra of the SQRT Nyquist filter obtained by performing the square roots of the center sequences. Again we see that both transitions pass through the 0.707 level at the rectangle filter's boundary. What we miss seeing in this figure is the levels of in-band and out-of-band side-lobe levels.

The top pair of subplots in Figure 4.9 shows the impulse response of the cosine and harris tapered SQRT Nyquist filters. At this scale, they appear to be essentially the same filters. The center subplot shows spectra of the two filters and, in particular, their different side-lobe levels in the stop-band region beyond frequency 0.625. The cosine taper filter is -30 dB at this frequency, while the harris taper filter is -60 dB. The additional 30 dB attenuation for the same length filter is quite impressive. It will be much easier to meet

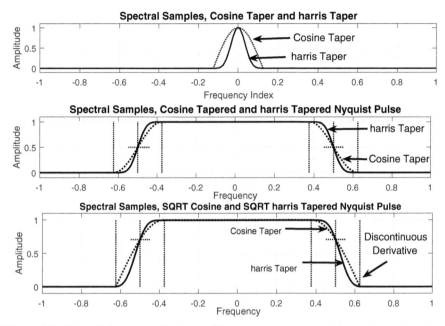

Figure 4.8 Top subplot: spectral cosine and harris tapers; center subplot: cosine and harris tapered Nyquist spectra; bottom subplot: cosine and harris tapered SQRT tapered spectra.

spectral band limitations with this additional attenuation. Finally, the bottom subplot shows the levels of in-band ripple for the two filters. Remember, in-band ripple is responsible for pre- and post-echoes, an objectionable signal degradation due to ISI. The ripple levels are seen to differ by an order of magnitude. The cosine taper level at its third side lobe is about 0.016 dB or a bit shy of 2 parts per thousand, while the harris taper level at the same frequency position is about 0.0016 dB or a bit shy of 0.2 parts per thousand. We would expect that the levels of pre- and post-echoes caused by in-band ripple are an order of magnitude smaller in the harris tapered filters than they are in the cosine-tapered filters.

Figure 4.10 presents the spectral side-lobe levels and the ISI time domain levels for both the cosine taper and the harris taper SQRT Nyquist filters, the responses obtained at the output of the cascade of shaping and matched filters. The top pair of subplots shows the in-band ripple levels of the two cascade filters. The shaping filters were implemented to operate at four samples per symbol. The sinc sequence sampled at four samples per symbol would have the amplitude of the maximum sample in its first side lobe of 0.2122. The

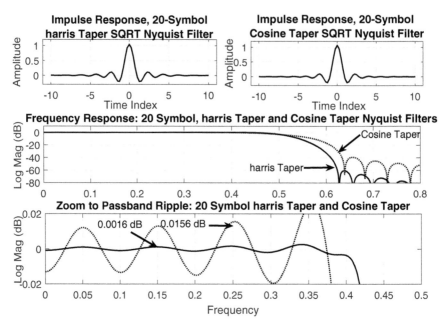

Figure 4.9 Top subplot: impulse responses of harris taper and cosine taper SQRT Nyquist shaping filters with 25% excess bandwidth; center subplot: spectra of shaping filters with same 25% transition bandwidth; bottom subplot: zoom to in-band ripple levels of the harris and cosine-tapered shaping filters.

measured value of the first side lobe in the cosine-tapered and harris tapered filters were 0.1857 and 0.2019. We would expect slightly lower values due to the equivalent time domain windows weighting of the underlying sinc sequence. Nothing unexpected yet! The lower subplots present a zoom to the neighborhood of the expected zero crossings of the same time series. In the ideal implementation of the windowed Nyquist filters, we would expect every fourth time sample to be zero-valued. We also expect that a subset of the zero-valued sample will not be zero but will be displaced by the additive pre- and post-echoes linked to the cosine shaped spectral in-band ripple of the shaping filters. When we plotted the time series, we placed "*circle*" markers on every fourth sample. Examining the cosine-tapered filter response, we see that the markers clearly show the pre- and post-echoes patterns centered at the offset indices ±10. The echo patterns span seven samples and have a peak amplitude of 0.0027. The RMS value of the set of echo samples was found to be 0.0044. The error vector

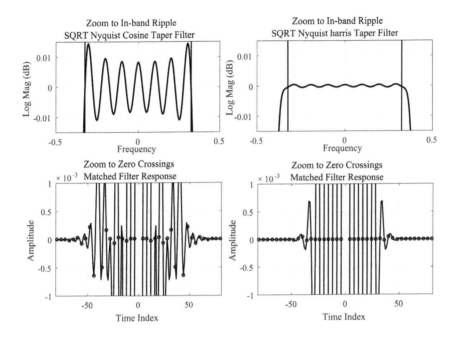

Figure 4.10 Frequency and time responses of cosine and harris tapered SQRT Nyquist filters. Top subplots: in-band ripple frequency responses. Bottom subplots: low-level ISI, deviations from nominal time response zero crossings.

magnitude (EVM) due to these echoes, for the unit length data samples, is a respectable −47 dB.

When we examine the harris tapered filter response, we see that the markers show reduced amplitude echo patterns centered at the offset indices ±12. These echo patterns also span seven samples and have a peak amplitude of 0.00013. This level is 1/20th of the cosine taper echo amplitude. The RMS value of the set of echo samples was found to be 0.00033. The EVM due to these echoes for the unit length data samples is a quite respectable −70 dB.

A second option that improves the shaping filter ISI is an iterative algorithm based on the Remez algorithm that will transform an initial low-pass filter to a SQRT Nyquist spectrum with the specified roll-off while preserving the ability to independently control pass-band ripple and stop-band ripple. Figure 4.11 illustrates the form of the algorithm by starting the

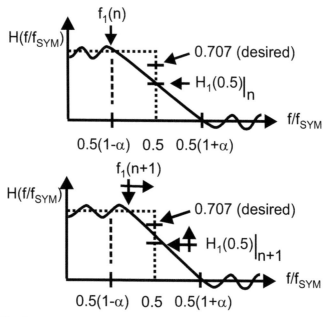

Figure 4.11   Spectrum of Remez algorithm output at iteration *n* matching band edges of Nyquist pulse and then increase of frequency f_1 for iteration $n + 1$ to raise $h_1(n)$ to $h_1(n + 1)$ and eventually to the desired −3 dB level.

Remez algorithm with pass-band and stop-band edges matched to the roll-off boundaries of the Nyquist spectrum. The resulting filter will cross the band edge ($f/f_{SYM} = 0.5$) with more attenuation than the desired −3 dB level. We can raise the attenuation level toward the desired −3 dB level by increasing the frequency of the pass-band edge. The algorithm performs the successive shifts to the right of the pass-band edge (frequency f_1) until the error between the desired 0.707 and the $H_1(0.5)$ level is reduced to zero by using the gradient descent method shown in Equation (4.22). A complete description of this technique, developed by harris and Moerder, can be found in the literature.

$$error(n) = \sqrt{2}/2 - abs(H_1(0.5)|_n)$$
$$f_1(n + 1) = f_1(n) \cdot [1 + mu \times error(n)].$$
(4.22)

A MATLAB script we called *nyq2* implemented the algorithm just described. Figure 4.12 presents the spectra of the h-M filter and compares its response to the standard cosine taper filter. The top and center subplots

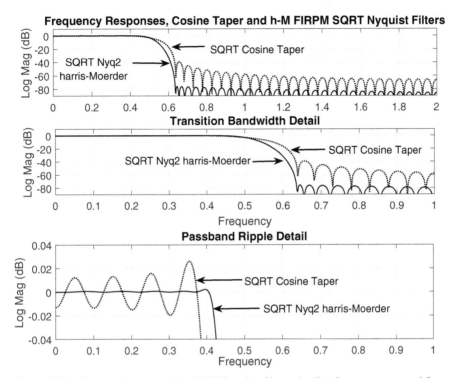

Figure 4.12 Spectra of converged and SQRT cosine filters, details of same spectra, and finer details of same spectra.

show that the h-M filter stop-band level is at −70 dB while the cosine taper level is −40 dB. The h-M filter offers 30 dB improved stop-band level. The lower subplot shows the in-band ripple level of the two filters. The h-M filter ripple is about 1.5 parts in 1000 (0.0015 dB) while the cosine taper filer is about 19 parts in 1000 (0.019 dB). This is more than an order of magnitude reduction. Remember that it is the in-band ripple level responsible for pre- and post-echoes, or ISI levels, in the filter time response. In both cases, the filter sample rate is four times the symbol rate with roll-off factor α equal to 0.25. The filters are both of length 20 symbols, the length suggested by *nyq2* in response to the input specifications.

Figure 4.13 shows the time domain echo levels associated with the matched filter responses of the two filters, the h-M, and the cosine taper. The first item of interest in the *nyq2* design is the side-lobe level obtained by the design. The peak responses of the matched filter are normalized to 1.0 and we zoomed to the low-level terms of the time response to see the ISI contributors.

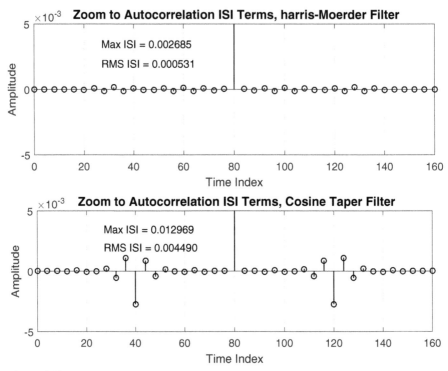

Figure 4.13   Details of ISI levels matched filter responses of harris–Moerder and cosine taper SQRT Nyquist filters.

We quite quickly can see the pre- and post-echoes of the cosine taper filters that are located 10 symbols on either sides of the peak which is at sample 20. By comparison, the h-M filter has many low-level offsets from level 0 but no obvious dominant echoes. We measured the peak and RMS levels of the ISI from the two filters and placed the numbers in the figures. The RMS level of ISI from the cosine taper is seen to be more than an order of magnitude larger than the ISI from the h-M filter. Note that by permitting the transition bandwidth to differ from the cosine taper of the standard SQRT Nyquist pulse, we obtain significant improvement in filter characteristics. The resulting pulse is an approximation to a SQRT Nyquist pulse, but the transition is no longer cosine tapered. When we examined the taper by a spectral derivative, we found that it has the appearance of a Gaussian spectral mass. A complete description of this design technique, developed by harris and Moerder (h-M) can be found in the literature.

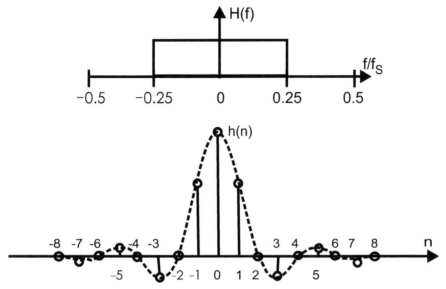

Figure 4.14   Impulse response and spectrum of ideal low-pass half-band filter.

4.4 Half-Band Filters

A half-band filter has a particularly attractive property that makes it uniquely desirable for use in multirate filters. We now identify that attribute. The frequency response and the impulse response of a zero-phase, hence nonrealizable, half-band filter is shown in Figure 4.14. We note in a normalized frequency axis that the pass band is the interval between the plus and minus quarter sample rate. The equation describing the impulse response is shown as follows:

$$h(n) = \frac{1}{2} \frac{\sin\left(\frac{n\pi}{2}\right)}{\left(\frac{n\pi}{2}\right)}. \tag{4.23}$$

The first thing we note about the impulse response is that, except for the sample value at the origin, all the even indexed sample values are zero. It is this property that makes the half-band filter interesting to us. The impulse response of the filter shown in Equation (4.23) extends over all integers, and our first modification to this filter is to apply a window as shown in Equation (4.24) to make it finite duration. The window, of course, causes a ripple in the pass band and in the stop band of the filter as well as induces an odd

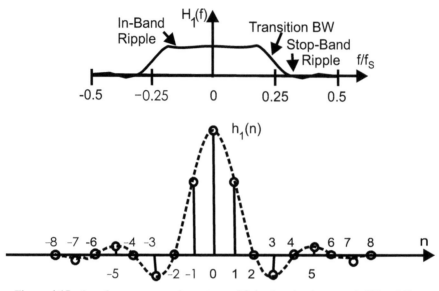

Figure 4.15   Impulse response and spectrum of finite duration low-pass half-band filter.

symmetric transition bandwidth that passes through the quarter sample rate with a gain of 0.5, the midpoint of the gain discontinuity. To control the amplitude of the in-band and out-of-band ripple, we apply a smooth window such as a Hann or Kaiser window. The time-limited version of the impulse response and its related finite transition bandwidth spectrum is shown in Figure 4.15.

$$h_{\text{LOW}}(n) = h(n) \cdot w(n). \tag{4.24}$$

The windowed impulse response continues to exhibit zero amplitude at all even indexed data samples as well as amplitude 0.5 for the data sample at index zero. The window only affects the odd indexed filter coefficients.

The half-band low-pass filter can be converted to a half-band high-pass filter by using the modulation theorem to translate the spectral center from DC, $\theta = 0$, to the half sample rate, $\theta = \pi$. This is shown in Equation (4.25). The impulse response and the frequency response of the translated filter are shown in Figure 4.16.

$$h_{\text{HIGH}}(n) = h_{\text{LOW}}(n) \cdot \cos(\pi n). \tag{4.25}$$

We note that the heterodyning cosine is alternately +1 and −1 and that the +1 occurs on the even indices and the −1 occurs on the odd indices. We

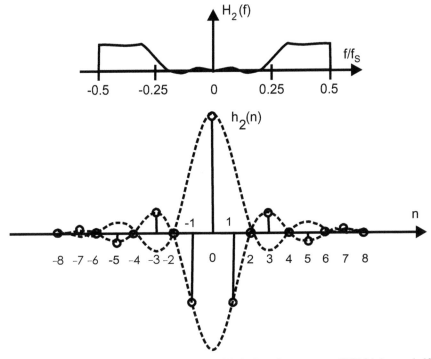

Figure 4.16 Impulse response and spectrum of finite impulse response (FIR) high-pass half-band filter.

already noted that except for the zero index, all the nonzero coefficients of the low-pass impulse response occur on the odd indices. Thus, the product formed in Equation (4.25) simply changes the sign of all the coefficients except the coefficient at index zero. As a result of sign changes just described, we can relate the even-indexed and odd-indexed coefficients of the two filters $h_{LOW}(n)$ and $h_{HIGH}(n)$ as shown in the following equation:

$$h_{HIGH}(2n) = h_{LOW}(2n)$$
$$h_{HIGH}(2n + 1) = -h_{LOW}(2n = 1). \tag{4.26}$$

It is convenient to identify the Z-transform of the even-indexed coefficients and the Z-transform of the odd-indexed coefficients of the low-pass filter. This is shown in Equation (4.27). The two units of delay between successive samples in the two sequences are accounted for in their Z-transforms by the Z^2 in the argument of the transform.

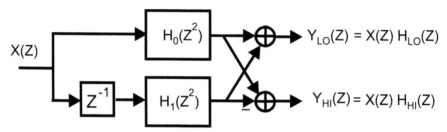

Figure 4.17   Polyphase partition of half-band filter pair.

$$H_0(Z^2) = \sum_n h_{\text{LOW}}(2n)Z^{-2n}$$

$$Z^{-1}H_1(Z^2) = \sum_n h_{\text{LOW}}(2n+1)Z^{-(2n+1)}. \tag{4.27}$$

Using the relationships identified in Equation (4.27) and the sign reversals described in Equation (4.26), we can form the transform of the low-pass and of the high-pass filters as a weighted sum of the sub-transforms as shown in the following equation:

$$H_{\text{LOW}}(Z) = H_0(Z^2) + Z^{-1}H_1(Z^2)$$

$$H_{\text{HIGH}}(Z) = H_0(Z^2) - Z^{-1}H_1(Z^2). \tag{4.28}$$

The block diagram of the filter structure suggested in Equation (4.28) is shown in Figure 4.17.

This form of filter is known as a *polyphase partition* of the prototype filter and since it offers both the low-pass and the high-pass version of the half-band filter, it is also known as a *quadrature mirror filter*.

If we form the sum of the two transforms identified in Equation (4.28), we obtain twice the transform of the even-indexed coefficients. But the even-indexed coefficients have only a single nonzero term, which is the index zero term. The resultant transform can be seen in Equation (4.29) to be zero everywhere except at index zero

$$H_{\text{LOW}}(Z) + H_{\text{HIGH}}(Z) = 2H_0(Z)$$

$$= 2h(0) = 1. \tag{4.29}$$

Equation (4.29) can be rearranged to form Equation (4.30), which demonstrates that $H_{\text{LOW}}(Z)$, the low-pass filter, and $H_{\text{HIGH}}(Z)$, the high-pass

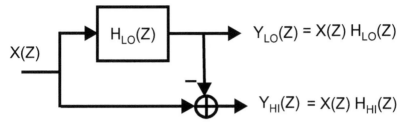

Figure 4.18 Complementary partition of nonrealizable half-band filter pair.

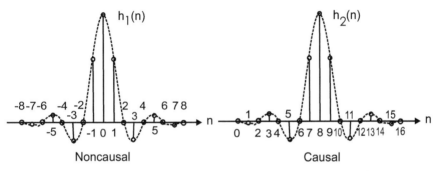

Figure 4.19 Noncausal and causal forms of half-band low-pass filter.

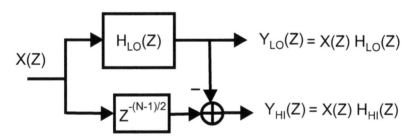

Figure 4.20 Complementary partitions of realizable half-band filter pair.

filter, are complementary

$$H_{LOW}(Z) = 1 - H_{HIGH}(Z). \tag{4.30}$$

An alternate representation of the quadrature mirror filter pair presented in Figure 4.16 is shown in Figure 4.18 as a complementary filter pair.

The filter presented in Figures 4.15 and 4.16 are noncausal. To make them causal, a delay of length $(N - 1)/2$ must be inserted in the time response so that the impulse response values for negative time indices are exactly zero. The redefinition of the origin for the impulse response of the half-band filter is shown in Figure 4.19. When the delay is inserted in the filter path to make it realizable, a matching delay has to be inserted in the second path containing the direct input to output connection. This change is shown in Figure 4.20.

References

Anderson, John. *Digital Transmission Engineering*, Englewood Cliff, NJ, Prentice-Hall, 1999. Crochiere, Ronald and Lawrence Rabiner. *Multirate Signal Processing*, Englewood Cliff, NJ, Prentice-Hall Inc., 1983.

Fliege, Norbert. *Digital Signal Processing: Multirate Systems, Filter Banks, Wavelets*, West Sussex, John Wiley & Sons, Ltd, 1994.

Ashrafi, Ashkan and harris, fred, *A novel Square Root Nyquist Filter Design with Prescribed ISI-Energy*, Elsevier Signal processing Journal, Signal Processing 93 (2013), pp 2626-2635.

harris, fred and Chis Dick, *An alternate Design Technique for Square Root Nyquist Shaping Filter,* SDR Wincomm-2015 Conference, San Diego, CA, 15-17 March 2015.

harris, fred, Chris Dick, and Sridhar Seshigari, *An improved Square-Root Nyquist Shaping Filter*, -2005, Anaheim, CA, 15-17 November 2015, Software Defined Radio Conference

harris, fred, Gurantz, Itzhak, and Tzukerman, Shimon, *Digital T/2 Low-Pass Nyquist Filter Using recursive All-Pass Polyphase Resampling Filter for Signaling rates 10-kHz to 10 MHz*, 26-th Annual Asilomar conference on Signals, Systems, and Computers, Pacific Grove, CA., 26-28 October 1992.

Harada Hiroshi, and Ramjee Prasad. *Simulation and Software Radio for Mobile Communications*, Norwood, MA ,Artech House, 2002.

Hentschel, Tim. *Sample Rate Conversion in Software Configurable Radios*, Norwood, MA, 2002.

Jovanovic-Dolecek, Gordana. *Multirate Systems: Design and Applications*, London, Idea Group, 2002. Mitra , Sanjit and James Kaiser. *Handbook for Digital Signal Processing*, New York, John Wiley & Sons, 1993.

Proakis, John. *Digital Communications*, 4th ed., New York, McGraw-Hill, 2001.

Problems

4.1 The following sequences all are time limited to an interval that permits modulation with independent peak samples every 10 output samples:

```
h1 = rectpuls([-0.5:0.1:0.5]);
h2 = tripuls([-0.5:0.1:0.5]);
h3 = cos(pi*(-0.5:0.1:0.5));
h4 = cos(pi*(-0.5:0.1:0.5)).^2;
h5 = exp(-20*(-0.5:0.1:0.5).^2);
```

On a set of subplots, plot the time response and the frequency response of each wave shape.
Comment on their respective bandwidths, side-lobe levels, and relative smoothness of the time series.

4.2 The following sequences all are time limited to an interval that permits modulation with overlapped but independent peak samples every 20 output samples:

```
h1 = tripuls([-1.0:0.1:1.0]);
h2 = cos(pi*(-1.0:0.1:1.0));
h3 = cos(pi*(-1.0:0.1:1.0)).^2;
h4 = exp(-5*(-1.0:0.1:1.0).^2);
h5 = sinc(-1.0:0.1:0.1);
```

On a set of subplots, plot the time response and the frequency response of each wave shape.
Comment on their respective bandwidths, side-lobe levels, and relative smoothness of the time series.

4.3 Use the MATLAB script file *rcosine* to generate samples of the root raised cosine Nyquist filter. The following MATLAB script will accomplish this:

```
hh1 = rcosine(1,8,'sqrt',0.5,6);
```

Convolve the sequence hh1 with itself and plot the two sequences hh1/max(hh1) and conv(hh1,hh1) and their spectra. Comment on the locations of the zero crossings of the two time series hh1/max(hh1) and conv(hh1,hh1). Comment on the amplitude response of the two spectra, particularly at the band edges, $\pm 1/16$ of the sample rate.

4.4 Use the MATLAB script file *rcosine* to generate samples of the root raised cosine Nyquist filter and then generate a time series or random modulated data and pass it through the receiver matched filter. Finally, form eye diagrams for the input and output of the matched filter. The following MATLAB script will accomplish this:

```
hh1=rcosine(1,8,'sqrt',0.25);
dat1=(floor(4*rand(1,1000))-1.5)/1.5;
dat2=reshape([dat1;zeros(7,1000)],1,8000)
dat3=conv(dat2,hh1)/max(hh1);
dat4=conv(dat3,hh1)
plot(0,0);
hold on
for nn=1:16:8000-16
plot(-1:1/8:1,dat3(nn:nn+16))
end
hold off
grid
% Repeat for dat4
```

Examine and comment on the two eye diagrams paying particular attention to the width and height of the eye opening.

4.5 Repeat problem 4.4 except change the excess bandwidth parameter from 0.25 to 0.125. The following MATLAB script will accomplish this:

```
hh1=rcosine(1,8,'sqrt',0.125);
```

4.6 Repeat problem 4.4 except change the excess bandwidth parameter from 0.25 to 0.50. The following MATLAB script will accomplish this:

```
hh1=rcosine(1,8,'sqrt',0.50);
```

4.7 Design a half-band low-pass FIR by windowing a sinc series filter with transition bandwidth 10% of the sample rate and with in-band and out-of-band ripple less than 0.001. Plot the impulse response and the frequency response. Plot the frequency response with linear and with log magnitude coordinates.

4.8 Design a half-band high-pass FIR by windowing a sinc series filter with transition bandwidth 10% of the sample rate and with in-band and out-of-band ripple less than 0.001. The high-pass is formed from the low-pass as a heterodyned filter or as a complementary filter. Try both! Plot the impulse response and the frequency response. Plot the frequency response with linear and with log magnitude coordinates.

4.9 Design a half-band low-pass FIR with the Remez algorithm. The filter has a transition bandwidth 10% of the sample rate and with in-band and out-of-band ripple less than 0.001. Plot the impulse response and the frequency response. Plot the frequency response with linear and with log magnitude coordinates.

4.10 Design a half-band high-pass FIR with the Remez algorithm. The filter has a transition bandwidth 10% of the sample rate and with in-band and out-of-band ripple less than 0.001. The high pass is formed from the low pass as a heterodyned filter, as a complementary filter, or by changing the gain vector in the Remez algorithm. Try all three. Are they the same? Plot the impulse response and the frequency response. Plot the frequency response with linear and with log magnitude coordinates.

5

Systems That Use Resampling Filters

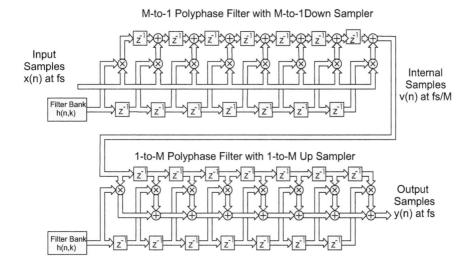

M-to-1 Polyphase Filter with M-to-1Down Sampler

Input
Samples
x(n) at fs

Filter Bank
h(n,k)

Internal
Samples
v(n) at fs/M

1-to-M Polyphase Filter with 1-to-M Up Sampler

Output
Samples
y(n) at fs

Filter Bank
h(n,k)

There are many applications for multirate filters that, when invoked, lead to reduced cost to implement the desired processing task. A common theme in many of these applications is that the filtering should occur at a rate matching the Nyquist rate. The first two examples presented next illustrate this concept.

5.1 Filtering With Large Ratio of Sample Rate to Bandwidth

A common application for multirate signal processing is the task of filtering to reduce the signal bandwidth without changing the sample rate. This might occur when the output of the filter is to be presented to a digital-to-analog converter (DAC) operating at a fixed output rate matching the input rate. Let us examine the specifications of a filter with a small bandwidth relative to

117

Table 5.1   Filter specifications.

Parameter	Specification
Sample rate	20.0 MHz
Pass-band frequency	±100 kHz
Stop-band frequency	±300 kHz
Pass-band ripple	0.1 dB
Stop-band ripple	80 dB
Rate of side-lobe attenuation	6 dB/Octave

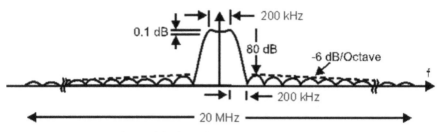

Figure 5.1   Specifications of low-pass filter.

sample rate. The specifications required for the filter are listed in Table 5.1 and are illustrated in Figure 5.1.

We determine that a 360-tap finite impulse response (FIR) filter is required to meet these specifications. The processing task implemented directly with this FIR filter is indicated in Figure 5.2. The computational workload required of this implementation is 360 ops per output or, since there is no sample rate change, 360 ops per input.

The Nyquist rate for this filter, equal to the filter two-sided bandwidth plus the transition bandwidth, is seen to be 400 kHz. This rate is 1/50th of the input sample rate. If we were to permit sample rate changes, we could reduce the sample rate by 50 and operate the filter at an output rate of 400 kHz. Let us examine this option. A 50-stage polyphase partition of the 360-tap filter is shown in Figure 5.3. It would be nice if the number of taps were a multiple of 50: we can drop the number to 350 or raise it to 400. We elected to raise it to 400 and obtain a filter design that outperforms the filter specifications.

Figure 5.2   Initial implementation of a filtering task.

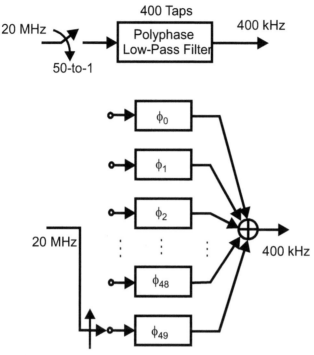

Figure 5.3 50-to-1 polyphase partition and down-sampling of low-pass filter.

By simple division, we determine that the average length of each polyphase filter stage is 360/50 or 7.2 taps. In reality, when loading the filter coefficients by successive columns of length 50, we would find that the first 10 subfilters contain eight taps and the remaining 40 subfilters contain seven taps. There is no problem with the subfilters being of different lengths. The hardware or software implementation of the filter would, in the interest of regularity, simply zero-extend all the subfilters to length 8. Since we would use all the coefficients in those zero-extended locations, we may just as well increase the filter length to 400 and obtain eight taps per path, design the filter with 400 taps, and obtain a filter that oversatisfies the specification. With the 400-tap version, the polyphase partition would require 400 ops per output but would only require 8 ops per input. This is significantly less than that required by the direct implementation suggested by Figure 5.2.

Our only problem at this point is we have not satisfied one of the filter specifications, namely that the output sample rate be the same as the input sample rate.

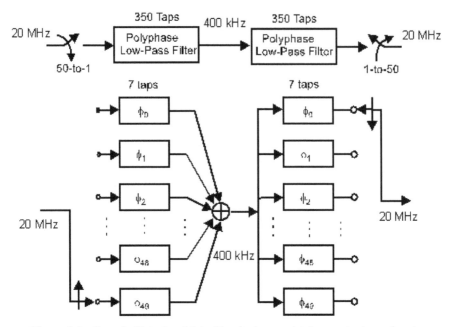

Figure 5.4 Cascade 50-to-1 and 1-to-50 polyphase maintains constant sample rate.

We respond to this last objection by following the down-sample filter by an up-sample filter. This cascade filter form is shown in Figure 5.4. In this form, we output one sample from the input process for every 50 input samples and then output 50 output samples for every intermediate input sample. This process generates one output sample for each input sample. We note that when the filter is implemented in this form, there are 8 ops per input and 8 ops per output for a total workload of only 16 ops per input–output pair. This compares with 350 ops per input–output pair in the direct implementation. Similarly, since only one arm of the polyphase filter is engaged at any one time, we only require 8 data registers for the input and 8 data registers for the output, a total of 16 registers which is certainly a reduction from the 400 data registers for the direct implementation. Figure 5.5 presents the same filter but reflects the reduced resources required for this filter. Compare Figure 5.5 with Figure 5.2. The lesson we have learned in this example is that a filter should be operated at its Nyquist rate. If the desired output sample rate is different from the Nyquist rate, consider that as a problem to be addressed in a second filter.

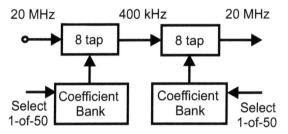

Figure 5.5 Reduced resource cascade 50-to-1 and 1-to-50 polyphase filter.

Examining Figure 5.4, in particular, the up-sampling half of the cascade, we can clearly see that a single-stage 8-tap FIR filter can perform the sequential processing tasks of the entire polyphase filter. This is obvious since all the stages of the filter store and process the same input series obtained from the previous stage. With this awareness, when we examine the input down-sampling structure, we face a quandary. The separate stages store and process different input data streams and it would appear that the resource sharing of the output stage is not applicable to the input stage. The quandary is that if we can share resources in the output filter, we must be able to do the same in the input filter since the two processes perform dual functions. The statement of the problem contains the solution to the problem. The two filter banks must be dual structures; they cannot both be tapped delay line implementations of the filter stages. We now examine the dual form or alternate form of the FIR filter.

5.1.1 Partial Sum Accumulator: The Dual Form

We start with the up-sampling filter to first establish a sequence of transformations that we then apply to the down-sampling filter. Figure 5.6 presents the polyphase partition of a three-stage up-sampling filter. The separate arms of the filter are implemented as the standard tapped delay line filter structure that supports the conventional inner product form or multiplier-accumulator (MAC) form of the FIR filtering process. We note that the data registers of the three filters all contain the same input samples. In Figure 5.7, we implement the filter with a single register set that is accessed by the three sets of MACs that feed the output commutator. We note that even though there are three sets of MACs, we use them sequentially, one at a time. Rather than using three MACs, we can time share one MAC and commutate filter weights to synthesize successive filter sections. This structure is shown in Figure 5.8. Note here that one polyphase filter segment and a commutator pointing to filter weights can form a filter with any number of polyphase arms. The process of presenting successive weight sets to the single stage can be likened to the operation of a Gatling gun in which sets of weights are successively rolled into the filtering task.

The sequence of transformations just applied to the up-sampling filter is now applied to the dual down-sampling filter. Figure 5.9 presents the polyphase partition of a 1-to-3 down-sampling filter. The separate stages are implemented in the conventional MAC structure.

Note that the three stages are combined as a dual of the three stages in the corresponding up-sampling process of Figure 5.6. Comparing the two figures, we see that the input node of Figure 5.6 has become the output-summing junction of Figure 5.9 and that the arrows indicating signal flow to and from the three stages and the commutator direction have been reversed. Since the tasks of up-sampling and down-sampling are dual processes, it is not

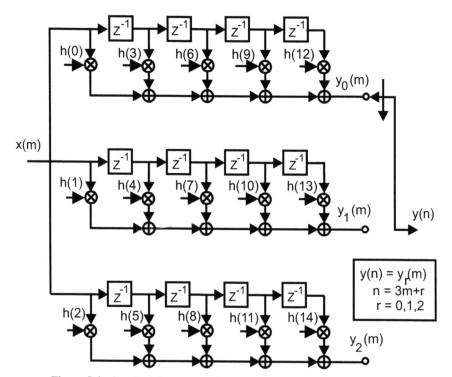

Figure 5.6   Standard three-stage, three-MAC, polyphase up-sampling filter.

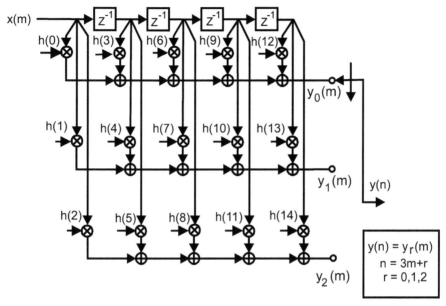

Figure 5.7 Shared registers of three-stage, three-MAC, polyphase up-sampling filter.

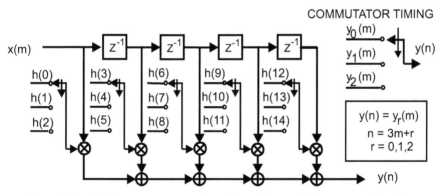

Figure 5.8 Minimum resource version of three-stage polyphase up-sampling filter.

surprising that the stages are connected in dual forms. What we now realize is that dual structures should also be applied to the stage implementations. This dual form with nodes replacing summing junctions, summing junctions replacing nodes, and reversed arrow directions and coefficient ordering is shown in Figure 5.10.

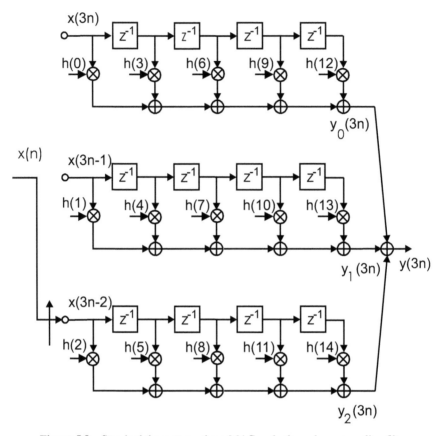

Figure 5.9   Standard three-stage, three-MAC, polyphase down-sampling filter.

This structure represents an alternate implementation of an FIR filter structure and is known as the partial sum accumulator form. Interestingly, the tapped delay line model performs an inner product and is the mental image of an FIR filter conjured by an imbedded system programmer. The partial sum accumulator model performs simultaneous parallel processing and is the mental image of an FIR filter conjured by an application-specific integrated circuit (ASIC) designer. Note that in the tapped delay line model, the registers contain input data samples. These samples are stored and accessed by shifting to apply the successive filter weights to form the sum for successive output samples. In the partial sum accumulator, the registers contain the sum of products. The products are formed between each input sample and all filter coefficients. The products are accumulated over successive inputs and then

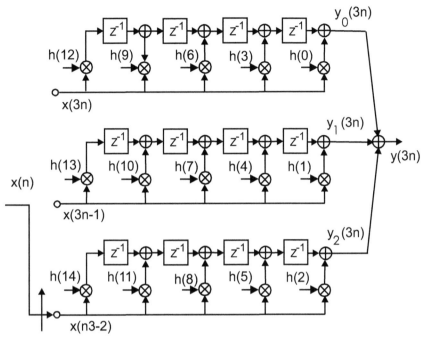

Figure 5.10   Dual form three-stage polyphase down-sampling filter.

shifted toward the filter output port to form the final sum for successive output samples.

In Figure 5.10, we note that the sum formed at the output of the three filters and at the output of the combined filter is in fact performed in the same accumulator. We also note that the outputs of the earlier accumulators eventually arrive at the output accumulator where they are combined to form the composite filter output. We can save memory by combining the outputs of the separate accumulators as a single accumulator. This combination is shown in Figure 5.11, which is seen to be the dual form of Figure 5.7.

We note here that even though there are three sets of multipliers to apply coefficients to the data that feed the common accumulators, we use them sequentially, one at a time. Rather than using three multiplier sets, we can time share one set of multipliers and commutate filter weights to synthesize successive filter sections. This structure is shown in Figure 5.12. Note here again that one polyphase filter segment and a commutator pointing to filter weights can form a filter with any number of polyphase arms. This process of presenting successive weight sets to the single stage can again be likened to

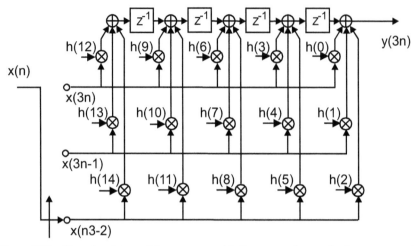

Figure 5.11   Shared registers of three-stage, multiplier-accumulator polyphase up- sampling filter.

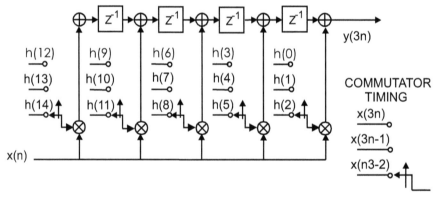

Figure 5.12   Minimum resource version of three-stage polyphase down-sampling filter.

the operation of a Gatling gun as was the dual process presented in Figure 5.8. Note, in particular, the dual structure of the filters shown in Figures 5.8 and 5.11. These two efficient forms of the polyphase filter are used in their appropriate position in the filter structure presented in Figure 5.5.

5.1.2 Generate Baseband Narrowband Noise

We now address a related signal-processing task: that of forming a baseband time series with a narrow bandwidth at a sample rate that is larger compared

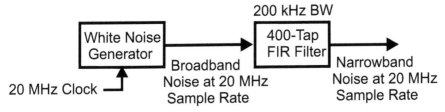

Figure 5.13 Block diagram of simple brute force narrowband noise generator.

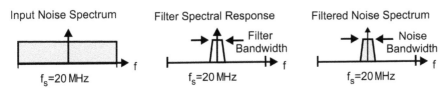

Figure 5.14 Input spectrum, filter spectral response, and output spectrum.

to the signal bandwidth. The specific example we examine is that of forming a digital narrowband noise source. The spectral characteristics of the filtered noise, by a remarkable coincidence, are the same as those presented in Table 5.1. The narrowband noise is to be sampled at 20 MHz, having a 2-sided bandwidth of 200 kHz, a transition bandwidth of 200 Hz, and a dynamic range of 80 dB. In a previous section, we learned that the FIR filter that meets these specifications has a length of 360 taps which we changed, for convenience, to 400 taps. Figure 5.13 presents a brute force solution to the processing task. We operate a long period pseudo noise (PN) noise generator at the desired 20 MHz output rate and deliver the noise sequence to the 400-tap filter. As shown in Figure 5.14, the output spectrum is equal to the product of the input spectrum and the power spectral response of the filter. The output time series of the filter will inherit the spectral properties of the filter and hence meets the spectral requirements placed on the processing task.

The problem with the noise generation process presented in Figure 5.13 is the high workload per output noise sample of 400 ops per output. Following the lesson taught in an earlier section, that the filter should operate at its Nyquist rate, we recast the solution in the following way. We operate the noise generator at the filter's Nyquist rate of 400 kHz and then use the polyphase partition of the 400-tap filter as a 1-to-50 up-sampler. This is shown in Figure 5.15. The polyphase partition is implemented as a single 8-tap filter

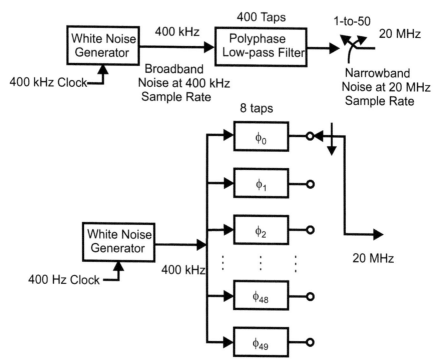

Figure 5.15   Block diagram of efficient narrowband noise generator.

with the 50 coefficient sets applied sequentially to the filter for each input sample to the filter. In this structure, the workload is 8 ops per output point, a rather significant improvement relative to the 400 ops per output point of the simple solution.

Conceptually, the input noise spectrum at 400 Hz sample rate has been up-sampled by 1-to-50 zero-packing which gives us access to the desired output sample rate of 20 MHz. The 400-tap filter processes the zero-packed data to extract the original baseband spectrum at the higher sample rate. Only the non-zero samples of the zero-packed input data contribute to the output samples. Tracking the position of these samples in the filter to determine which coefficients interact with the known data samples is equivalent to the polyphase partition of the filter. The zero-packing is introduced to help visualize the process, not to offer implementation instructions. The spectrum of the input signal, of the zero-packed input signal, of the filter, and of the filter output is shown in Figure 5.16.

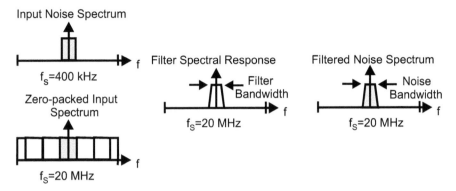

Figure 5.16 400-Hz input spectrum, 1-to-50 zero-packed 20 kHz input spectrum, filter spectral response, and filtered output spectrum.

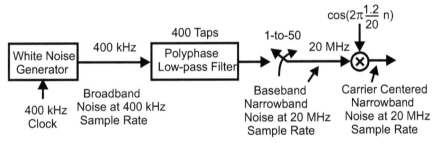

Figure 5.17 Translate baseband signal formed by polyphase filter.

5.1.3 Generate Narrowband Noise at a Carrier Frequency

A variant of the previous example is that we want the same narrowband noise sampled at 20 MHz rate but want it centered at 1200 kHz. We might want such a noise signal to add to a narrowband communication signal to test its susceptibility to additive noise. One way to obtain the narrowband noise centered at the 1200 kHz carrier is to upconvert the baseband signal generated by the previous multirate up-sampling process. The implementation of this option is shown in Figure 5.17.

The center frequency selected for the upconversion process is 1.2 MHz, a frequency that happens to be a multiple of the input sample rate of 400 kHz. In our model of the process, the input signal was zero-packed to raise the sample rate to 20 MHz. In the zero-packed model, copies of the spectrum are distributed over the 20 MHz interval at multiples of the 400 kHz input sample rate. What we accomplished was to place a spectrum at baseband,

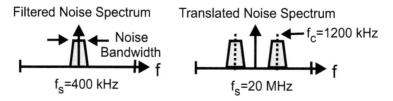

Figure 5.18   Baseband spectrum and translated spectrum at 20 kHz sample rate.

zero-pack it to access spectral copies at multiples of 400 kHz, filter out all the copies, and translate the surviving baseband spectrum up to 1200 kHz. In an alternate process, we can translate the spectral response of the polyphase filter and have it extract the spectral copy of the input signal already residing at 1200 kHz.

The effect of translating a baseband prototype up-sampling polyphase filter to a multiple of the input rate can be easily described. Equation (5.1) presents the relation between the baseband prototype and its polyphase partition, while Equation (5.2) presents the same relationship for the translated version of the same filter.

$$
\begin{aligned}
\text{Prototype} & \quad h\,(n) \\
\text{Polyphase Partition} & \quad h_r\,(n) \;=\; h\,(r\,+nM)
\end{aligned}
\tag{5.1}
$$

$$
\begin{aligned}
\text{Translated Filter} & \quad g(n,k) \;=\; h(n)\,\cos(n\,\theta_k) \\
\text{Polyphase Partition} & \quad g_r(n,k) \;=\; h_r(n)\,\cos[(r+nM)\,\theta_k].
\end{aligned}
\tag{5.2}
$$

When we substitute the center frequency θ_k defined in Equation (5.3) into Equation (5.2), we obtain Equation (5.4).

$$
\theta_k \;=\; \frac{2\pi}{M}k
\tag{5.3}
$$

$$
\text{Polyphase Partition} \quad g_r(n,k) \;=\; h_r(n)\,\cos(\frac{2\pi}{M}r\,k).
\tag{5.4}
$$

Here we see that when the heterodyne is applied to the polyphase partition and the heterodyne corresponds to a multiple of the input sample rate, the heterodyne defaults to a scalar for each path that depends only on the index r, the polyphase arm, and the index k, the selected frequency offset. The scalar can be distributed through the polyphase coefficients as an offline operation or can be applied as an online scaling term to the input data as it enters each polyphase filter arm. The former makes sense if the filter is to service a

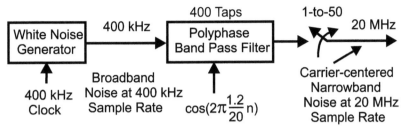

Figure 5.19   Band pass signal formed by translated polyphase filter.

single frequency, and the latter makes sense if the filter is to service multiple frequencies.

Applying the heterodyne to the filter coefficients allows the polyphase filter to accomplish three operations simultaneously: sample rate change, bandwidth control, and center frequency translation. This filter requires only 8 ops per input sample to accomplish all three processing tasks. The form of the translated polyphase filter that accomplishes these three tasks is shown in Figure 5.19.

5.2 Workload of Multirate Filter

It is common practice to describe a filter's specifications relative to its sample rate. For instance, a filter may have a bandwidth that is one-fourth of the sample rate. It is also common to describe specifications relative to the filter's bandwidth. As an example, a Nyquist or square-root Nyquist filter with a roll-off of 0.20 has an excess bandwidth equal to 20% of the symbol rate. Describing filter parameters of a multirate filter relative to its sample rate presents an interesting quandary. Which sample rate, input or output? What does one do when the filter supports arbitrary resampling ratios? We now address a perspective that proves to be useful when specifying filter specifications relative to filter sample rate.

Figure 5.20 shows the sample rates associated with a multirate filter, while Figure 5.21 shows the spectral description of a multirate filter at the input rate and output sample rates. We use the two versions of the spectral description to examine a connection between filter length N and the resampling ratio M.

Equation (5.5) is a statement of how the transition bandwidth Δf is related to filter length N and filter pass-band and stop-band ripple parameters δ_1 and δ_2.

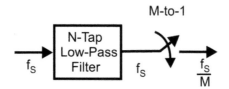

Figure 5.20   Multirate filter indicating input and output sample rates.

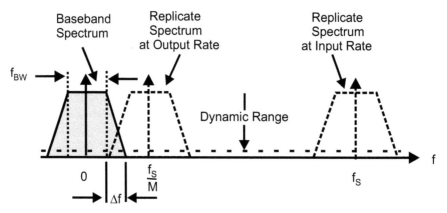

Figure 5.21   Spectra of low-pass signal at input and output sample rates.

$$\Delta f = K(\delta_1, \delta_2)\frac{f_S}{N}. \tag{5.5}$$

The filter bandwidth, described relative to output sample rate, is indicated in Equation (5.6) where the parameter α represents the fractional bandwidth of the pass-band relative to the output sample rate.

$$f_{\mathrm{BW}} = \alpha\frac{f_S}{M}. \tag{5.6}$$

As shown in Equation (5.7), we satisfy the Nyquist criterion when the output sample rate is equal to the sum of the filter's two-sided bandwidth and the transition bandwidth.

$$\frac{f_S}{M} = \alpha\frac{f_S}{M} + \Delta f. \tag{5.7}$$

Substituting Equation (5.5) in Equation (5.7), we obtain Equation (5.8) which, when rearranged, leads to Equation (5.9).

$$\frac{f_S}{M} = \alpha\frac{f_S}{M} + K(\delta_1, \delta_2)\frac{f_S}{N} \tag{5.8}$$

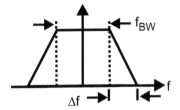 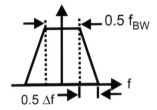

Figure 5.22 Spectra of two low-pass filters of different bandwidth but with fixed ratio of bandwidth to transition bandwidth.

$$N = M \frac{K(\delta_1, \delta_2)}{(1 - \alpha)}. \tag{5.9}$$

We now interpret the relationship derived in Equation (5.9). We know that the filter length N is proportional to the stop-band and pass-band ripple controlled parameter $K(\delta_1, \delta_2)$ and the ratio of sample rate to transition bandwidth. This relationship is shown in Equation (5.10) with respect to the input sample rate f_s. $N = K(\delta_1, \delta_2)$.

We find conflicting definitions of the estimate of filter length when comparing Equations (5.10) and (5.9). Equation (5.9), strangely enough, seems to couple the filter length to the resampling ratio that follows the filter, while Equation (5.10) does not. How can this be? We first resolve this apparent inconsistency.

The inconsistency is related to the fact that transition bandwidth is not normally related to filter bandwidth but is related in the multirate filter. Figure 5.22 presents the spectra of two low-pass filters with different bandwidths but with a fixed ratio of filter bandwidth to transition bandwidth. Note that as the filter bandwidth is reduced, so is the transition bandwidth. This means that the transition bandwidth is proportional to the filter bandwidth and as the filter bandwidth is reduced, so is the transition bandwidth. Filters with a fixed ratio between bandwidth and transition bandwidth are said to have a *constant form factor*. Analog filters have this property. Hence, as the filter bandwidth and the transition bandwidth become smaller, we apply additional down-sampling or we increase M, the down-sampling ratio. We now conclude that a filter with a constant form factor inserted in a multirate filter has a transition bandwidth that varies inversely with the down-sample ratio. The relationship that illustrates this is found by rearranging Equation (5.7) to obtain Equation (5.10).

$$\Delta f = (1 - \alpha) \frac{f_S}{M}. \tag{5.10}$$

We now can solve for the ratio of sample rate to transition bandwidth as shown in the following equation:

$$\frac{f_S}{\Delta f} = M\frac{1}{(1-\alpha)}. \tag{5.11}$$

Now solving for filter length with the transition bandwidth coupled to filter bandwidth, we have the results shown in Equation (5.12), which matches the result derived in Equation (5.9).

$$N = K(\delta_1, \delta_2)\frac{f_S}{N} = M\frac{K(\delta_1, \delta_2)}{(1-\alpha)}. \tag{5.12}$$

Rearranging Equation (5.12), we have Equation (5.13), a form that offers interesting insight into multirate filters.

$$\frac{N}{M} = \frac{K(\delta_1, \delta_2)}{(1-\alpha)}. \tag{5.13}$$

We first examine the units of the ratio N/M. N, the filter length, has units of ops/output, while M, the resampling ratio, has units of inputs/output. Thus, the units of N/M are seen in Equation (5.14) to be ops/input.

$$\frac{N(\text{ops/output})}{M(\text{inputs/output})} = \frac{N}{M}(\text{ops/input}). \tag{5.14}$$

The quantity N/M comes about naturally as we partition a length N prototype filter into M paths. The length of each path is N/M, which is seen as the amount of work required to perform each path in the polyphase filter partition. From Equation (5.13), we learn that ops/input is a constant dependent on the filter quality, where quality is defined by pass-band and stop-band ripples δ_1 and δ_2, respectively, and the fractional bandwidth $(1-\alpha)$ allocated to the transition band. Typical values of $K(\delta_1, \delta_2)$ are on the order of 3-to-4, while typical values of $1/(1-\alpha)$ are on the order of 2-to-5. From these typical values, we determine that N/M spans the range of 6-to-20 ops/input. The example demonstrated in the narrowband noise generator required an N/M ratio equal to 8. Of course, if we reduce the filter pass-band or stop-band ripple or if we reduce the filter transition bandwidth, the quality of the filter is increased, and the N/M ratio is increased appropriately.

The important concept presented here is that in any data-resampling task, the workload per data point is defined entirely by the filter quality. If we are performing a 10-to-1 down-sampler, the workload per input sample N_1/M_1

is a fixed constant, say 8. If we elect to perform a 20-to-1 down-sampler, the filter length doubles but so does the number of stages so that the ratio N_2/M_2 is still 8. The workload per input N/M is fixed independently of sample rate change even though the filter length N is proportional to the sample rate change. As the sample rate ratio changes, the filter length changes but the number of stages changes proportionally so that the ratio N/M is fixed, fixed to the ratio defined by the filter quality.

References

Crochiere, Ronald and Lawrence Rabiner, *Multirate Signal Processing*, Englewood Cliff, NJ, Prentice-Hall Inc., 1983.

Fliege, Norbert. *Multirate Digital Signal Processing: Multirate Systems, Filter Banks, Wavelets*, West Sussex, John Wiley & Sons, Ltd, 1994.

Harada, Hiroshi and Ramjee Prasad, *Simulation and Software Radio for Mobile Communications*, Norwood MA, Artech House, 2002.

Hentschel, Tim, *Sample Rate Conversion in Software Configurable Radios*, Norwood, MA, Artech House, 2002.

Jovanovic-Dolecek, Gordana. *Multirate Systems: Design and Applications*, London, Idea Group, 2002. Mitra, Sinjit and James Kaiser, "Handbook for Digital Signal Processing", New York, John Wiley & Sons, 1993.

Problems

5.1 Design a 5-to-1 five-path polyphase down-sampling filter for the following performance requirements:

Input Sample Rate	100 kHz		
Pass Band	0−8 kHz	In-Band Ripple	0.1 dB
Stop Band	12−50 kHz	Stop-Band Atten.	60 dB
Output Sample Rate	20 kHz		

First verify that the design meets the specifications. Then partition the filter into a five-path filter and verify that it successfully rejects out-of-band signals by having it process 1000 samples of out-of-band signal, a sinusoid in the first Nyquist zone: specifically the sequence shown here.

$x_1 = \exp(j*2*\mathrm{pi}*(0{:}999)*21/100);$

Plot the real part of the time domain response and the windowed log magnitude fast Fourier transform (FFT) response of the polyphase filter output response to the signal x_1.

Generate a signal containing an in-band and an out-of-band component as shown next.

$$x_2 = x_1 + \exp(j*2*pi*(0:999)*1.5/100);$$

Plot the real part of the time domain response and the windowed log magnitude FFT response of the polyphase filter output response to the signal x_2.

5.2 Design a 1-to-4 four-path polyphase up-sampling filter for the following performance requirements:

Input Sample Rate	25 kHz		
Pass Band	0–10 kHz	In-Band Ripple	0.1 dB
Stop Band	15–50 kHz	Stop-Band Atten.	60 dB
Output Sample Rate	100 kHz		

First verify that the design meets the specifications. Then partition the filter into a four-path filter and verify its operation by performing an impulse response test. Plot the time domain response and the log magnitude FFT response of the polyphase filter output response to the impulse.

Now have the filter up-sample the in-band sequence shown next.

$$x_1 = \exp(j*2*pi*(0:199)*3/20);$$

Plot the real part of the time domain response and the windowed log magnitude FFT response of the polyphase filter output response to the signal x_1.

5.3 Design a 1-to-20 20-path polyphase up-sampling filter for the following performance requirements:

Input Sample Rate	10 kHz		
Pass Band	0–4 kHz	In-Band Ripple	0.1 dB
Stop Band	6–100 kHz	Stop-Band Atten.	60 dB
Output Sample Rate	200 kHz		

First verify that the design meets the specifications. Then partition the filter into a 20-path filter and verify that its operation by performing an impulse response test. Plot the time domain response and the log magnitude FFT response of the polyphase filter output response to the impulse.

Now have the filter up-sample following the in-band sequence: x_1 = randn(1,200);

Plot the time domain response and the windowed log magnitude FFT response of the polyphase filter output response to the signal x_1.

5.4 Design a 1-to-20 20-path polyphase up-sampling filter for the following performance requirements:

Input Sample Rate	10 kHz		
Pass Band	0–4 kHz	In-Band Ripple	0.1 dB
Stop Band	6–100 kHz	Stop-Band Atten.	60 dB
Output Sample Rate	200 kHz		

Tune the filter by heterodyning the impulse response to the second Nyquist zone. This is shown here

$$h_2 = h_1.*cos(2*pi*(0:length(h_1)-1)*20/200)$$

First verify that the heterodyned design meets the specifications but offset to 20 kHz, the center of the second Nyquist zone. Then partition the filter into a 20-path filter and verify its operation by performing an impulse response test. Plot the time domain response and the log magnitude FFT response of the polyphase filter output response to the impulse.

Now have the filter up-sample the in-band sequence noise sequence shown next. x_1 = randn(1,200);

Plot the time domain response and the windowed log magnitude FFT response of the polyphase filter output response to the signal x_1.

6

Polyphase FIR Filters

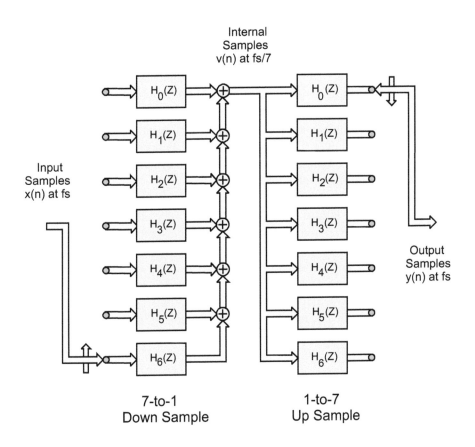

Internal
Samples
v(n) at fs/7

Input
Samples
x(n) at fs

Output
Samples
y(n) at fs

7-to-1
Down Sample

1-to-7
Up Sample

Digital filters are employed in various signal processing systems. The filter can often perform more than its intended primary filtering task. It can absorb many of the secondary signal processing tasks of a signal processing system.

In this light, a filter should always be designed from a system viewpoint rather than as an isolated component in a system. There are so many ways of folding other functions into the filtering process we hardly know where to start. We thus use this section to demonstrate the great versatility to be had by performing multiple functions in a single resampling filter. This approach gives the reader an idea of how the material developed in later sections can be applied to specific applications. This path differs from the traditional approach of first presenting the theoretical tools and then presenting applications. The material presented here is expanded upon in later sections.

6.1 Channelizer

The problem we address here to introduce the fundamental concepts is that of down-converting a single frequency band or channel located in a multichannel Frequency division multiplexing (FDM) input signal. The spectrum of the input signal is a set of equally spaced, equal bandwidth FDM channels as shown in Figure 6.1. The input signal has been band limited by analog filters and has been sampled at a sufficiently high sample rate to satisfy the Nyquist criterion for the full FDM bandwidth.

The standard single processing process to down-convert a selected channel is shown in Figure 6.2. This structure performs the standard operations of down-conversion of a selected frequency band with a complex heterodyne, low-pass filtering to reduce bandwidth to the channel bandwidth, and down sampling to a reduced rate commensurate with the reduced bandwidth. The structure of this processor is seen to be a digital signal processing (DSP) implementation of a prototype analog I-Q down-converter. We mention that the down sampler is commonly referred to as a decimator, a term which means to destroy every tenth one. Since nothing is destroyed,

Figure 6.1 Spectrum of multichannel input signal, processing task: extract complex envelope of selected channel.

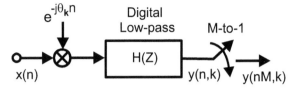

Figure 6.2   Standard single channel down-converter.

and nothing happens in tenths, we prefer, and will continue to use, the more descriptive name, down sampler.

The spectrum we would observe at various points in the processing chain of Figure 6.2 is shown in Figure 6.3. Here we see the spectra at the input and output of the complex heterodyne, the spectral response of the base-band filter, the now oversampled spectrum at the output of the filter, and finally the spectrum at the output of the resampler.

The expression for $y(n,k)$, the time series output from the down-converted kth channel, prior to resampling, is a simple convolution as shown in the following equation:

$$y(n, k) = [x(n) \cdot e^{-j\theta_k n}] * h(n)$$

$$= \sum_{r=0}^{N-1} x(n - r) \cdot e^{-j\theta_k (n-r)} \cdot h(r). \tag{6.1}$$

The output data from the complex mixer is complex and, hence, is represented by two time series, $I(n)$ and $Q(n)$. The filter with real impulse response $h(n)$ is implemented as two identical filters, each processing one of the quadrature time series. The convolution process is often performed by a simple digital filter that performs the multiply and add operations between data samples and filter coefficients extracted from two sets of addressed memory registers. In this form of the filter, one register set contains the data samples while the other contains the coefficients that define the filter impulse response. This structure is shown in Figure 6.4.

We can rearrange the summation of Equation (6.1) to obtain a related summation reflecting the *equivalency theorem*. The equivalency theorem states that the operations of down-conversion followed by a low-pass filter are totally equivalent to the operations of a band-pass filter followed by a down-conversion. The block diagram demonstrating this relationship is shown in Figure 6.5, while the rearranged version of Equation (6.1) is shown in Equation (6.2).

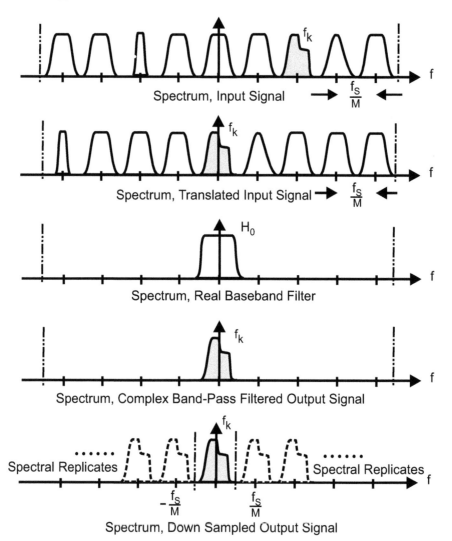

Figure 6.3   Spectra observed at various points in processing chain of standard down-converter.

Note, here, that the up-converted filter, $h(n)\,\exp(j\theta_k n)$, is complex and as such its spectrum resides only on the positive frequency axis without a negative frequency image. This is not a common structure for an analog prototype because of the difficulty of forming a pair of analog quadrature filters exhibiting a 90° phase difference across the filter bandwidth. The

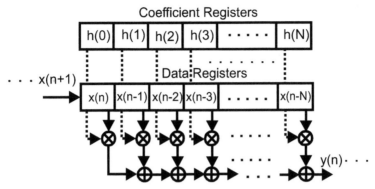

Figure 6.4 Tapped delay line digital filter: coefficients and data registers, multipliers, and adders.

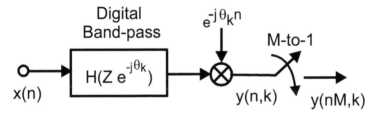

Figure 6.5 Band-pass filter version of down-converter.

closest equivalent structure in the analog world is the filter pair used in image-reject mixers and even there, the phase relationship is maintained by a pair of complex heterodynes.

$$
\begin{aligned}
y(n, k) &= \sum_{r=0}^{N-1} x(n-r) \cdot e^{-j\theta_k(n-r)} \cdot h(r) \\
&= \sum_{r=0}^{N-1} x(n-r) \cdot e^{-j\theta_k n} \cdot h(r) \cdot e^{+j\theta_k r} \qquad (6.2) \\
&= e^{-j\theta_k n} \cdot \sum_{r=0}^{N-1} x(n-r) \cdot h(r) \cdot e^{+j\theta_k r}.
\end{aligned}
$$

Applying the transformation suggested by the equivalency theorem to an analog prototype system does not make sense since it doubles the required hardware. We would have to replace a complex scalar heterodyne (two mixers) and a pair of low-pass filters with a pair of band-pass filters,

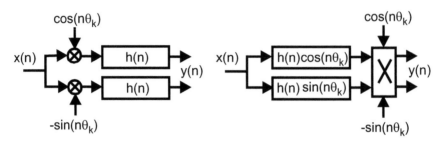

Figure 6.6   Block diagrams illustrating equivalency between operations of heterodyne and base-band filter with band-pass filter and heterodyne.

containing twice the number of reactive components, and a full complex heterodyne (four mixers). If it makes no sense to use this relationship in the analog domain, why does it make sense in the digital world? The answer is found in the fact that we define a digital filter as a set of weights stored in coefficient memory. Thus, in the digital world, we incur no cost in replacing the pair of low-pass filters $h(n)$ required in the first option with the pair of band-pass filters $h(n) \cos(n\theta_k)$ and $h(n) \sin(n\theta_k)$ required for the second one. We accomplish this task by a simple download to the coefficient memory. The filter structures corresponding to the two sides of the equivalency theorem are shown in Figure 6.6. Note that the input signal interacts with the complex sinusoid as a product at the filter input in the first option or through products in the convolution with the filter weights in the second option.

An interesting historical perspective is worth noting here. In the early days of wireless, radio tuning was accomplished by sliding a narrow-band filter to the center frequency of the desired channel to be extracted from the FDM input signal. These radios were known as tuned radio frequency (TRF) receivers. The numerous knobs on the face of early radios adjusted reactive components of the amplifier chain to accomplish this tuning process. Besides the difficulty in aligning multiple tuned stages, shifting the center frequency of the amplifier chain often initiated undesired oscillation due to the parasitic coupling between the components in the radio. Edwin Howard Armstrong, an early radio pioneer, suggested that rather than moving the filter to the selected channel, move the selected channel to the fixed frequency filter. This is known as the *superheterodyne principle,* a process invented by Armstrong in 1918 and quickly adopted by David Sarnoff of the Radio Corporation of America (RCA) in 1924. Acquiring exclusive rights to Armstrong's single tuning dial radio invention assured the commercial dominance of RCA in

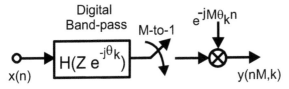

Figure 6.7 Down sampled band-pass down-converter.

radio broadcasting as well as the demise of hundreds of manufacturers of TRF radio receivers. It seems that we have come back full circle. We inherited from Armstrong a directive to move the desired spectrum to the filter and we have readily applied this legacy to DSP-based processing. We are now proposing that, under appropriate conditions, it makes sense to move the filter to the selected spectral region.

We still have to justify the full complex heterodyne required for the down-conversion at the filter output rather than at the filter input. Examining Figure 6.5, we note that following the output down-conversion, we perform a sample rate reduction in which we retain one sample in every M samples. Recognizing that there is no need to down-convert the samples we discard in the down sample operation, we choose to down sample only the retained samples. This is shown in Figure 6.7. Here, we note that when we bring the heterodyne to the low data rate side of the resampler, we are, in fact, also down sampling the time series of the complex sinusoid. The rotation rate of the sampled complex sinusoid is θ_k and $M\theta_k$ radians per sample at the input and output, respectively, of the M-to-1 resampler.

This change in the observed rotation rate is due to aliasing. When aliased, a sinusoid at one frequency or phase slope appears at another phase slope due to the resampling. We now invoke a constraint on the sampled data center frequency of the down-converted channel. We choose center frequencies θ_k which will alias to DC (zero frequency) as a result of the down sampling to $M\theta_k$. This condition is assured if $M\theta_k$ is congruent to 2π, which occurs when $M\theta_k = k\,2\pi$ or, more specifically, when $\theta_k = k\,2\pi/M$. The modification to Figure 6.7 to reflect this provision is seen in Figure 6.8. The constraint that the center frequencies be limited to integer multiples of the output sample rate assures aliasing to base band by the sample rate change. When a channel aliases to base band by the resampling operation, the resampled related heterodyne defaults to a unity-valued scalar, which consequently is removed from the signal processing path. If the center frequency of the aliased signal is offset by $\Delta\theta$ rad/sample from a multiple of the output sample rate, the aliased

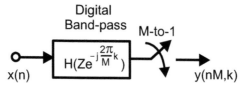

Figure 6.8   Band-pass down-converter aliased to base band by down sampler.

signal will reside at an offset of $\Delta\theta$ rad/sample from zero frequency at base band. A complex heterodyne or base-band converter will shift the signal by the residual $\Delta\theta$ offset. This base-band mixer operates at the output sample rate rather than at the input sample rate for a conventional down-converter. We can consider this required final mixing operation a post-conversion task and allocate it to the next processing block.

The operations invoked by applying the equivalency theorem to the down-conversion process guided us to the following sequence of maneuvers:

1) slide the input heterodyne through the low-pass filters to their outputs;
2) doing so converts the low-pass filters to a complex band-pass filter;
3) slide the output heterodyne to the downside of the down sampler;
4) doing so aliases the center frequency of the oscillator;
5) restrict the center frequency of the band-pass signal to be a multiple of the output sample rate;
6) doing so assures the alias of the selected pass band to base band by the resampling operation;
7) discard the now unnecessary heterodyne.

The spectral effect of these operations is shown in Figure 6.9. The savings realized by this form of the down-conversion is due to the fact we no longer require a quadrature oscillator or the pair of input mixers to effect the frequency translation.

6.1.1 Transforming the Band-Pass Filter

Examining Figure 6.8, we note that the current configuration of the single channel down-converter involves a band-pass filtering operation followed by a down sampling of the filtered data to alias the output spectrum to base band. Following the idea developed in the previous section that led us to down-convert only those samples retained by the down sampler, we similarly conclude that there is no need to compute the output samples from the pass-band filter that will be discarded by the down sampler. Conceptually, we

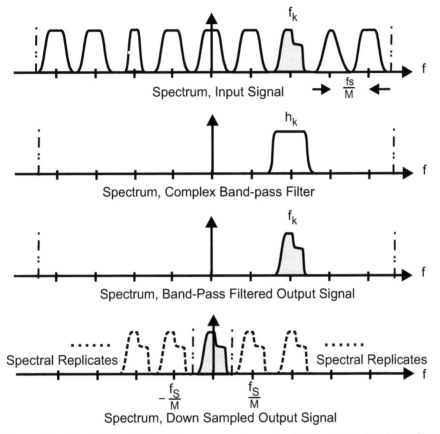

Figure 6.9 Spectral description of down-conversion realized by a complex band-pass filter at a multiple of output sample rate aliased to base band by output resampling.

accomplish this in the following manner: after computing an output sample from the finite impulse response (FIR) filter, we shift the data in the filter register M positions and wait till M new inputs are delivered to the filter before computing the next output sample. In a sense, by moving input data through the filter in stride of length M, we are performing the resampling of the filter at the input port rather than at the output port. Performing the resampling at the filter input essentially reorders the standard operations of the filter followed by a down sample with the operations of down sample followed by the filter. The formal description of the process that accomplishes this interchange is known as the noble identity, which we now describe in detail.

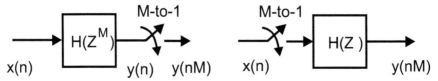

Figure 6.10   Statement of noble identity: a filter processing every math input sample followed by an output M-to-1 down sampler is the same as an input M-to-1 down sampler followed by a filter processing every input sample.

The noble identity is compactly presented in Figure 6.10 which we describe with similar conciseness by "the output from a filter $H(Z^M)$ followed by an M-to-1 down sampler is identical to an M-to-1 down sampler followed by the filter $H(Z)$." The Z^M in the filter impulse response tell us that the coefficients in the filter are separated by M samples rather than the more conventional one sample delay between coefficients in the filter $H(Z)$. We must take care to properly interpret the operation of the M-to-1 down sampler. The interpretation is that the M-to-1 down sampled time series from a filter processing every Mth input sample presents the same output by first down sampling the input by M-to-1 to discard the samples not used by the filter when computing the retained output samples and then operating the filter on only the retained input samples. The noble identity works because M samples of delay at the input clock rate is the same interval as one-sample delay at the output clock rate.

We might ask, "under what condition does a filter manage to operate on every Mth input sample?" We answer by rearranging the description of the filter to establish this condition so that we can invoke the noble identity. This rearrangement starts with an initial partition of the filter into M parallel filter paths. The Z-transform description of this partition is presented in Equations (6.3)−(6.6), which we interpret in Figures 6.11−6.13. For ease of notation, we first examine the base-band version of the noble identity and then trivially extend it to the pass-band version.

$$
\begin{aligned}
H(Z) &= \sum_{n=0}^{N-1} h(n)Z^{-n} \\
&= h(0) + h(1)Z^{-1} + h(2)Z^{-2} + h(3)Z^{-3} \\
&\quad + \cdots + h(N-1)Z^{-(N-1)}.
\end{aligned} \tag{6.3}
$$

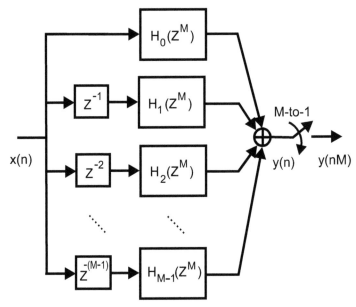

Figure 6.11 *M*-path partition of prototype low-pass filter with output resampler.

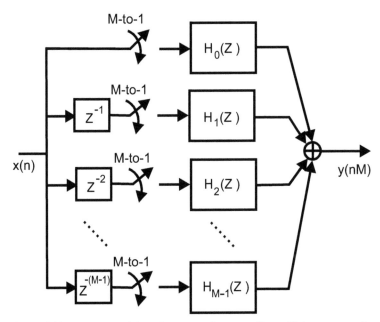

Figure 6.12 *M*-path partition of prototype low-pass filter with input resamplers.

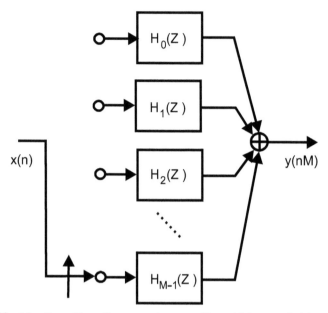

Figure 6.13   *M*-path partition of prototype low-pass filter with input path delays and *M*-to-1 resamplers replaced by input commutator.

Anticipating the M-to-1 resampling, we partition the sum shown in Equation (6.3) to a sum of sums as shown in Equation (6.4). This partition maps a one-dimensional array of weights (and index markers Z^{-n}) to a two-dimensional array. This mapping is sometimes called lexicographic, for natural order, a mapping that occurs in the Cooley–Tukey fast Fourier transform. In this mapping, we load an array by columns but process the array by rows. In our example, the partition forms columns of length M containing M successive terms in the original array and continues to form adjacent M-length columns until we account for all the elements of the original one-dimensional array.

$$
H(Z) = \begin{array}{llllll}
h(0) & + & h(M+0)Z^{-M} & + & h(2M+0)Z^{-2M} & + \cdots \\
+h(1)Z^{-1} & + & h(M+1)Z^{-(M+1)} & +h(2M+1)Z^{-(2M+1)} & + \cdots \\
+h(2)Z^{-2} & + & h(M+2)Z^{-(M+2)} & +h(2M+2)Z^{-(2M+2)} & + \cdots \\
+h(3)Z^{-3} & + & h(M+3)Z^{-(M+3)} & +h(2M+3)Z^{-(2M+3)} & + \cdots \cdot \\
\quad\vdots & & \quad\vdots & \quad\vdots & \quad\vdots \cdots \\
+h(M-1)Z^{-(M-1)} & +h(2M-1)Z^{-(2M-1)} & +h(3M-1)Z^{-(3M-1)} & + \cdots
\end{array}
$$

$$(6.4)$$

We note that the first row of the two-dimensional array is a polynomial in Z^M, which we will denote as $H_0(Z^M)$, a notation to be interpreted as an addressing scheme to start at index 0 and increments in stride of length M. The second row of the same array, while not a polynomial in Z^M, is made into one by factoring the common Z^{-1} term and then identifying this row as $Z^{-1} H_1(Z^M)$. It is easy to see that each row of Equation (6.4) can be described as $Z^{-r} H_r(Z^M)$ so that Equation (6.4) can be rewritten in a compact form as shown in the following equation:

$$
\begin{aligned}
H(Z) = {} & H_0(Z^M) + Z^{-1}H_1(Z^M) + Z^{-2}H_2(Z^M) + Z^{-3}H_3(Z^M) \\
& + Z^{-4}H_4(Z^M) + \cdots + Z^{-(M-2)}H_{(M-2)}(Z^M) \\
& + Z^{-(M-1)}H_{(M-1)}(Z^M).
\end{aligned}
\tag{6.5}
$$

We rewrite Equation (6.5) in the traditional summation form as shown in Equation (6.6), which describes the original polynomial as a sum of delayed polynomials in Z^M. The block diagram reflecting this M-path partition of a resampled digital filter is shown in Figure 6.11. The output formed from the M separate filter stages representing the M separate paths is the same as that obtained from the nonpartitioned filter. We have not yet performed the interchange of filter and resampling.

$$
\begin{aligned}
H(Z) &= \sum_{r=0}^{M-1} Z^{-r} H_r(Z^M) \\
&= \sum_{r=0}^{M-1} Z^{-r} \sum_{n=0}^{\frac{N}{M}-1} h(rnM) Z^{-nM}.
\end{aligned}
\tag{6.6}
$$

We first pull the resampler through the output summation element and down sample the separate outputs by performing the output sum only for the retained output sample points. With the resamplers now at the output of each filter, which operates on every Mth input sample, we are prepared to invoke the noble identity and pull the resampler to the input side of each filter stage. This is shown in Figure 6.12.

The input resamplers operate synchronously, all closing at the same clock cycle. When the switches close, the signal delivered to the filter on the top path is the current input sample. The signal delivered to the filter one path down is the content of the one sample delay line, which, of course, is the previous input sample. Similarly, as we traverse the successive paths of the

M-path partition, we find upon switch closure that the kth path receives a data sample delivered k samples ago.

We conclude that the interaction of the delay lines in each path with the set of synchronous switches can be likened to an input commutator that delivers successive samples to successive legs of the M-path filter. This interpretation is shown in Figure 6.13.

We now complete the final steps of the transformation that changes a standard mixer down-converter to a resampling M-path down-converter. We note and apply the frequency translation property of the Z-transform. This property is illustrated and stated in Equation (6.7). Interpreting the relationship presented in Equation (6.7), we note that if $h(n)$, the impulse response of a base-band filter, has a Z-transform $H(Z)$, then the sequence $h(n)e^{+j\theta n}$, the impulse response of a pass-band filter, has a Z-transform $H(Z\,e^{-j\theta n})$. Simply stated, we can convert a low-pass filter to a band-pass filter by associating the complex heterodyne terms of the modulation process of the filter weights with the delay elements storing the filter weights.

$$\text{If } H(Z) = h(0) + h(1)Z^{-1} + h(2)Z^{-2} + \cdots + h(N-1)Z^{-(N-1)}$$

$$= \sum_{n=0}^{N-1} h(n)Z^{-n}$$

$$\text{and } G(Z) = h(0) + h(1)e^{j\theta}Z^{-1} + h(2)e^{j2\theta}Z^{-2} + \cdots$$
$$+ h(N-1)e^{j(N-1)\theta}Z^{-(N-1)}$$

$$= h(0) + h(1)[e^{-j\theta}Z]^{-1} + h(2)[e^{-j2\theta}Z]^{-2} + \cdots$$
$$+ h(N-1)[e^{-j(N-1)\theta}Z]^{-(N-1)}$$

$$= \sum_{n=0}^{N-1} h(n)[e^{-j\theta}Z]^{-n}$$

$$\text{then } G(Z) = H(Z)|_{Z \Rightarrow e^{-j\theta}Z} = H(e^{-j\theta}Z).$$

(6.7)

We now apply this relationship to Equation (6.2) or, equivalently, to Figure 6.11 by replacing each Z with $Z\,e^{-j\theta}$, or perhaps more clearly, replacing each Z^{-1} with $Z^{-1}\,e^{j\theta}$, with the phase term θ satisfying the congruency constraint of the previous section, that is, $\theta = k\,(2\pi/M)$. Thus, Z^{-1} is replaced with $Z^{-1}\,e^{jk\,(2\pi/M)}$, and Z^{-M} is replaced with $Z^{-M}\,e^{jkM}$ $(2\pi/M)$. By design, the kMth multiple of $2\pi/M$ is a multiple of 2π for which

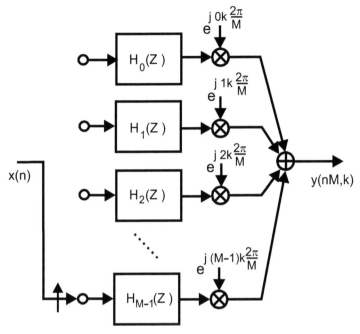

Figure 6.14 Resampling *M*-path down-converter.

the complex phase rotator term defaults to unity or, in our interpretation, aliases to base band (DC or zero frequency). The default to unity of the complex phase rotator occurs in each path of the *M*-path filter shown in Figure 6.14. The nondefault complex phase angles are attached to the delay elements on each of the *M* paths. For these delays, the terms Z^{-r} are replaced by the terms $Z^{-r} e^{jkr}(2\pi/M)$. The complex scalar $e^{jkr(2\pi/M)}$ attached to each path of the *M*-path filter can be placed anywhere along the path, and, in anticipation of the next step, we choose to place the complex scalar after the down sampled path filter segments $H_r(Z)$. This is shown in Figure 6.14.

The modification to the original partitioned *Z*-transform of Equation (6.6) to reflect the added phase rotators of Figure 6.14 is shown in the following equation:

$$H(Ze^{-j\frac{2\pi}{M}k}) = \sum_{r=0}^{M-1} Z^{-r} e^{-j\frac{2\pi}{M}rk} H_r(Z). \qquad (6.8)$$

The computation of the time series obtained from the output summation in Figure 6.14 is shown in Equation (6.9). Here the argument nM reflects the down sampling operation which increments through the time index in stride of length M, delivering every Mth sample of the original output series. The variable $y_r(nM)$ is the nMth sample from the filter segment in the rth path, and $y(nM,k)$ is the nMth time sample of the time series from the kth center frequency. Remember that the down-converted center frequencies located at integer multiples of the output sample frequency are the frequencies that alias to zero frequency under the resampling operation. Note that the output $y(nM,k)$ is computed as a phase coherent summation of the M output series $y_r(nM)$.

$$y(nM, k) = \sum_{r=0}^{M-1} y_r(nM)\, e^{-j\frac{2\pi}{M} r k} \tag{6.9}$$

The M-path down-converter structure is known as a polyphase filter. We will shortly describe the property of the filter partition that is responsible for the name polyphase. We first take a side trip to examine an amazing property of the filter partition and the application of the noble identity. We have already noted that when we down sample by a factor of M-to-1, every multiple of the output sample rate aliases to base band. In Figure 6.8, we used the band-centered digital filter to suppress all but one of the frequency bands that would alias to the base band. After successfully cleaning house, as it were, we were free to down sample and let the only surviving band-pass spectra alias to base band without concern that signal components in other spectral bands would alias on top of our spectral band. During our development of the resampled M-path filter, we blithely used the noble identity to interchange the order of filtering and down sampling. By interchanging the order of filtering and resampling to resampling and filtering, we have reduced the sample rate prior to the bandwidth reduction and have caused M-fold aliasing of the input spectrum. It seems that we have started the process by violating the Nyquist criterion. We have indeed violated the Nyquist criterion and, in fact, have violated it M times at each of the M commutator ports. We will show that the M different realizations of the M-fold aliasing represent M distinct equations with M distinct aliasing terms and hence contain sufficient information to separate and remove the aliases. The polyphase filter structure performs the processing required to separate the aliases formed by the input resampling.

6.2 Separating the Aliases

Each of the M time series observed at the ports of the input commutator represents M-fold aliased data. The aliases are the spectral terms from each multiple of the output sample rate, which, due to the resampling by the factor M-to-1, appear at base band centered at DC. The alias from any particular center frequency exhibits a unique phase profile across the set of M aliased time series. Prior to aliasing, the signal component at the kth center frequency ($k \, 2\pi/M$) exhibits an arbitrary phase angle, and at each successive data sample, the kth center frequency experiences a phase rotation of $k \, 2\pi/M$. In the same way a heterodyned signal preserves the phase of an input sinusoid, the spectral terms that alias to base band when down sampled also preserve their phase. Thus, as we progress up the successive ports of the input commutator, the kth alias exhibits successive phase increments of $k \, 2\pi/M$. Starting at the top of the commutator and counting down to the rth port, we would find a phase angle for the kth center frequency equal to $-r \, k \, 2\pi/M$. The phase rotators shown in Equation (6.9) and in Figure 6.14 are seen to be the phase shift required to de-rotate and align the phase shifts from the kth center frequency seen at successive ports of the input commutator. When the sum of the phase aligned terms are formed in Equation (6.9), the remaining aliasing terms, residing on the M roots of unity, sum to zero and are destructively canceled during the summation. The destructive cancelation of the aliased terms is the process that separates the aliases formed by the resampling process that occurs at the input commutator.

This phase coherent sum is, in fact, a discrete Fourier transform (DFT) of the M-path outputs, which can be likened to beam-forming the output of the path filters. The beam-forming perspective offers interesting insight into the operation of the resampled down-converter system we have just examined. The reasoning proceeds as follows: the commutator delivering consecutive samples to the M input ports of the M-path filter performs a down sampling operation. Each port of the M-path filter receives data at one-Mth of the input rate. The down sampling causes the M-to-1 spectral folding, effectively translating the M multiples of the output sample rate to the base band. The alias terms in each path of the M-path filter exhibit unique phase profiles due to their distinct center frequencies and the time offsets of the different down sampled time series delivered to each port. These time offsets are, in fact, the input delays shown in Figure 6.11 and in Equation (6.10). Each of the aliased center frequency experiences a phase shift shown in Equation (6.10), equal to the product of its center frequency and the path time delay.

$$\phi(r, k) = -\omega_k \Delta T_r$$

$$= -2\pi \frac{fs}{M} k \, r \, T_S \tag{6.10}$$

$$= -2\pi \frac{fs}{M} k \, r \, \frac{1}{fs} = -\frac{2\pi}{M} k \, r.$$

The phase shifters of the DFT perform phase coherent summation, very much like that performed in narrow-band beam-forming, extracting from the myriad aliased time series, the alias with the particular matching phase profile. This phase sensitive summation aligns contributions from the desired alias to realize the processing gain of the coherent sum while the remaining alias terms, which exhibit rotation rates corresponding to the M roots of unity, are destructively canceled in the summation.

The inputs to the M-path filter are not narrowband, and phase shift alone is insufficient to effect the destructive cancelation over the full bandwidth of the undesired spectral contributions. Continuing with our beam-forming perspective, to successfully separate wideband signals with unique phase profiles due to the input commutator delays, we must perform the equivalent of time-delay beam-forming. The M-path filters, obtained by M-to-1 down sampling of the prototype low-pass filter, supply the required time delays. The M-path filters are approximations to all-pass filters, exhibiting, over the channel bandwidth, equal ripple approximation to unity gain and the set of linear phase shifts that provide the time delays required for the time delay beam-forming task. The inputs to the polyphase paths are delivered sequentially with time offsets. The separate time series must be time aligned at the output of the polyphase filters in order to perform the summation required to compute the output samples. Figure 6.15 illustrates the task required of the path filters of the polyphase partition, while Figure 6.16 illustrates the time alignment of the various path sequences for a commutated input series.

The filter achieves this time alignment process by virtue of the way we partitioned the low-pass prototype. As shown in Equation (6.6), we formed each of the M-path filters, filter $h_r(n)$ for instance, with weights $h(r + nM)$ starting at an initial offset of r samples and then incrementing by stride of M samples. In the mapping of the polyphase partition, the r sample offset is suppressed, as the rth sample becomes the initial weight for the rth filter path. These initial offsets, unique to each path, are the source of the different linear phase shift profiles. It is for this reason, the different linear phase profiles, that the filter partition is known as a *polyphase* filter. The phase shift and group

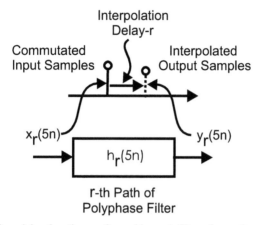

Figure 6.15 Time delay function performed by path filter of m-path polyphase partition.

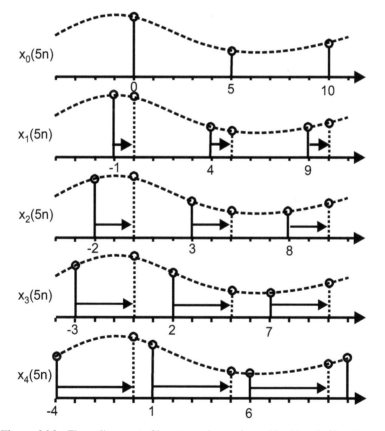

Figure 6.16 Time alignment of input samples performed by *M*-path filter stages.

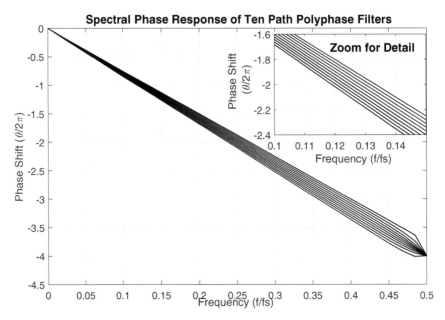

Figure 6.17   Phase profiles for 10 paths of 10-path polyphase partition.

delay profiles for the filter weights of a 10-path polyphase filter are shown in Figures 6.17 and 6.18. Figures 6.19 and 6.20 present the phase profile of the weights $h(r + nM)$ formed at the initial step of the polyphase partition. Here we have not yet dropped the $(r - 1)$ leading zeros or the $(M - 1)$ intersample zeros, so the weights still share a common time origin and offer the spectral replicates of the zero packing. We clearly see that in the successive Nyquist zones, the 10-phase profiles have the same slope but exhibit phase offsets that are multiples of the Nyquist zone number k times $2\pi/10$. The phase rotators after the polyphase rotators access a selected Nyquist zone by canceling the phase offsets of that zone. Figure 6.20 is a set of subplots presenting zooms of the phase profiles showing the phase transition between the zones as well as the fixed phase offsets unique to that zone. These figures are part of the output suite of figures formed by the MATLAB *m*-file *filter_ten* accompanying this text. This file synthesizes a 10-stage polyphase channelizer and presents input and output time series and spectra of the channelizers as well as various figures demonstrating characteristics of the 10-path filter.

A useful perspective is that the phase rotators following the filters perform phase alignment of the band center for each aliased spectral channel or

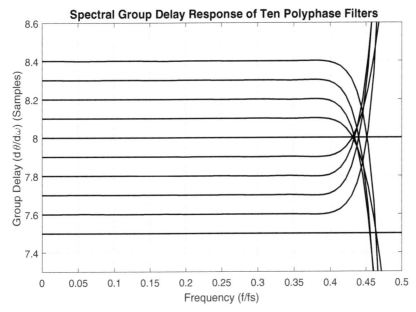

Figure 6.18 Group delay profiles for 10 paths of 10-path polyphase partition.

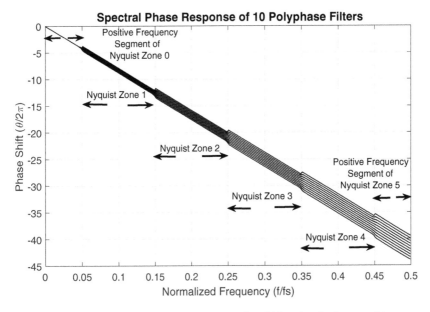

Figure 6.19 Phase profiles of 10 separate paths of 10-path polyphase partition.

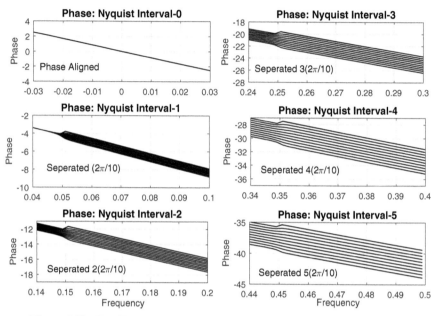

Figure 6.20   Details of phase profiles of 10 paths of 10-path polyphase partition.

Nyquist band, while the polyphase filters perform the required differential phase shift across these same channel bandwidths to compensate for the offset time origins of the commutated data streams delivered to each path. When the polyphase filter is used to down-convert and down sample a single channel, the phase rotators are implemented as external complex products following each path filter. When a small number of channels are being down-converted and down sampled, appropriate parallel sets of phase rotators can be applied to the filter stage outputs and summed to form each channel output. We take a different approach when the number of channels becomes sufficiently large. Here sufficiently large means on the order of $\log_2(N)$. Since the phase rotators following the polyphase filter stages are the same as the phase rotators of a DFT, we can use the DFT to simultaneously apply the phase shifters for all of the channels we wish to extract from the aliased signal set. This is the structure shown in Figure 6.21. This is reminiscent of phased array beam-forming. For computational efficiency, the fast Fourier transfer (FFT) algorithm implements the DFT. Additional material in this vein will be found in the section describing multichannel receivers.

It is instructive to compare the conventional mixer down-converter and the resampled polyphase down-converter. The input to either process can

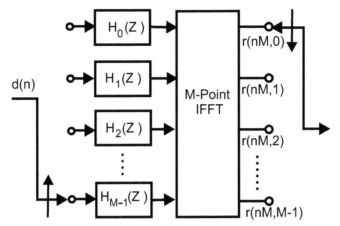

Figure 6.21 Polyphase filter bank: *M*-path polyphase filter and *M*-point FFT.

be real or complex. In the mixer down-converter model, a separate mixer pair and filter pair must be assigned to each channel of the channelizer, and these mixers and filters all operate at the high input data rate, prior to down sampling. By way of contrast, in the resampled polyphase, there is only one low-pass filter required to service all the channels of the channelizer, and this single filter accommodates all the channels as co-occupying alias contributors of the base-band bandwidth. This means that all the processing performed in the resampled polyphase channelizer occurs at the low output sample rate. When the input signal is real, there is another significant difference between the two processes. In the mixer down-converter model, the signal is made complex by the input mixers as we enter the filtering process, which means that the low-pass filtering task requires two filters, one for each of the quadrature components, while in the resampling channelizer, the signal is made complex by the phase rotators as we leave the process. Consequently, we require only one partitioned low-pass filter to process the real input signal.

Let us summarize what we have accomplished to this point. The commutator performs an input sample rate reduction by commutating successive input samples to selected paths of the *M*-path filter. Sample rate reduction occurring prior to any signal processing causes spectral regions residing at multiples of the output sample rate to alias to base band. This desired result allows us to replace the many down-converters of a standard channelizer, implemented with dual mixers, quadrature oscillators, and bandwidth reducing filters, with a collection of trivial aliasing operations performed in a single partitioned and resampled filter.

The partitioned *M*-path filter performs the task of aligning the time origins of the offset sampled data sequences delivered by the input commutator to a single common output time origin. This is accomplished by the all-pass characteristics of the *M*-path filter sections that apply the required differential time delay to the individual input time series. The DFT performs the equivalent of a beam-forming operation: the coherent summation of the time-aligned signals at each output port with selected phase profiles. The phase coherent summation of the outputs of the *M*-path filters separate the various aliases residing in each path by constructively summing the selected aliased frequency components located in each path, while simultaneously destructively canceling the remaining aliased spectral components.

This section introduced the structure of a process that can perform down-conversion of one or more channels using a polyphase filter. A similar exposition can be mounted for a process that up-converts one or more channels with a polyphase filter. We discuss both options in detail in a later chapter. For now, it suffices to state that the up-converter process is the dual process of the down-converter process. The dual process simply reverses all signal flow of the original process. In the dual structure, we enter the channel process at the phase rotators (or DFT) and leave the process by the polyphase commutator. Reversing the signal flow results in a process that up samples and up-converts rather than one that down-converts and down samples.

References

Crochiere, Ronald, and Lawrence Rabiner. *Multirate Signal Processing*, Englewood Cliff, NJ: Prentice-Hall Inc, 1983.

Fliege, Norbert. *Multirate Digital Signal Processing: Multirate Systems, Filter Banks, Wavelets*, West Sussex: John Wiley & Sons, Ltd, 1994.

Harada, Hiroshi and Ramjee Prasad. *Simulation and Software Radio for Mobile Communications*, Norwood, MA: Artech House, 2002.

Hentschel, Tim. *Sample Rate Conversion in Software Configurable Radios*, Norwood, MA: Artech House, 2002.

Jovanovic-Dolecek, Gordana, *Multirate Systems: Design and Applications*, London: Idea Group, 2002.

Lutovac, Miroslav, Dejan Tošić and Brian Evans. *Filter Design for Signal Processing: Using MATLAB and Mathematica*, Upper Saddle River, NJ: Prentice Hall Inc, 2001.

Mitra, Sanjit and James Kaiser. *Handbook for Digital Signal Processing*, New York: John Wiley & Sons, 1993.

Mitra, Sanjit *Digital Signal Processing: A Computer-Based Approach*, 2nd ed., New York:, McGraw- Hill, 2001.

Paul Nahin. *The Science of Radio*, 2nd ed. New York: Springer-Verlag, 2001.

Vaidyanathan, P. P. *Multirate Systems and Filter Banks*, Englewood Cliff, NJ, Prentice-Hall Inc., 1993.

6.3 Problems

6.1 Design a low-pass FIR filter satisfying the following specifications:

Sample Rate	100 kHz		
Pass Band	0–4 kHz	In-Band Ripple	0.1 dB
Stop Band	6–50 kHz	Stop-Band Atten.	60 dB

Form an input sequence as 2000 samples of a real sinusoid at frequency 32.0 kHz with random phase. Use a complex heterodyne centered at 30 kHz to down-convert the input signal and present the down-converted data to the base-band filter designed previously. Form subplots of the 10-to-1 down sampled time series from the filter pair and the log magnitude windowed spectrum of the same time series.

Now up-convert the base-band low-pass filter designed earlier to form a complex band centered filter centered at 30 kHz. Pass the input signal through the complex filter and down-convert the complex output with the 30 kHz heterodyne. Form subplots of the 10-to-1 down sampled time series from the sequence following the heterodyne and the log magnitude windowed spectrum of the same time series.

Finally, form a 10-to-1 down sampled time series without the complex heterodyne. Form subplots of this 10-to-1 down sampled time series and the log magnitude windowed spectrum of the same time series.

Compare the three time series and their corresponding spectra. They should be the same signal and should have the same spectra. We are using this exercise to illustrate the equivalency theorem.

6.2 Consider a trivial filter corresponding to the Z-transform shown here:

$$H(Z) = 1 + Z^{-10} + Z^{-20} + Z^{-30} + Z^{-40}.$$

Form subplots of the impulse response and magnitude spectrum of this filter.

Now down sample the impulse response by 10-to-1 and form subplots of the down sampled impulse response series and its magnitude spectrum.

Apply 500 samples of a real sinusoid at normalized frequency 0.11 to the original filter and then plot the time series of the 10-to-1 down sampled output. Also pass the 10-to-1 down sampled input series through the 10-to-1 down sampled filter and plot the output time response. Compare the two time series and comment on the results.

6.3 Design a low-pass FIR filter satisfying the following specifications with the filter length increased to the next multiple of 10:

Sample Rate	100 kHz		
Pass Band	0–4 kHz	In-Band Ripple	0.1 dB
Stop Band	6–50 kHz	Stop-Band Atten.	60 dB

Perform a 10-path polyphase partition of this impulse response and form and plot on a single subplot the magnitude frequency response of all 10 paths. Also form and plot on a single subplot the unwrapped phase (angle) response of all 10 paths.

Comment on what the similarity is and what the difference is about the 10 overlaid magnitude and phase response plots.

6.4 Design a low-pass FIR filter satisfying the following specifications with the filter length increased to the next multiple of 10:

Sample Rate	100 kHz		
Pass Band	0–4 kHz	In-Band Ripple	0.1 dB
Stop Band	6–50 kHz	Stop-Band Atten.	60 dB

Perform a 10-path polyphase partition of this impulse response and convolve each path of the partition with the Gaussian sequence exp(5*(−0.5:0.1:0.5).*(−0.5:0.1:0.5)).

On 10 subplots, plot the input and output series from the 10 convolutions. The successive time series represent the time-shifted samples of the input signal. Can you explain the amount of time shift exhibited by each output time series?

6.5 Design a low-pass FIR filter satisfying the following specifications with the filter length increased to the next multiple of 10:

Sample Rate	100 kHz		
Pass Band	0–4 kHz	In-Band Ripple	0.1 dB
Stop Band	6–50 kHz	Stop-Band Atten.	60 dB

Perform a 10-path polyphase partition of this filter and follow the path outputs with a 10-point FFT. Generate a 2000 sample input time series comprising the sum of two complex sinusoids at frequency 10.5 kHz and at –22 kHz. Feed the polyphase filter with 10 input samples per input cycle and compute the 10 separate outputs to form the 10 time series from the 10 polyphase paths. Transform the time series from two paths and identify the spectral aliases. Now perform the FFT on the polyphase vector sequences to form the 10 phase corrected time series that correspond to the channelized filter bank. Plot the real part and the imaginary part of each time series in a set of subplots. Identify the transients associated with the unoccupied channels and the time responses of the two occupied channels. Also form and plot the windowed FFT of the 10 time series. Identify the spectral lines that correspond to the two input spectral lines in the occupied channels as well as the suppressed copies of these lines in the unoccupied channels.

7

Resampling Filters

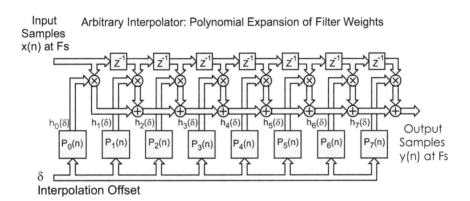

Input Samples x(n) at Fs

Arbitrary Interpolator: Polynomial Expansion of Filter Weights

δ

Interpolation Offset

In this chapter, we discuss variants of the interpolation process and present a number of applications dependent on the process. The simplest such process is a 1-to-M up-sampler where M is an arbitrary integer. Many applications require up-sampling by a small integer in the range $4-10$, but the integer M may be in the range of hundreds to thousands. When M is a large nonprime integer, the up-sampling is performed by a sequence of low-order up-sampling filters. The next most common interpolation process is interpolation by a rational ratio P/Q, where both P and Q can be arbitrary integers but most often are small integers. A common interpolation task is one that requires arbitrary resampling including slowly varying time varying resampling. The later examples are very common in modern digital receivers. The filters participate in the timing recovery process by interpolating input sample values to offset output time samples. We will present these examples and others as well as present mechanisms for control of the time varying interpolation process.

167

7.1 Interpolators

My Webster's Second Collegiate Dictionary lists, in its third entry, a math definition of *interpolate* as: "To estimate a missing functional value by taking a weighted average of known functional values at neighboring points." Not bad, and that certainly describes the processing performed in a multirate filter. Interpolation is an old skill that many of us learned before the advent of calculators and key strokes replaced tables of transcendental functions such as $\log(x)$ and the various trigonometry functions. Take, for example, the NBS Applied Mathematics Series, AMS-55 *Handbook of Mathematical Functions with Formulas, Graphs, and Mathematical Tables* by Abramowitz and Stegan. This publication contains tables listing functional values of $\sin(\theta)$ for values of θ equal to ... 40.0, 40.1, 40.2,..., etc. Interpolation is required to determine the value of $\sin(\theta)$ for the value of θ equal to 40.137. Interpolation was such an important tool in numerical analysis that three pages in the introduction of the handbook are devoted to the interpolation process. Interpolation continues to be an important tool in signal processing and we now present the digital signal processing (DSP) description of the interpolation process.

7.1.1 Simple 1-to-M Interpolator

A very common interpolator task is raising the sample rate of a sampled data sequence that is already oversampled by a factor of 2 or by a factor of 4. This happens, for instance, when the original data is the output of a shaping filter in a modulator in which the sample rate is increased relative to the symbol rate to accommodate the excess bandwidth of the shaping filter. The interpolator following the original shaping filter is used to raise the sample rate so that the first spectral translation to an intermediate frequency is performed in the DSP domain rather than in the analog domain. To compensate for sinc response of the output digital-to-analog converter (DAC), the up converter is followed by a $[\sin(x)/(x)]^{-1}$ predistortion filter and then an analog smoothing filter. A block diagram of such a system is shown in Figure 7.1.

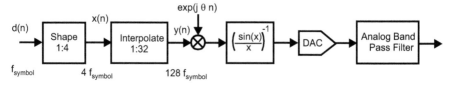

Figure 7.1   Example of interpolator following initial time series generator.

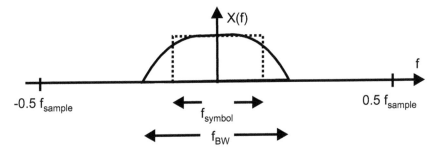

Figure 7.2 Spectrum at output of shaping filter with embedded 1-to-4 up-sampler.

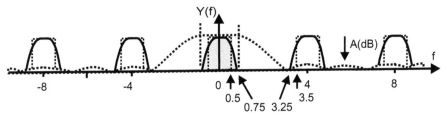

Figure 7.3 Periodic input spectrum and filter response of 1-to-32 interpolator filter.

Figure 7.2 presents a typical output spectrum at the output of the shaping filter. In this example, we can assume a square root Nyquist filter with 50% roll-off with side-lobe levels held to –60 dB. We will use normalized frequency, with the reference being the symbol rate which we can call "1" or if we require a typical value, we will use 100 kHz. The two-sided bandwidth in this example is 1.5 or 150 kHz, and the sample rate is 4 or 400 kHz. We now examine the design, properties, and performance of the 1-to-32 up-sampler that is to raise the sample rate to 128 or 12.8 MHz.

Figure 7.3 presents the periodic spectrum of the input sampled data signal with an overlay approximating the frequency response of the interpolating filter that is designed to suppress the spectral replicates. The performance specifications required of this filter are listed in Table 7.1.

An estimate of the number of filter taps required to implement the 1-to-32 interpolating filter is obtained from standard estimate rules or algorithms. One such rule, shown in Equation (7.1), leads to the estimate shown in Equation (7.2).

$$N \cong \frac{f_S}{\Delta f} K(\delta_1, \delta_2) \approx \frac{f_S}{\Delta f} \frac{\text{attn(dB)}}{22} \qquad (7.1)$$

Table 7.1   Parameters required for interpolating filter.

Parameters	Values
Pass-band ripple	0.1 dB
Stop-band attenuation	60 dB
Pass-band frequencies	0–0.75 (0–75 kHz)
Stop-band frequencies	3.25–64 (325 kHz to 6.4 MHz)
Sample frequency	128 (12.8 MHz)

$$N \cong \frac{128}{(3.25 - 0.75)} \frac{60}{22} = 139.6 \Rightarrow 140. \tag{7.2}$$

An estimate for the filter length obtained by a call to the MATLAB script file *firpmord* is shown next. The help response to *firpmord* suggests that the estimates are understated which we have also observed.

```
NN=firpmord([0.75 3.25],[1 0],[0.01 0.001],128)
```

The response to this call is

$$NN = 130.$$

The two estimates of filter length are comparable, but the actual filter length must be verified by applying the *firpm* algorithm to the design task. The filter length required to meet the ripple specifications was found, in fact, to be 140. We have the option of modifying the pass-band and stop-band frequency edges to widen the transition band and hopefully meet the requirements with a 128-point filter. We are permitted to do this because the input spectra at the stop-band edge has significantly less energy than at the band center and thus does not require as much attenuation to meet the desired spectral mask. Similarly, the filter rolls off very slowly at the pass-band edge and since the input signal has very little spectral energy at this edge, there is essentially no signal degradation if the pass-band edge overlaps the input signal's transition band edge.

It would be desirable to have the filter length be 128 taps so that the 32-polyphase stages would all be of length 4 taps. This requirement assures that all stages are of the same length. To allow a length 128 filter to meet the attenuation specifications, the band edge specifications of the filter were modified slightly. The next length chosen would be 32 times 5 or 160 taps, but we stayed with the 128. This entailed allowing overlap of the transition band and the first spectral replicate. By trial and error, we found that widening the transition band by moving the stop-band edge from 3.3 to 3.5, the 128-point filter would meet the required stop-band attenuation. The filter specifications

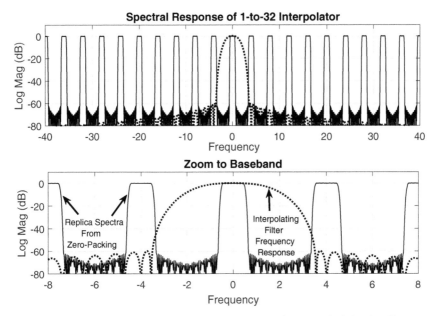

Figure 7.4 Spectrum of interpolator and spectrum of zero-packed shaping filter.

were also modified to obtain a nonequiripple stop band by the use of a frequency-dependent penalty sequence in the penalty vector of the *firpm* algorithm.

Figure 7.4 presents the spectrum of the 1-to-32 interpolating filter overlaid on the replicated spectrum of the shaping filter over the full bandwidth and a zoomed bandwidth to show the stop-band detail. Here we plainly see the overlap of the filter's transition band with the adjacent channel. Figure 7.5 presents the result of the interpolation process with this interpolation filter. Here we see that the adjacent spectral copies are suppressed by at least 60 dB and that the spectra residing in the overlapped transition band is also attenuated by 60 dB.

Using an alternate strategy, we can redefine the interpolating filter to have multiple stop bands of width ±0.7 bracketing the periodic input spectra centered at multiples of 4. We treat the intervals between the stop bands as do not care regions. These regions are permitted since we know that they contain no significant spectral content by design of the oversampled-by-4 shaping filter. Since there is no need to suppress the spectral interval between the periodic spectra, we can release the stop-band zeros normally assigned to this

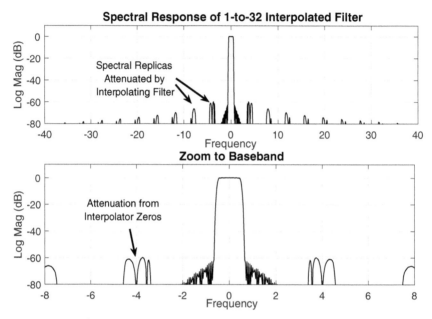

Figure 7.5   Spectrum of 1-to-32 interpolated shaping filter.

region and permit the *firpm* algorithm to use them in the desired stop-band regions. This filter exhibits additional attenuation and smaller in-band ripple than the previous filter design with full stop-band attenuation. The spectra of the filter designed in this manner and of the filtered data obtained with this filter are shown in Figures 7.6 and 7.7.

Figures 7.4 and 7.6 presented the frequency responses of the interpolator filter designed with a single stop band and then with multiple stop bands interspersed with do-not-care bands. As mentioned, use of the do-not-care bands permitted the design routine to obtain improved suppression in the spectral regions containing the spectral replicates. An unexpected consequence of the design option using the do-not-care regions is the change in the filter's impulse response. Figure 7.8 presents the impulse response of the two versions of the interpolator filter whose spectra appeared in Figures 7.4 and 7.6. We call the impulse response of the second design the Tibetan Hat filter. The outlier samples that appear in this impulse response are reminiscent of the end-point outliers generated by the *firpm* algorithm designs for filters with equiripple stop band. These outliers present no problem to the implementation of the interpolator. They simply map to the first or last

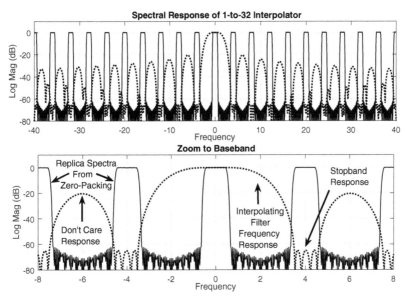

Figure 7.6 Spectrum of alternate interpolator and periodic spectrum of zero-packed shaping filter.

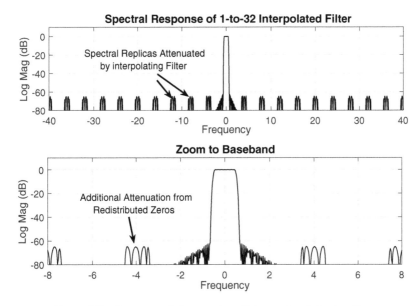

Figure 7.7 Spectrum of alternate 1-to-32 interpolated shaping filter.

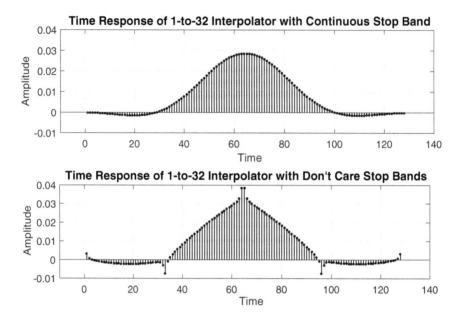

Figure 7.8 Impulse response of interpolator filter with single stop band and with multiple stop bands with do-not-care band.

segments of the 32-stage polyphase partition. We only called attention to the impulse response because of its unusual shape.

7.2 Interpolator Architecture

We now examine the architectural structure of the interpolator derived in the previous section. The initial model of the 1-to-32 interpolator is a 1-to-32 up-sampler followed by an appropriate low-pass filter. In this configuration, the zero packed up-sampler redefines the Nyquist interval, the observable frequency span, from the input sample rate to a span 32 times as wide, the output sample rate. The 1-to-32 up-sampler zero-packs the input series, effectively decreasing the distance between input samples without modifying the spectral content of the series. The wider Nyquist interval, spanning 32 input Nyquist intervals, presents 32 spectral copies of the input spectrum to the finite impulse response (FIR) filter. The amplitude of each of the 32 copies is 1/32 of the amplitude of the input signal spectrum. When an FIR low-pass filter is scaled so that the peak coefficient is unity, the filter exhibits

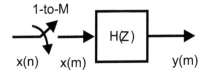

Figure 7.9   Initial structure of 1-to-32 interpolator.

a processing gain inversely proportional to its fractional bandwidth. Thus, as the filter eliminates the 31 spectral copies, reducing the bandwidth by a factor of 1/32, the filter gain precisely compensates for the attenuation of the input spectra due to the zero-packing of the input series. This structure is shown in Figure 7.9.

7.2.1 Polyphase Partition

The zeros of the zero-packed input time series do not contribute to the weighted sums formed at the filter output. Since they do not contribute, there is no need to perform the product sum from the input data registers containing the known zero-valued samples. Since only the nonzero-packed samples contribute to the filter output, we can track their location in the filter and perform the weighted sum only from the register locations containing these samples. These locations are separated by 32 sample values and their position shifts through the filter as each new zero-valued input is presented to the input of the filter. Keeping track of the coefficient stride and the position of the each coefficient set is automatically performed by the polyphase partition of the filter as shown next. Equation (7.3) is the Z-transform of the standard FIR filter structure representing a set of delayed filter coefficients. Equation (7.4) is the polyphase partition of the same filter representing the filter as a sum of successively delayed sub-filters with coefficients separated by stride of M samples. Finally, Equation (7.5) is a compact representation of Equation (7.4) where the rth stage $H_r(Z^M)$ of the polyphase filter is formed by the coefficient set that starts at index r and increments in steps of length M.

$$H(Z) = \sum_{n=0}^{N-1} h(n)\, Z^{-n} \tag{7.3}$$

$$H(Z) = \sum_{r=0}^{M-1} Z^{-r} \sum_{n=0}^{\frac{N}{M}-1} h(r + nM)\, Z^{-nM} \tag{7.4}$$

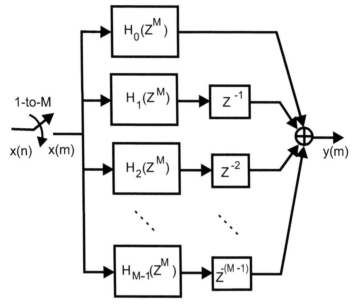

Figure 7.10   Initial structure of 1-to-*M* polyphase interpolator.

$$H(Z) = \sum_{r=0}^{M-1} Z^{-r} H_r(Z^M). \qquad (7.5)$$

The form of the filter in the *M*-path polyphase partition is shown in Figure 7.10. This structure enables the application of the noble identity in which we slide the resampler through the filter and replace the *M* units of delay at the output clock rate with one unit of delay at the input clock rate. This form of the filter is shown in Figure 7.11. Note that the resampler cannot slide through the delays Z^{-r} following each filter segment H_r.

The resamplers following the separate filter stages up-sample each time series by a factor of *M* and the delays in each arm shift each resulting time series a different time increment so that only one nonzero time sample is presented to the summing junction at each output time. Thus, rather than performing the sum with multiple zeros, we can simply point to the arm that sequentially supplies the nonzeros sample. The output commutator in Figure 7.12 performs this selective access.

We note that the polyphase arms contribute their outputs one at a time as the commutator points to successive output ports. We also note that the separate filters all contain the same input data and differ by only their unique coefficient sets. We can replace the *M*-path version of the polyphase

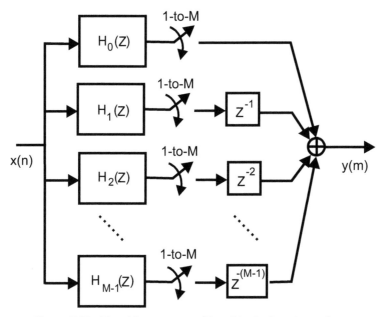

Figure 7.11 Transition structure of 1-to-*M* polyphase interpolator.

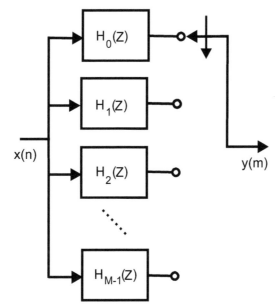

Figure 7.12 Standard structure of 1-to-*M* polyphase interpolator.

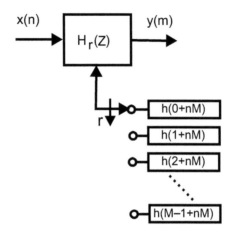

Figure 7.13   Efficient implementation structure of 1-to-*M* polyphase interpolator.

filter with a single stage filter with M coefficient sets that are sequentially presented to the filter to compute successive outputs. This structure is shown in Figure 7.13. What we have accomplished here is, first, move the input commutator that performed the input zero-packing to the output of the filter where it selected successive filter outputs and, second, move the commutator to coefficient memory where it selected successive coefficient sets rather than filtering outputs.

7.3 Band-Pass Interpolator

This option reduces system cost by using a single DAC and band-pass filter to replace the standard baseband process requiring matched DACs, matched low-pass filters, matched balanced mixers, and a quadrature oscillator to form the first IF frequency band. For completeness, the baseband version of the modulator structure comparison is shown in Figure 7.14 and the digital IF version is shown in Figure 7.15.

The first model of the interpolating process zero-packed the input data to give us access to 32 spectral copies of the input time series. The spectral copies reside at multiples of the input sample rate, at normalized frequencies 0, 4, 8, 12, ..., 60, and 64. The low-pass filter rejected the spectral copies, retrieving the baseband copy at frequency 0 that was then passed on to the digital up converter for translation to the desired center frequency. It is possible to perform the spectral translation as part of the interpolation process

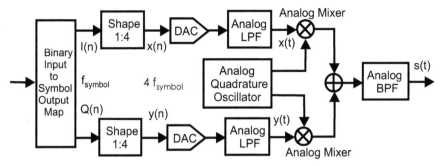

Figure 7.14 Baseband modulator with dual analog components to be replaced by digital components in digital IF modulator.

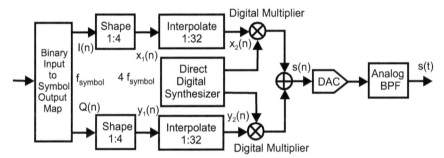

Figure 7.15 Digital baseband and IF modulator with minimum analog components.

when the desired center frequency coincides with one of the multiples of the input sample rate. Rather than extracting the spectral copy at baseband from the replicated set of spectra, we can directly extract one of the spectral translates by using a band-pass filter as opposed to a low-pass filter. The band-pass filter is simply an up converted version of the low-pass filter with weights shown in Equation (7.6). Here the center frequency is interpreted as the kth multiple of $1/M$th of the output sample rate.

$$
\begin{aligned}
&\text{Low Pass}: \quad h(n) \\
&\text{Band Pass}: \quad g(n,k) = h(n)\exp\left(j\tfrac{2\pi}{M}nk\right).
\end{aligned}
\tag{7.6}
$$

The Z-transform of the two filters, the low pass, and the band pass are shown in the following equation:

$$
\text{Low Pass}: \quad H(Z) = \sum_{n=0}^{N-1} h(n)\,Z^{-n}
$$

$$\text{Band Pass}: \quad G_k(Z) = \sum_{n=0}^{N-1} h(n)\exp(j\frac{2\pi}{M}nk)\,Z^{-n}. \qquad (7.7)$$

The Z-transform of the polyphase versions of the two filters is shown in Equation (7.8). Here we see that M-length stride in coefficient index due to the 1-to-M resampling aliases the phase rotators in the polyphase filter stages to DC and hence have no effect on the polyphase weights. The phase rotator does however have a contribution related to the delay associated with each arm of the partition. The output commutator process absorbs the delays while the phase rotators are applied to each arm to obtain the spectral translation as part of the interpolation.

Low Pass :

$$H(Z) = \sum_{r=0}^{M-1} Z^{-r} \sum_{n=0}^{\frac{N}{M}-1} h(r+nM)\,Z^{-nM}$$

Band Pass :

$$G_k(Z) = \sum_{r=0}^{M-1} Z^{-r} \exp\left(j\frac{2\pi}{M}rk\right) \sum_{n=0}^{\frac{N}{M}-1} h(r+nM) \qquad (7.8)$$

$$\times \exp\left(j\frac{2\pi}{M}nMk\right)\,Z^{-nM}$$

$$= \sum_{r=0}^{M-1} Z^{-r} \exp(j\frac{2\pi}{M}rk) \sum_{n=0}^{\frac{N}{M}-1} h(r+nM)\,Z^{-nM}.$$

The filter that accomplishes the task of simultaneous interpolation and up conversion to the kth Nyquist zone is shown in Figure 7.16.

Figure 7.17 presents the spectrum of the 1-to-32 interpolating filter with the phase rotators tuned to the third Nyquist zone. This spectrum is overlaid on the replicated spectrum of the shaping filter over the full bandwidth and a zoomed bandwidth to show stop-band detail. Here we plainly see the phase rotators have positioned the spectrum of the filter to recover the third Nyquist zone centered at normalized frequency 12. Figure 7.18 presents the result of the interpolation process with this interpolation filter. Here we see that the band passed by the filter is the third Nyquist zone and that the channels adjacent to this zone are suppressed by the filtering action. We have an interest only in the real part of the time series associated with the spectrum shown in

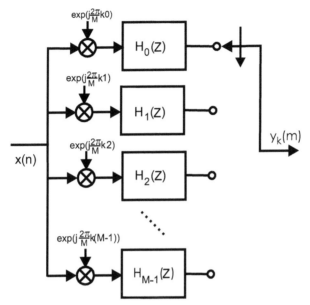

Figure 7.16 Structure of 1-to-*M* polyphase interpolator with phase rotators for selecting spectrum centered in *M*th Nyquist.

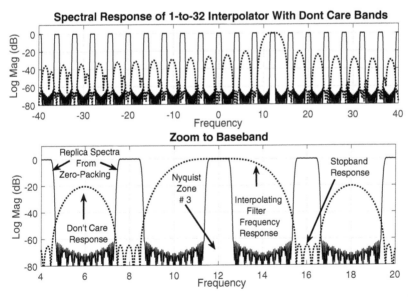

Figure 7.17 Spectrum of translated interpolator and periodic spectrum of zero-packed shaping filter.

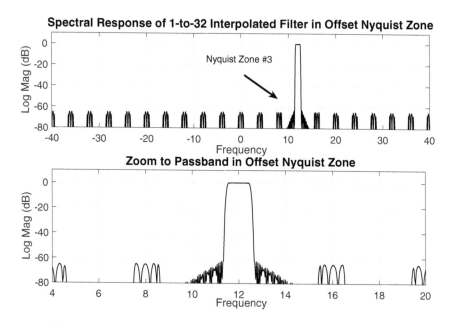

Figure 7.18   Spectrum of 1-to-32 interpolated and translated shaping filter.

Figure 7.18. The spectrum shown in Figure 7.18 represents the most general case of processing complex input data with a complex filter requiring four convolutions as shown in Equation (7.9). The real part is generated with only two convolutions as shown in Equation (7.10), which is also indicated in the block diagram of Figure 7.15.

$$
\begin{aligned}
d_{\mathrm{IF}}(n) &= [x_{\mathrm{in}}(n) + j\, y_{\mathrm{in}}(n)] \cdot [h_{\mathrm{RL}}(n) + j\, h_{\mathrm{IM}}(n)] \\
&= [x_{\mathrm{in}}(n) \cdot h_{\mathrm{RL}}(n) - y_{\mathrm{in}}(n) \cdot h_{\mathrm{IM}}(n)] \\
&\quad + j[x_{\mathrm{in}}(n) \cdot h_{\mathrm{IM}}(n) + y_{\mathrm{in}}(n) \cdot h_{\mathrm{RL}}(n)]
\end{aligned}
\tag{7.9}
$$

$$
\mathrm{Real}[d_{\mathrm{IF}}(n)] = [x_{\mathrm{in}}(n) \cdot h_{\mathrm{RL}}(n) - y_{\mathrm{in}}(n) \cdot h_{\mathrm{IM}}(n)].
\tag{7.10}
$$

7.4 Rational Ratio Resampling

We know how to up-sample a time series by any integer. This integer is usually small, say up 4 or up 8, and in the example examined in the previous section, up 32. There are occasions that the desired resampling ratio is a ratio of integers, say up P and down Q for an increase in sample rate of P/Q. While

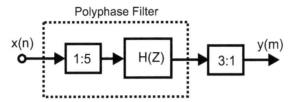

Figure 7.19 Straightforward approach to 5/3 interpolator.

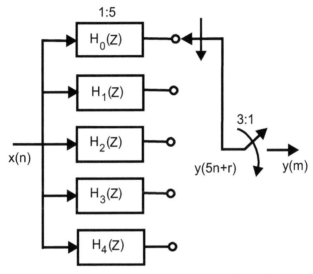

Figure 7.20 Detailed presentation of 5/3 interpolator.

there is no restriction on the values of the two integers P and Q, they are often selected to small integers such as up 5 and down 3. We will use the 5/3 ratio as an example to demonstrate rational ratio resampling. The resampling process can be visualized as a two-step process. In the first step, we raise the sample rate by a factor of 5 by a standard five-path polyphase filter. In the second step, we reduce the new sample rate by a factor of 3-to-1 with a resampling switch. In this process, for every three input samples, we extract five output samples. This straightforward approach is shown conceptually in Figure 7.19.

Figure 7.20 presents the polyphase partition of the 1-to-5 interpolator with the output port commutating between successive output samples at the intermediate output rate of 5 times the input sample rate. Since the 3-to-1 resampler following the 1-to-5 interpolator discards two out of three samples, it would be foolish to compute the output points from the interpolator that are destined to be discarded.

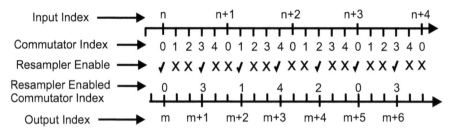

Figure 7.21   Time indices for input series, for output commutator series, save and discard indicator for 3-to-1 resampler, saved commutator indices, and output series.

Table 7.2   Indices for input, for commutator, and for output series.

Input index	Commutator index	Output index
n	0, 3	$m, m + 1$
$n + 1$	1, 4	$m + 2, m + 3$
$n + 2$	2	$m + 4$

Figure 7.21 shows successive indices of the input time series with respect to an arbitrary starting index n as well as the indices of the commutated output series from the 1-to-5 interpolator. The 3-to-1 resampler following the commutator saves one sample in three. We indicate the saved sample location by a check mark and the discarded samples with an X. The successive indices of the output commutator matching the check marked samples are shown as final output samples. The output samples are labeled with successive sample numbers with respect to an arbitrary starting index m. As expected, we observe that when the input index has stepped through three input samples, the output index steps through five output samples.

We note that the preserved samples from the resampled commutator indexes in stride of 3 with an address overflow implied by a modulo-5 operation on successive increments. An input sample is delivered to the polyphase filter each time the modulo-5 operation is invoked. For instance, starting at input n, we access commutator 0, then $0 + 3 = 3$. The next port is 3 $+ 3 = 6$ or 6 mod-5 $= 1$, so we insert the next sample $n + 1$ and access port 1 then $1 + 3 = 4$. Again the next port is $4 + 3 = 7$ or 7 mod-5 $= 2$, so we insert the next sample and access port 2. The next port is $2 + 3 = 5$ or 5 mod-5 $= 0$ and the pattern continues to repeat. The indexing process is shown in Table 7.2 and a simple state machine can track and control the input–output sequencing process.

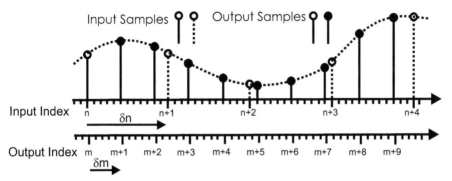

Figure 7.22 Resampling an input sequence by accessing every fifth output port of a 12-stage polyphase filter.

7.5 Arbitrary Resampling Ratio

We demonstrated in the previous section that we could use a P-stage polyphase filter to up-sample by P and down-sample by Q to obtain any rational ratio resampling of the form P/Q. Thus, for instance, if we have a 12-stage polyphase filter, we can obtain 12 possible output rates formed by taking an output sample from each output port (12/1), from every second output port (12/2), or from every third output port (12/3) and so on. A visualization of a possible resampling option is to be seen in Figure 7.22 where an input sequence is resampled by a factor of 12/5. This is accomplished by accessing the output commutator in stride of length 5 with commutator address computed modulo 12.

In general, the interpolation process computes output samples with fractional spacing between samples of αT with T being the input sample interval. When the parameter α is a rational ratio of the form Q/P, we can obtain that value precisely with a P-stage polyphase filter. This filter partitions the interval between successive input samples into P increments and forms successive output samples from every Qth commutator port. The number of stages, P, in the polyphase filter defines the time granularity of the interpolating process. Increasing the number of available stages in the polyphase filter will satisfy a requirement for smaller granularity. An increased number of stages does not increase the computational complexity of the interpolator but rather increases the number of coefficient sets available to compute output points. The increased number of coefficient sets requires additional memory space that leads to practical limits on the time granularity of this form of the interpolation process.

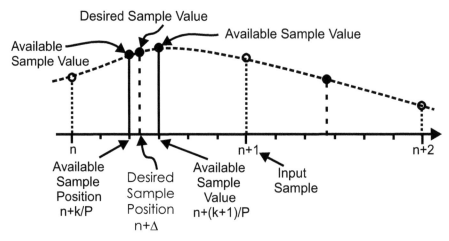

Figure 7.23 Interpolating to a position between available output points in a *P*-stage interpolator.

With a sufficiently large number of polyphase partitions, the time granularity may be fine enough to use the outputs from the nearest available sample positions to form an acceptable approximation to the output at any arbitrary time location. The task of forming a sample value at a location that does not correspond to an available output location of a *P*-stage interpolator is illustrated in Figure 7.23.

7.5.1 Nearest Left Neighbor Interpolation

One option for arbitrary resampling with a polyphase interpolator is to assign the sample value obtained from the nearest neighbor sample location. This nearest neighbor replacement is shown in Figure 7.24. In practice, nearest neighbor can be replaced with the neighbor to the immediate left. This is equivalent to truncating the desired offset value rather than rounding the desired offset value.

The amplitude error in the interpolation process is related to an error in sample position. The sample value errors formed over successive output samples are modeled as timing jitter errors. If the resulting amplitude errors are smaller than the errors due to amplitude quantization of the sample values, the timing jitter errors do not degrade the quality of the interpolated samples. A simple way to estimate the size of the timing jitter errors is to describe the errors due to nearest neighbor replacement as an equivalent linear process

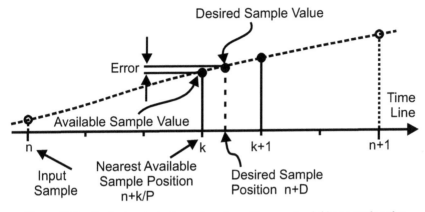

Figure 7.24 Replacing desired sample value with nearest neighbor sample value.

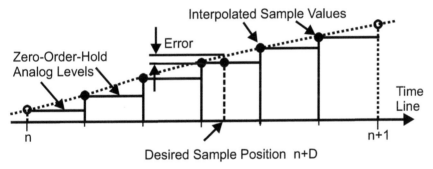

Figure 7.25 Timing jitter model: sampling of virtual DAC output for maximum interpolated sample values.

and then estimate the errors in that process. We do this by imagining that we fully interpolate the input series to the maximum output sample rate using every output port, form an analog signal from these samples with a perfect DAC or zero-order hold, and then resample this virtual analog waveform at the desired time locations. This model is shown in Figure 7.25.

We can upper bound the errors due to the timing jitter by examining the spectrum of the signal obtained by maximally up-sampling and then converted to an analog signal by the virtual DAC. We model the original sampled signal as uniformly occupying unity bandwidth, such as a Nyquist shaped spectrum in a communication system. A signal might originally be formed with a sample rate that is 4 times the input bandwidth and then up-sampled to a new sample rate N times its Nyquist rate with an $N/4$ stage

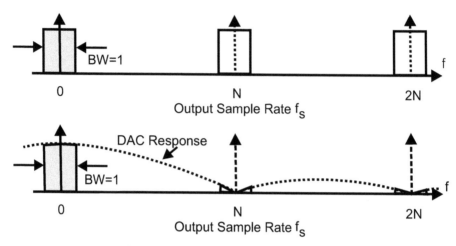

Figure 7.26   Spectrum of up-sampled signal at input and output of DAC.

interpolator. The important parameter here is the output sample rate, not how it was obtained. The spectrum of the up-sampled signal at the input and output of the DAC processed signal is shown in Figure 7.26. The frequency response of the DAC is the standard sin(x)/x, shown in Equation (7.11), with zeros located at multiples of the output sample rate. These zeros suppress the spectral copies centered at multiples of the sample rate but leave a residual spectrum in the neighborhood of the spectral zero.

The DAC spectral response is shown in Equation (7.11), and its first derivative is shown in Equation (7.12). Evaluating Equation (7.12) at the first zero crossing of the spectrum, at $f = 1/T$, we obtain the results shown in Equation (7.13).

$$H(f) = \frac{\sin\left(2\pi f\frac{T}{2}\right)}{\left(2\pi f\frac{T}{2}\right)} \tag{7.11}$$

$$\frac{d}{df}H(f) = \frac{\left(2\pi f\frac{T}{2}\right)\left(2\pi\frac{T}{2}\right)\cos\left(2\pi f\frac{T}{2}\right) - \left(2\pi\frac{T}{2}\right)\sin\left(2\pi f\frac{T}{2}\right)}{\left(2\pi f\frac{T}{2}\right)^2}$$

$$\tag{7.12}$$

$$\frac{d}{df}H(f)\bigg|_{f=\frac{1}{T}} = -T = -\frac{1}{f_S}. \tag{7.13}$$

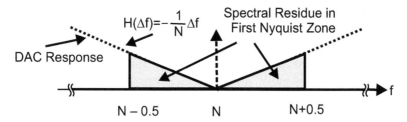

Figure 7.27 Frequency response in neighborhood of the DAC's first spectral null.

The Taylor series expansion of the DAC's spectral response at the first zero crossing is shown in the following equation:

$$H(\Delta f) = -\frac{1}{f_S}\Delta f. \tag{7.14}$$

Substituting the sample rate indicated in the normalized frequency axis presented in Figure 7.26, we obtain the local Taylor series shown in the following equation:

$$H(\Delta f) = -\frac{1}{N}\Delta f. \tag{7.15}$$

A zoom to the spectral response of the DAC in the neighborhood of the first spectral zero is shown in Figure 7.27.

The maximum amplitude of the residual spectrum centered about the first spectral null, obtained by substituting 1/2 for Δf, is seen in the following equation:

$$|H(\Delta f)|_{\text{Max}} = \frac{1}{N}\Delta f\bigg|_{\Delta f=\frac{1}{2}} = \frac{1}{2N}. \tag{7.16}$$

The smallest resolvable signal level of a b-bit quantizer is 2^{-b}. If the residual spectral levels at the output of the DAC are below this level, the error attributed to the timing jitter of the nearest neighbor interpolator is below the quantization noise level of quantized signal samples. To assure this condition, the maximum spectral level of the residual spectrum must satisfy the condition shown in the following equation:

$$\frac{1}{2N} < \frac{1}{2^b} \quad \text{or} \quad N > 2^{(b-1)}. \tag{7.17}$$

The virtual analog signal described by the up-sampling condition satisfying Equation (7.17) can now be sampled at any output rate that satisfies the Nyquist criterion for the input bandwidth. The aliasing terms that fold into the primary Nyquist zone are the multiple residual spectra residing at the successive spectral nulls of the DAC interpolator. The amplitudes of the aliasing terms inherit the alternating signs and $1/M$ gain terms of the DACs $\sin(x)/x$. In general, the spectral terms do not alias to the same frequency, so we are not concerned with a build-up of their contributions. If the amplitudes of the successive alias terms do stack up, they form a geometric series with alternating signs or with the same signs. If we up-sample by N and down-sample by Q, the worst case cumulative gain due to the folding is $(Q/(Q-1)$, which results in worst case folded spectral levels of $Q/(Q-1) * 1/(2N)$. In practice, the collective aliases of all the residual spectra from the DAC's spectral zero crossings to the Nyquist interval of the new sample rate do not rise above the highest level of $1/(2N)$.

By way of example, consider interpolating a time series represented by 8-bit samples. For this case, if we operate the interpolator to obtain a maximum output sample rate 128 times the signal bandwidth, the noise spectrum due to nearest neighbor interpolation will be below the noise caused by the signal quantization process. If the input signal is originally oversampled by a factor of 4, the interpolator must make up the additional factor of 128/4 or 32. Thus, a 32-stage interpolator can resample an 8-bit input signal with jitter-related noise levels below the -48 dB dynamic range noise level of the 8-bit quantized data. Figure 7.28 presents the time and spectral response of a 45-tap filter response that is initially oversampled by a factor of 4. Note the spectrum is scaled for two-sided 3-dB bandwidth of 1 for which the sample rate of 4 is presented on an axis of ± 2.

Figure 7.29 presents the time and frequency response obtained by applying the interpolation process to resample the time response of Figure 7.28. The sample rate change shown here is 32/6.4, a sample rate change of 5. This sample rate change is obtained by stepping through the 32 output commutator port indices by the integer part of an accumulator that is incremented in steps of 6.4. The number of output samples formed by the interpolation process is seen to be 245 points. The spectrum obtained by the interpolation process clearly shows the four spectral regions at multiples of the input rate that have been suppressed by the $\sin(x)/x$ response of the virtual DAC employed by the nearest neighbor interpolator. The level of suppression

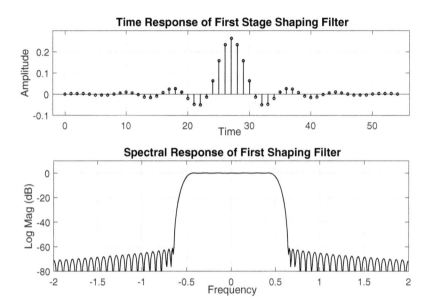

Figure 7.28 Shaping filter: time and frequency response four times oversampled.

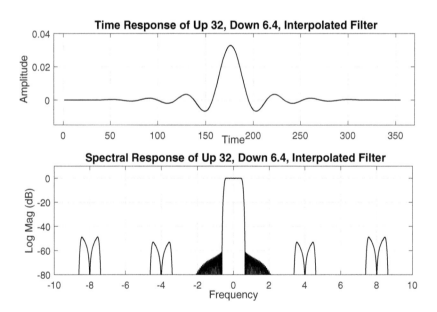

Figure 7.29 Time and frequency response of 32/6.4 nearest neighbor interpolator.

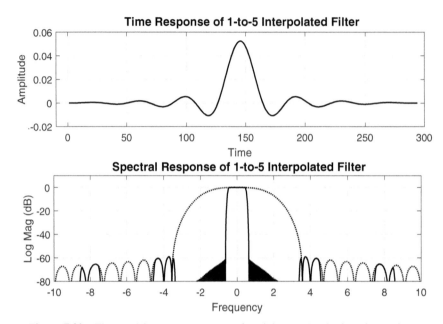

Figure 7.30   Time and frequency response of straight 1-to 5 polyphase interpolator.

is 50 dB, 2 dB more than the maximum 48 dB level estimated in Equation (7.14). This apparent excess attenuation is due to the fact that the spectral width of the input signal was less than unity bandwidth due to its transition roll-off. The reduced spectral occupancy lowers the Δf in Equation (7.13) and results in a reduced level estimate for the maximum amplitude spectral residue.

Note that the spectrum shown in Figure 7.29 differs slightly from the spectrum obtained by a straight 1-to-5 interpolator performing the same up-sampling task. A spectrum obtained from a 1-to-5 up-sampler is shown in Figure 7.30. Note that here the residual spectra are filtered versions of the spectral replicates as opposed to the folded $\sin(x)/x$ suppressed spectral images of Figure 7.29. Both filters were designed to suppress spectral copies by 60 dB and different effects form the spectral residues we observe in Figures 7.29 and 7.30. Both filters satisfy the design requirements with the residual spectra submerged under the quantizing noise of the 8-bit representation of the input signal.

For comparison, Figure 7.31 shows the time and spectral response when the same 32-stage interpolator is incremented in steps of 6.37 to obtain a

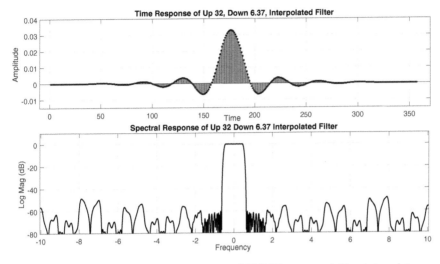

Figure 7.31   Time and frequency response of 32/6.37 nearest neighbor interpolator.

sample rate change of 32/6.37 or 5.0235. The slightly higher sample rate presents 246 instead of the 245 output samples of the previous example. We note here that the aliases no longer fold to four common frequencies and now appear over the entire span of the new Nyquist interval of width 4* 5.0235 or 20.094.

The nearest neighbor indexing is usually implemented as a truncation-indexing scheme. A block diagram of the input and output index processing is shown in Figure 7.32. The process can be visualized with the aid of a pair of saw-tooth waveforms representing the input and output clocks shown in Figure 7.33. The process operates as follows: a new input data sample is shifted into the filter register. The integer part of the accumulator content selects one of the 32 filter weights that are used to compute and output a sample point. The output clock then directs the accumulator process to add the desired increment delta to the modulo-32 accumulator and increment the output index m. This process continues until the accumulator overflows at one of the output clock increments. Upon this overflow, the input index is incremented, a new input data point is shifted into the filter, and the process continues as before.

A segment of MATLAB code that performs the alignment of output clock index m with filter weight index *pntr_k* and input clock index n is shown next. The interpolating filter weights are stored in *hh* and filtering is performed as

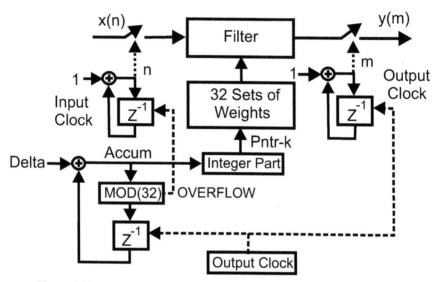

Figure 7.32   Block diagram of input, output, and filter weight set index control.

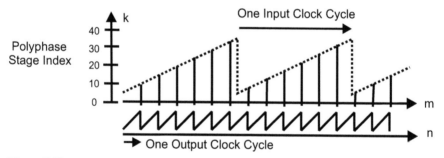

Figure 7.33   Visualization of relationship between input clock, output clock, and polyphase filter index pointer.

the inner product of the four-tap register named *reg* and the selected filter weights *wts(pntr_k,:)*. Figure 7.34 displays a segment of the sequence of the *pntr_k* indices obtained using this MATLAB code with *del* = 3.14. Note the precession of the start and stop index in successive cycles through the 32-stage index selection.

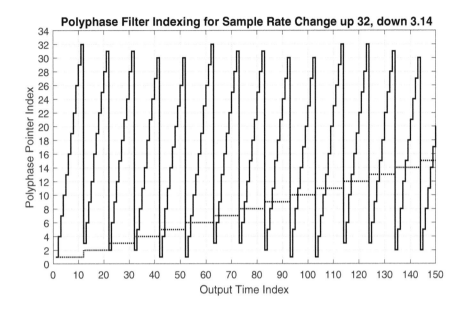

Figure 7.34 Example of indexing sequence formed by address control in resampling routine.

Segment of MATLAB Code to Control Input and Output Indexing

```
del=5.3;    %del=32*f_in/f_out;
accum=1;
mm=1;       % Output Clock
nn=1;       % Input Clock
reg=zeros(1,4);
wts=reshape(hh,32,4);
    % append zeros to flush filter
register
xx=[xx zeros(1,4)];
while nn<=length(xx)
    reg=[xx(nn) reg(1,3)];
        while accum < 33
            pntr_k=floor(accum);
            yy(mm)=reg*wts(pntr_k,:)';
            mm=mm+1;
        end
    accum=accum-32;% on overflow get new input
    nn=nn+1;
end
```

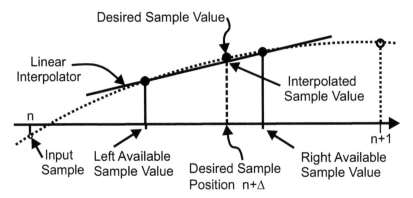

Figure 7.35   Linear interpolation between available left and right sample values.

7.5.2 Two-Neighbor Interpolation

Equation (7.17) provided an estimate of the amount of oversampling required to keep spectral artifacts in nearest neighbor interpolation below the quantizing noise level of a b-bit representation of the data samples. When the number of bits is large, the required oversample rate becomes excessive. For instance, to match the −96 dB noise level of a 16-bit data set, we require an oversample rate of 65,536 which in turn requires a polyphase filter of 16,384 stages. We must improve on the nearest neighbor approximation. One option that leads to a smaller number of stages is linear interpolation between the two available adjacent neighbors that bracket the desired sample position. This option is shown in Figure 7.35.

The equation for linear interpolation between two samples at k and $k + 1$ is shown in the following equation:

$$x(k + \Delta) = x(k) \cdot (1 - \Delta) + x(k + 1) \cdot \Delta. \qquad (7.18)$$

In a manner similar to the DAC process being equivalent to convolving a time sequence with a rectangle pulse, the linear interpolator can be modeled as convolving a time sequence with a triangle pulse. This interpretation is shown in Figure 7.36. The arithmetic required to compute the linearly interpolated sample is shown in the following equation:

$$\begin{aligned} x(k + \Delta) &\cong x(x) + [x(k + 1) - x(k)] \cdot \Delta \\ &= x(x) + \dot{x}(k) \cdot \Delta. \end{aligned} \qquad (7.19)$$

The triangle pulse can be formed as the convolution of the rectangle with a second identical rectangle. Since convolution in time is equivalent to a

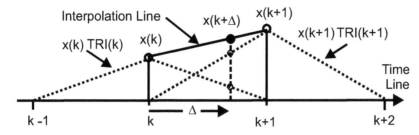

Figure 7.36 Linear interpolation as convolution of sample points with triangle pulse.

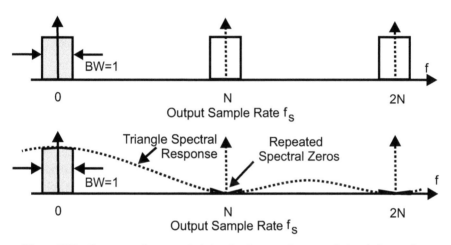

Figure 7.37 Spectrum of up-sampled signal at input and output of triangle interpolator.

product in frequency, the Fourier transform of the triangle is the product of the transform of the rectangle with itself. The transform of the triangle is shown in the following equation:

$$H(f) = \left[\frac{\sin(2\pi f \frac{T}{2})}{(2\pi f \frac{T}{2})} \right]^2 . \tag{7.20}$$

The spectral response of the up-sampled data set subjected to the linear interpolator is a $[\sin(x)/x]^2$ weighting applied to periodic spectra of the up-sampled sequence formed by the interpolator. This is shown in Figure 7.37.

Note that the repeated zeros of the $[\sin(x)/x]^2$ offer additional suppression of the spectral replicates. We seek the first term nonzero term in the Taylor

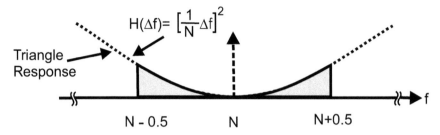

Figure 7.38   Frequency response in neighborhood of triangle first spectral null.

series at the first repeated zero of the $[\sin(x)/x]^2$. This is simply the square of the first nonzero term of the Taylor series of $\sin(x)/x$. This is shown in Equation (7.21) and is illustrated in Figure 7.38.

$$H(\Delta f) = \left[\frac{1}{N}\Delta f\right]^2 \tag{7.21}$$

As in the previous section, the errors due to the linear interpolation will not be observable if the residual spectral level is below the noise level attributed to the quantization noise of the signal. Since the error due to the b-bit quantized signal is 2^{-b}, we require the residual spectral levels to be below this level. Substituting the maximum frequency offset Δf of 1/2 in Equation (7.21) and comparing this level to the quantization noise level, we obtain the condition shown in the following equation:

$$\left[\frac{1}{2N}\right]^2 < \frac{1}{2^b} \quad \text{or} \quad N > 2^{(b-2)/2}. \tag{7.22}$$

Thus, for this example, if we have a requirement to interpolate a 16-bit data set, we can keep the spectral artifacts below the quantization noise level if N, the oversample rate, is greater than 128. Continuing with the assumption of the previous example that the input signal is oversampled by a factor of 4, the number of stages required in the polyphase interpolator is 128/4 or 32. Figure 7.39 shows the time and frequency response of a 4 times oversampled shaping filter with –96 dB side lobes. Figure 7.40 shows the time and frequency response formed by a 32-stage interpolating filter using linear interpolation between available output samples bracketing the desired time samples. Note that the levels of the spectral errors seen in this figure are at the expected –96 dB level. In order to obtain the desired 96 dB attenuation, the length of the interpolating filter had to be increased from four taps per

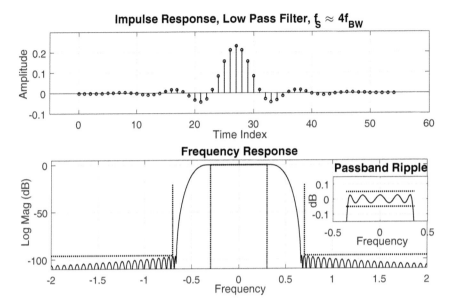

Figure 7.39 Shaping filter: time and frequency response four times oversampled.

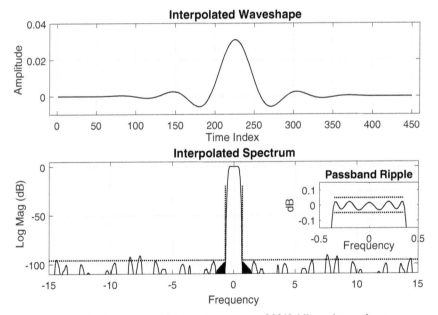

Figure 7.40 Time and frequency response of 32/6.4 linear interpolator.

stage, used for the previous example, to five taps per stage. There is a slight spectral droop of the in-band spectrum caused by the main lobe curvature of the $[\sin(x)/x]^2$. We can precompensate for this droop in the design of the input shaping filter or we can correct for the anticipated droop with a compensating filter prior to use of the interpolator.

7.6 Farrow Filter

The Farrow filter offers two alternative approaches to the arbitrary resampling problem. In the first approach, coefficients of the polyphase filter stages are computed on the fly from low-order piecewise polynomials. These polynomials form approximations to segments of the impulse response of the original interpolating filter prior to the polyphase partition. In the second alternative, the coefficients of the approximating polynomials are rearranged and are applied directly to the input data to form data-dependent, locally valid, polynomial approximations to the input data as opposed to the filter. The data polynomial is in turn evaluated at the desired sample points.

7.6.1 Classical Interpolator

Arbitrary resampling filters are designed and implemented as polyphase P-path filters. Each path provides a delay equal to an integer multiple of $1/P$th of the input sample interval for an up-sampler and of $1/P$ of the output sample interval for a down-sampler. Down-sampling is implemented concurrently with the up-sampling by stepping through the commutator output ports in increments of Q to realize a rational ratio sample rate change of P/Q. In efficient implementations, the filter is a single stage with the commutation process performed by a pointer in coefficient (memory) space. Figure 7.41 shows the functional structure of a 1-to-P up-sampler with an embedded Q-to-1 down-sampler performed by stepping in stride of Q through P commutated stages.

When used for arbitrary resampling, P, the number of stages in the filter, is selected to be sufficiently large so that phase jitter artifacts due to selecting the nearest neighbor to the desired interpolation point are sufficiently small. This time jitter is shown in Figure 7.42 where the desired output sample point is located between sample points k and $k + 1$. The output port is normally selected from the left stage to avoid the problem of stage index overflow for the $P - 1$ stage of sample n and the 0 stage of sample $n + 1$.

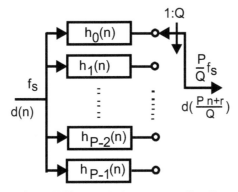

Figure 7.41 A polyphase up-sampling filter.

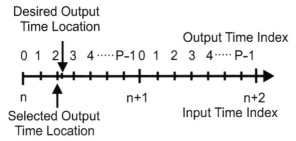

Figure 7.42 Input and output sample location for *P*-stage resampling filter.

We examine the polyphase partition or coefficient mapping of the prototype filter to understand how and where to apply the polynomial approximation. The weights $h_r(n)$ of the *r*th stage of a *P*-stage polyphase filter are obtained from the prototype filter weights $h(n)$ by the mapping $h_r(n) = h(r + nP)$. This mapping can be easily visualized as mapping the *N* taps of the prototype filter into a two-dimensional array with *P* rows and *N/P* columns by filling successive columns in natural order. Thus, the first column contains the first *P* points, the second column contains the next *P* points, and so on. The rows of this two-dimensional array are the polyphase filter sets with the *r*th set holding the weights to compute the *r*th interpolated output point located *r/P* fraction of the input intersample interval. This two-dimensional mapping is shown in Figure 7.43. Each of the rows of the partition corresponds to a sample point in the interpolated data stream. Row *k* would compute a sample $y(n + k/P)$ and row *k* + 1 would compute a sample $y(n + (k + 1)/P)$. Shown in Figure 7.43 is a pointer pointing to a nonexisting filter at sample $y(n + (k + \Delta)/P)$. Our traditional response to computing an output point at this position

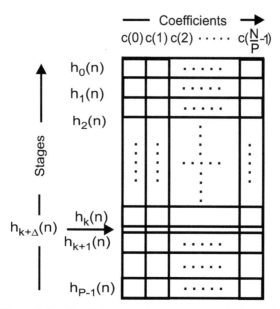

Figure 7.43   Coefficient mapping of *P*-stage polyphase filter.

is to use nearest neighbor $y(n + k/P)$ or compute the two outputs $y(n + k/P)$ and $y(n + (k + 1)/P)$ and linearly interpolate by distance Δ between them.

Figure 7.44 shows the time and frequency response of a 250-tap interpolating filter designed for 1-to-50 interpolation of data samples with 12-bit or 72 dB dynamic range. Note that the filter coefficients are samples of a smooth continuous filter.

The filter in Figure 7.44 is partitioned as shown in Figure 7.43 into five columns of length 50, with each row of the partition forming one path of the 50-path interpolator. Figure 7.45 presents the five 50-point sequences, which are seen to be contiguous segments of the prototype filter. A weight vector containing the first coefficient in each column is the first row of the polyphase filter; similarly, a vector containing the second coefficient in each column is the second row of the filter, and so on. Suppose we wanted a filter coefficient set to compute an output between any two filter sets, say filter set 25 and filter set 26. If we examine the coefficients for this pair of filters, we would recognize that they are very similar. If we interpolate between these weights, we would form a new set of weights that would indeed compute the output sample at the time offset corresponding to the interpolated distance between the weights.

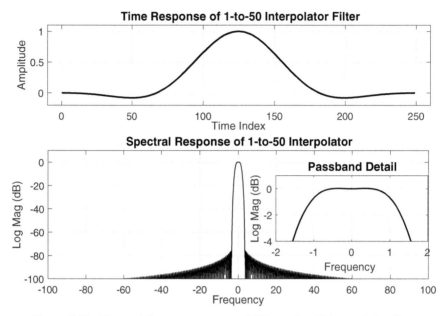

Figure 7.44 Time and frequency response of 300-tap, 1-to-50 interpolating filter.

The change to interpolating between filter weights rather than between sample outputs is seen in the following manipulations. Equation (7.23) presents the linear interpolation required to obtain an output distance Δ between the output from sub-filter r and sub-filter $r + 1$.

$$y(n + \frac{r + \Delta}{P}) = y(n + r) \cdot (1 - \Delta) + y(n + r + 1) \cdot \Delta. \qquad (7.23)$$

The outputs of these sub-filters are obtained as an inner product of the input samples in the filter register and the two sets of weights previously denoted by $h_r(n)$ and $h_{r+1}(n)$. The linear interpolation of the inner products is shown in the following equation:

$$y(n + \frac{r + \Delta}{P}) = (1 - \Delta) \sum_{k=0}^{5} x(n - k) \, h_r(k) + \Delta \sum_{k=0}^{5} x(n - k) \, h_{(r+1)}(k). \qquad (7.24)$$

Since the same data samples are used in the two inner products, the order of summation and inner product can be interchanged as in Equation (7.25).

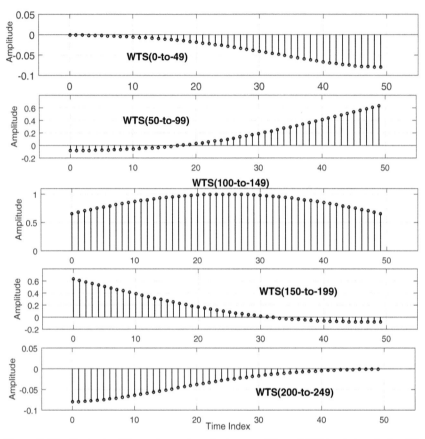

Figure 7.45   Contents of five columns in polyphase partition of 250-tap interpolating filter.

Here we see that the output sample has been computed at the desired sample location with a filter weight set designed specifically for that output location $r + \Delta$.

$$
\begin{aligned}
y(n + \frac{r + \Delta}{P}) &= \sum_{k=0}^{5} x(n - k)[(1 - \Delta)\, h_r(k) + \Delta\, h_{(r+1)}k] \\
&= \sum_{k=0}^{5} x(n - k)\, h_{(r+\Delta)}(k).
\end{aligned}
\tag{7.25}
$$

Examining the factored form shown in Equation (7.24), we can arrive at an alternate version of the interpolated filter set shown in Equation (7.26). Here we recognize that the filter weight, formed as the difference between adjacent filter sets, is a filter that forms the derivative output series.

$$y(n + \frac{r+\Delta}{P}) = \sum_{k=0}^{5} x(n-k)\, h_r(k) + \Delta \sum_{k=0}^{5} x(n-k)\, [h_{r+1}(k) - h_r(k)].$$

(7.26)

We can use the second term in this factored form as a second set of weights that computes the signal derivative to accompany the original interpolator set. We then perform two inner products to form the left sample at position $n + r/P$ and the derivative of the left sample at the same location. This option is shown in the following equation:

$$y(n + \frac{r+\Delta}{P}) = \sum_{k=0}^{5} x(n-k) \cdot h_r(k) + \Delta \sum_{k=0}^{5} x(n-k) \cdot \dot{h}_r(k). \quad (7.27)$$

We then use the local Taylor series to interpolate to the desired position $n + (r + \Delta)/P$. This local Taylor series option is shown in the following equation:

$$y(n + \frac{r+\Delta}{P}) = y(n + \frac{r}{P}) + \Delta\, \dot{y}(n + \frac{r}{P}). \quad (7.28)$$

We have just examined two options to perform fine-grain interpolation of an input time series. These were the interpolation of the filter weights or the forming of a Taylor series for the envelope of the output time sequence. These two options are the core of the Farrow filter that we examine next.

7.6.2 Polynomial Approximation

We return to the coefficient columns of the polyphase partition presented in Figure 7.45. We noted that the coefficients of the rth stage are located in the rth position of the successive columns of the two-dimensional representation of the prototype filter. Similarly, the coefficients of the $(r + 1)$th stage are in the $(r + 1)$th position of each column. We also noted that if we required a filter stage for an output sample between stages r and $r + 1$, say at $r + \Delta$, we could interpolate filter coefficients, an option presented in Equation (7.25).

We now take another tack and model the coefficient set in each column, as a P-sample wide section of the prototype filter, which can be considered

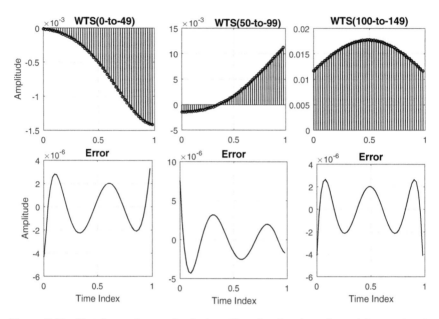

Figure 7.46   First three columns of polyphase filter, fourth-order polynomial approximation, and errors between columns and approximation.

to be samples of a smooth continuous function. We can approximate these sections with low-order polynomials and can compute the coefficients for a filter stage at any position by evaluating these polynomials. Tchebyschev or equal ripple approximations to the column entries are easy to compute and the maximum error can be controlled by selection of the degree of the approximating polynomials.

A useful rule for implementing FIR filters is that out-of-band side-lobe levels are bounded by 5 dB per bit. Thus, if we design an FIR filter with minimum side-lobe attenuation of 72 dB, we must represent the finite length coefficient set with at least 15 bits. Since leading zeros in a coefficient reduce arithmetic precision, we require that the coefficients be scaled to set the maximum coefficient to 1 (i.e. left justified in coefficient space). If we design a filter with a minimum of 72 dB attenuation, we require 15 bit precision of the coefficient list which translates to errors less than 2/32,768 or approximately 6×10^{-5}. This is the measure we use to specify the acceptable error between the polynomial approximation and the entries in the polyphase columns.

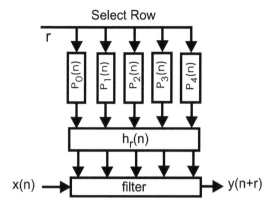

Figure 7.47 Polynomial form of arbitrary resampling filter.

Figure 7.46 presents the 50 samples of the first three columns of the polyphase partition of the prototype filter along with an approximating curve formed by a fourth-order polynomial. Also shown is the error between the 50 samples and the fourth-order approximating polynomial evaluated at these 50 points. Note that the error is everywhere less than 10^{-5}, better by a factor of 6 than required by the approximation requirement.

Figure 7.47 presents the structure of an arbitrary position resampling filter. Rather than using the control mechanism to select the desired sub-filter from a polyphase partition, this form of the filter uses the polynomials to compute the filter coefficients corresponding to the desired sample location r. Note that the original 50-stage polyphase filter required storage of 250 coefficients plus a mechanism to interpolate between output sample points. In this alternate form, the filter is defined by the five coefficients of each fourth-order polynomial, and since there are five polynomials, the filter is defined by 25 coefficients and a processor to evaluate the five polynomials at the correct time offset to form the filter coefficients $h_r(n)$, where r is not limited to an integer divided by P.

Figure 7.48 presents the time and frequency response of the polyphase filter impulse response formed by samples computed by the five fourth-order polynomials described earlier. Note that the side-lobe levels are maintained at or below the desired -72 dB level.

A word of caution: the *firpm* algorithm designs filters with equiripple pass-band and stop-band errors. We know that when the stop-band spectrum of filter exhibits constant level side lobes (or zero-decay rate), the impulse response almost always contains observable impulses at the two ends of the

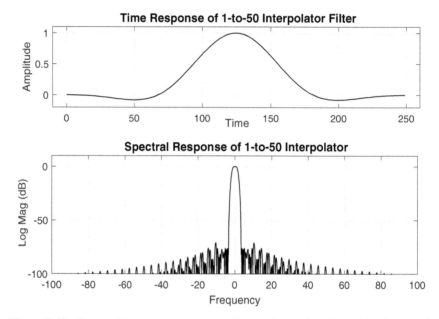

Figure 7.48   Time and frequency response of 50-stage interpolator formed by fourth-order polynomial approximations to five columns of polyphase partition.

time series. The presence of this impulse precludes the use of a low degree polynomial approximation to the section containing the impulse. We must either design the impulse out of the time response by use of a modified *firpm* algorithm, or in desperation, we simply cut the offending impulse down to the value determined by its two adjacent neighbors (to obtain continuous slope). The sample slicing may result in a 6 dB increase in nearby side-lobe levels and side-lobe decay rate of 6 dB per octave. Simply design the filter with additional 6 dB attenuation and proceed with the slicing operation. Having a filter with falling side lobes offers an incidental system benefit of reduced *integrated side lobe (ISL) levels*. Lower ISL results in improved filtering performance when the side lobes alias back into band due to the concurrent down-sampling when operating in the arbitrary resampling mode.

7.6.3 Farrow Structure

The *m*th coefficient of the interpolation polynomial for the offset Δ is computed by evaluating the polynomial $P_m(x)$ at $x = \Delta$. The polynomial P_m is the approximation assigned to the mth column of the polynomial interpolator. The form of this polynomial is shown in the following equation:

$$P_m(x) = \sum_{\ell=0}^{4} b(\ell, m) x^{\ell}. \tag{7.29}$$

The output of the filter using the coefficients from Equation (7.29) is shown in the following equation:

$$y(n + \Delta) = \sum_{m=0}^{4} P_m(\Delta) \, x(n - m). \tag{7.30}$$

We can substitute Equation (7.29) in Equation (7.30) and obtain the double sum of the following equation:

$$y(n + \Delta) = \sum_{m=0}^{4} \sum_{\ell=0}^{4} b(\ell, m) \, \Delta^{\ell} \, x(n - m). \tag{7.31}$$

Reordering the summations in Equation (7.31), we obtain the form shown in the following equation:

$$y(n + \Delta) = \sum_{\ell=0}^{4} \Delta^{\ell} \sum_{m=0}^{4} b(\ell, m) \, x(n - m). \tag{7.32}$$

Note here that the coefficients $b(l,m)$ are no longer applied as a Taylor series for the filter weights but now are coefficients of a filter applied to the data to form a new set of data-dependent coefficients denoted as $c(n,l)$ as in the following equation:

$$c(n, \ell) = \sum_{m=0}^{4} b(\ell, m) \, x(n - m). \tag{7.33}$$

We substitute the coefficients $c(n,l)$ in Equation (7.32) to obtain the following equation:

$$y(n + \Delta) = \sum_{\ell=0}^{4} c(n, \ell) \, \Delta^{\ell}. \tag{7.34}$$

Examining Equation (7.34), we recognize the form as the Taylor series representation of the output sequence. Consequently, the terms $c(n,l)$ of

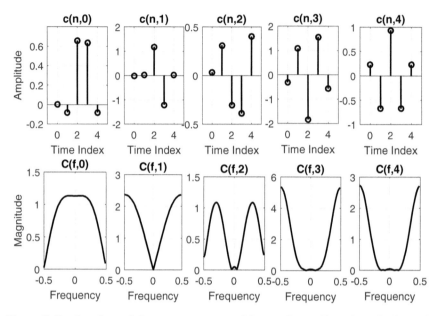

Figure 7.49   Impulse and frequency response of farrow filters: filters form Taylor series coefficients of input time series.

Equation (7.33) are successive local derivatives of the input series and hence of the output series. Equation (7.30) formed the desired output by evaluating the polynomial expansion of the filter weights and then applying these weights to the filter data. By rearranging the double summation, the coefficients of the polynomial expansion have become filters applied to the data to form the local data-dependent coefficients for the polynomial expansion of the data. These polynomial filters, along with their frequency responses, are shown in Figure 7.49. Here we see that the first filter estimates the local DC term, while the next filter estimates the local first derivative and so on.

Table 7.3 lists the coefficients of the polynomial interpolators for the 50-stage filter presented in the previous section. The $P_k(\Delta)$ are the polynomial coefficients of the filter that processes the input data samples $x(n - k)$ to compute the output $y(n - \Delta)$. Equation (7.35) shows how the coefficients are generated from the table for a specific Δ and then applied to the input data to compute the desired output sample.

Table 7.3 Polynomial filter for output at Δ.

	$\Delta4$	$\Delta3$	$\Delta2$	$\Delta1$	$\Delta0$
$P_0(\Delta)$	0.2012	−0.2536	−0.0198	−0.0090	−0.0018
$P_1(\Delta)$	−0.6648	1.0550	0.3287	0.0225	−0.0828
$P_2(\Delta)$	0.9231	−1.8092	−0.3134	1.1759	0.6587
$P_3(\Delta)$	−0.6648	1.5511	−0.4007	−1.2035	0.6347
$P_4(\Delta)$	0.2012	−0.5350	0.3940	0.0210	−0.0827

Table 7.4 lists the rearranged coefficient set that forms the Taylor series expansion of the desired output time series. The C_l (x) are the polynomial coefficients of the filter that processes the input data samples $x(n - k)$ to compute the Taylor series coefficients. The Taylor series coefficients are then evaluated at Δ to obtain the output $y(n - \Delta)$. Equation (7.36) shows how the coefficients, generated for a specific set of input samples, are evaluated at a specific Δ to compute the desired output sample. It is useful to select a sample $x(n)$ and see how it intersects with both sets of coefficients to contribute to an output sample.

$$y(n + \Delta) = p_0(\Delta)x(n) + p_1(\Delta)x(n - 1) + p_2(\Delta)x(n - 2)$$
$$+ p_3(\Delta)x(n - 3) + p_4(\Delta)x(n - 4). \tag{7.35}$$

$$y(n+\Delta) = c_0(x)+c_1(x) \cdot \Delta+c_2(x) \cdot \Delta^2+c_3(x) \cdot \Delta^3+c_4(x) \cdot \Delta^4. \tag{7.36}$$

The output samples of the interpolator are computed by evaluating the Taylor series terms of Equation (7.36) for each value of desired output Δ. Note that the coefficients are not updated until a new input sample is delivered to the coefficient engine. By comparison, a new polynomial must be formed and applied to the input data when the polynomial form of the interpolator is used. The Taylor series is most efficiently evaluated using the Horner Rule for evaluating a polynomial. This is shown in the following equation:

Table 7.4 Coefficients to compute Taylor series expansion of input time series.

	x(n)	x(n-1)	x(n-2)	x(n-3)	x(n-4)
$C_0(x)$	−0.0018	−0.0828	0.06587	0.6347	−0.0827
$C_1(x)$	−0.0090	0.0225	1.1759	−1.2035	0.0210
$C_2(x)$	−0.0198	0.3287	−0.3134	−0.4007	0.3940
$C_3(x)$	−0.2536	1.0550	−1.8092	1.5511	−0.5350
$C_4(x)$	0.2012	−0.6648	0.9231	−0.6648	0.2012

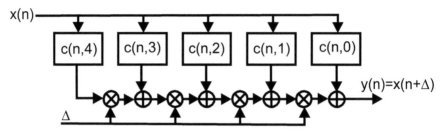

Figure 7.50   Signal flow diagram of Horner's rule for evaluating polynomial.

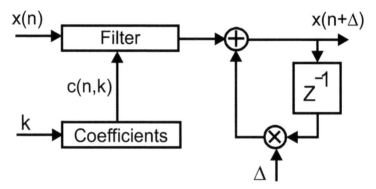

Figure 7.51   Efficient hardware version of Horner's rule processor.

$$y(\Delta) = c_0 + c_1 \cdot \Delta + c_2 \cdot \Delta^2 + c_3 \cdot \Delta^3 + c_4 \cdot \Delta^4$$
$$= c_0 + \Delta \cdot (c_1 + \Delta \cdot (c_2 + \Delta \cdot (c_3 + \Delta \cdot c_4))). \tag{7.37}$$

A processor flow diagram that matches Horner's rule is shown in Figure 7.50 and an efficient hardware version of the same process is shown in Figure 7.51.

References

Crochiere, Ronald, and Lawrence Rabiner. *Multirate Signal Processing*, Englewood Cliff, NJ, Prentice-Hall, 1983.

Farrow C. W., "A Continuously Variable Digital Delay Element", Proc. IEEE Int. Symp. Circuits Systems, (ICAS-88), Vol.3, pp 2642-2645, Espoo, Finland, June 6-9, 1988.

Fliege, Norbert, *Multirate Digital Signal Processing: Multirate Systems, Filter Banks, Wavelets*, West Sussex, John Wiley & Sons, Ltd., 1994.

Gardner, Floyd. M. "Interpolation in Digital Modems: Part-I: Fundamentals", IEEE Trans. Comm., Vol. 41, No. 3, pp. 502-508, Mar. 1993.

harris, fred. "Forming Arbitrary Length Windows or Filter Sequences From a Fixed Length Reference Table", 18-th Annual Asilomar Conf. on Signals Systems, and Computers, Pacific Grove, CA., Nov. 1984.

harris, fred. "Performance and Design of Farrow Filter Used for Arbitrary Resampling", 31st Asilomar Conference on Signals, Systems, and Computers, Pacific Grove, CA., Nov 1997.

Hentschel, Tim. *Sample Rate Conversion in Software Configurable Radios*, Norwood, MA, Artech House, Inc., 2002.

Jovanovic-Dolecek, Gordana. *Multirate Systems: Design and Applications*, London, Idea Group, 2002.

Laakso, Timo, Vesa Välimäki, Matti Karjalainen, and Unto Laine,, "Splitting the Unit Delay: Tools for Fractional Delay Filter Design", IEEE Signal Processing Magazine, Jan 1996, Vol. 13, No.1, pp. 30-60.

Mitra, Sanjit. *Digital Signal Processing: A Computer-Based Approach*, 2nd ed., New York, McGraw- Hill, 2001.

Mitra, Sanjit and James Kaiser. *Handbook for Digital Signal Processing*, New York, John Wiley & Sons, 1993.

Sable, Les and fred harris, "Using the High Order Nyquist Zones in the Design of Efficient Multi- channel Digital Upconverters", IEEE Personal, Indoor and Mobile Radio Conference, September 1-4 1997, Helsinki, Finland.

Vaidyanathan, P. P. *Multirate Systems and Filter Banks*, Englewood Cliff, NJ, Prentice-Hall, 1993.

Problems

7.1 Design an FIR filter to be used as a 1-to-20 up-sampler for an input signal initially oversampled by a factor of 4. The specifications for this design are as follows:

Sample Rate	400 kHz		
Pass Band	0−4 kHz	Pass-Band Ripple	0.1 dB
Stop Band	16−200 kHz	Stop-Band Atten.	60 dB,
			slope −6dB/octave

Plot the impulse response and log magnitude frequency response of the prototype filter to verify that it meets design specifications.

Form an input signal as the impulse response of a Nyquist filter, with the following specifications:

Sample Rate	20 kHz
Symbol Rate	5 kHz
Roll-Off Factor α	0.2
Pass-Band Ripple	0.1 dB
Stop-Band Attenuation	60 dB

Zero-pack the input signals $1-20$ and form and plot its log magnitude spectrum along with an overlay of the log magnitude spectrum of the prototype interpolating filter. In a sub-plot, zoom to the spectral region at the edge of the interpolating filter's transition bandwidth to view the rejection level of the replica spectra.

Convolve the zero-packed input signal with the prototype interpolating filter and display in sub-plots the time response and log magnitude spectrum of the interpolated signal. Verify that the residual spectral levels in the spectra of the interpolated signal have been suppressed at least 60 dB.

Partition the prototype interpolator into a 20-path polyphase filter and up-sample the input signal using the polyphase filter structure. Display in sub-plots the time response and log magnitude spectrum of the interpolated signal. Verify that the signal obtained in the polyphase form of the filter is identical to that obtained by the zero-packed and processed form of the process.

7.2 Repeat problem 7.1 except design the prototype low-pass filter with multiple stop bands and interleaved do-not-care bands. Keep the filter length the same for both designs. Be sure to note the difference in the replica attenuation in this design as well as the difference in the shape of the filter impulse response.

7.3 Repeat problem 7.1 except this time, cosine heterodyne the impulse response of the prototype interpolating filter to the center of the third Nyquist zone, at 60 kHz. The filter will now simultaneously up-sample, reject undesired spectral replicates, and translate.

7.4 Design an FIR filter to be used as a 1-to-40 up-sampler for an input signal initially oversampled by a factor of 4. The specifications for this design are as follows:

Sample Rate	800 kHz		
Pass Band	0−4 kHz	Pass-Band Ripple	0.1 dB
Stop Band	16−400 kHz	Stop-Band Atten.	60 dB, slope −6dB/octave

Plot the impulse response and log magnitude frequency response of the prototype filter to verify that it meets design specifications.

Form an input signal as the impulse response of a Nyquist filter, with the following specifications:

Sample Rate	20 kHz
Symbol Rate	5 kHz
Roll-Off Factor α	0.2
Pass-Band Ripple	0.1 dB
Stop-Band Attenuation	60 dB

Zero-pack the input signals 1−40 and form and plot its log magnitude spectrum along with an overlay of the log magnitude spectrum of the prototype interpolating filter. In a sub-plot, zoom to the spectral region at the edge of the interpolating filter's transition bandwidth to view the rejection level of the replica spectra.

Convolve the zero-packed input signal with the prototype interpolating filter and display in sub-plots the time response and log magnitude spectrum of the interpolated signal. Verify that the residual spectral levels in the spectra of the interpolated signal have been suppressed at least 60 dB.

Partition the prototype interpolator into a 40-path polyphase filter and up-sample the input signal using the polyphase filter structure. Display in sub-plots the time response and log magnitude spectrum of the interpolated signal. Verify that the signal obtained in the polyphase form of the filter is identical to that obtained by the zero-packed and processed form of the process.

7.5 Repeat the filter design of Problem 7.4 except this time, we will arrange to up-sample by 40 and down-sample by 12. We accomplish this by incrementing through the polyphase filter paths in stride of 40/12. Here, we form an output sample from every 12th filter path (modulo 40). This accomplishes a sample rate change of 10/3 or 3.333, a rational ratio. Plot the time response and the log magnitude frequency response of the interpolated output time series.

7.6 Repeat the filter design of problem 7.4 except this time, we will arrange to up-sample by 3.14159 (look familiar?). We accomplish this by incrementing through the polyphase filter paths in stride of 40/3.14159 or 12.73239..., using the integer part of the index accumulator (modulo 40) to select the nearest neighbor interpolating path. Plot the time response and the log magnitude frequency response of the interpolated output time series.

7.7 Repeat the filter design of Problem 7.4 except this time, we will arrange to up-sample by 127. We accomplish this by incrementing through the polyphase filter paths in stride of 40/127 or 0.31496..., using the integer part of the index accumulator (modulo 40) to select the nearest neighbor interpolating path. Because the increment is less than 1, there will be three and occasionally four repeats of a given output sample. Plot the time response and the log magnitude frequency response of the interpolated output time series. Be sure to zoom in on a segment of the interpolated time series to see the repeated output samples.

7.8 Repeat the filter design of Problem 7.4 and then the polyphase partition of the prototype interpolating filter. Use the MATLAB script *polyfit* to form fourth-order polynomial approximations to the individual columns of the partition. Use the vector [0:1/40:1-1/40] to define the values of x for the corresponding values of y, the elements of the columns. To test the quality of the approximation, use the MATLAB script *polyval* to evaluate the polynomials over the same input vector values and take the difference between the sample values and the polynomial approximants. Plot the error values.

Now use the polynomials that represent the columns of the original polyphase filter to form a new set of column values over the input grid $x2 = [0\ 1/80:1-1/80]$. This forms a filter with impulse response twice the length of the original design. Plot the time response and log magnitude frequency response to see the quality of the polynomial approximation process. Note the levels of the spectral artifacts. Is the order of the polynomial approximation sufficient for this filtering process?

7.9 This problem is a repeat of problem 7.8 except that here we will use fifth-order polynomials to approximate the columns of the polyphase filter. Repeat the filter design of problem 7.4 and then the polyphase partition of the prototype interpolating filter. Use the MATLAB script *polyfit* to form fifth-order polynomial approximations to the individual columns of the partition. Use the vector [0:1/40:1-1/40] to define the values of x for the corresponding values of y, the elements of the columns. To test the quality of the approximation, use the MATLAB script *polyval* to evaluate the polynomials over the same input vector values and take the difference between the sample values and the polynomial approximants. Plot the error values. Is the order of the polynomial approximation sufficient for this filtering process?

7.10 This problem is a repeat of problem 7.8 except that here we will use third-order polynomials to approximate the columns of the polyphase filter. Repeat the filter design of problem 7.4 and then the polyphase partition of the prototype interpolating filter. Use the MATLAB script *polyfit* to form third-order polynomial approximations to the individual columns of the partition. Use the vector [0:1/40:1-1/40] to define the values of x for the corresponding values of y, the elements of the columns. To test the quality of the approximation, use the MATLAB script *polyval* to evaluate the polynomials over the same input vector values and take the difference between the sample values and the polynomial approximants. Plot the error values. Is the order of the polynomial approximation sufficient for this filtering process?

7.11 Use the polynomial approximations formed in problem 7.8 as a replacement for the polyphase interpolator as shown in Figure 7.47 and use this process to up-sample by 3.14, the impulse response formed in Problem 7.4. We accomplish this by using an increment of 1/3.14 or 0.31847 (modulo 1) of the incrementing accumulator. Plot the time response and the log magnitude frequency response of the time series obtained from the polynomial-based polyphase interpolator.

8

Half-Band Filters

Four Stages of 2-to-1 Down Sample

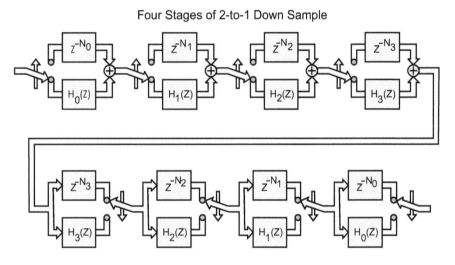

Four Stages of 1-to-2 Up Sample

Polyphase M-path filters can be designed with any number of paths. A particularly attractive M-path filter is to be had when the filter has two paths. Many of the properties of multirate filters are easily demonstrated with half-band filters, and, as such, the half-band filter is often used to introduce the reader to multirate filters. From the perspectives and insights gathered from the half-band filter, the M-path filter is presented as a reasonable extension. The half-band filter is used in many configurations including cascade resampling stages, iterated filter stages, and Hilbert transform filters.

219

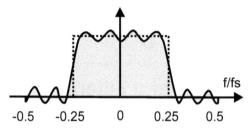

Figure 8.1   Spectral characteristics of half-band low-pass filter.

8.1 Half-Band Low-Pass Filters

The prototype half-band low-pass filter has the spectral characteristics shown in Figure 8.1.

The filter is designed for a pass-band bandwidth between the $\pm 1/4$ sample rate for a two-sided bandwidth equal to half the sample rate. The finite length implementation of the filter has a transition bandwidth that passes through the midpoint, 0.5, of the discontinuity at $\pm$ quarter sample rate.

The impulse response of the ideal non-casual continuous filter with two-sided bandwidth $f_s/2$ is shown in the following equation:

$$h_{\mathrm{LP}}(t) = \frac{f_S}{2} \frac{\sin\left(\dfrac{2\pi f_S/2}{2}t\right)}{\left(\dfrac{2\pi f_S/2}{2}t\right)}. \tag{8.1}$$

When sampled at multiples on nT or n/f_s, we obtain, after scaling by f_s, the sample data form of the ideal half-band filter as shown in Equation (8.2) and after canceling terms results in the form shown in Equation (8.3).

$$h_{\mathrm{LP}}(n) = \frac{f_S/2}{f_S} \frac{\sin\left(\dfrac{2\pi f_S/2}{2}\dfrac{n}{f_S}\right)}{\left(\dfrac{2\pi f_S/2}{2}\dfrac{n}{f_S}\right)} \tag{8.2}$$

$$h_{\mathrm{LP}}(n) = \frac{1}{2} \frac{\sin\left(\dfrac{n\pi}{2}\right)}{\left(\dfrac{n\pi}{2}\right)}. \tag{8.3}$$

The impulse response of the half-band filter formed by the sampling operation contains one sample at the point of symmetry with matching left and right samples about the symmetry point as shown in Figure 8.2.

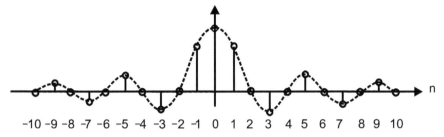

Figure 8.2   Impulse response of non-causal half-band low-pass filter.

The impulse response samples of the half-band filter are seen to be zero at the even index offsets from the center point of the filter. The coefficients with odd index offset are seen to exhibit even symmetry about the filter center point. These two properties permit the non-polyphase form of the filter of length $2N + 1$ to be implemented with only $N/2$ multiplications per output sample. When partitioned into a two-path filter and used for down-sampling, the $N/2$ multiplications are distributed over two input samples for a workload of $N/4$ multiplications per input point in the $2N + 1$ length filter.

8.2 Half-Band High-Pass Filter

The half-band high-pass filter has the spectral form shown in Figure 8.3.

The impulse response of the high-pass half-band filter is obtained by heterodyning the low-pass impulse response to the half sample rate. This is shown in the following equations:

$$h_{\text{HP}}(n) = h_{\text{LP}}(n)\, e^{j\, n\, \pi} \tag{8.4}$$

$$h_{\text{HP}}(n) = \frac{1}{2} \frac{\sin\left(\dfrac{n\,\pi}{2}\right)}{\left(\dfrac{n\,\pi}{2}\right)} \cos(n\,\pi). \tag{8.5}$$

The heterodyne indicated in Equation (8.5) is a sequence of unit-valued samples with alternating signs. The unit-valued samples with a positive sign correspond to the even offset index samples of the low-pass filter, which are zero except for the sample at the symmetry point. The unit-valued samples with a negative sign correspond to the odd offset index samples of the low-pass filter. The heterodyne to the half sample rate has the effect of reversing the signs of all the samples of the low-pass filter except for the

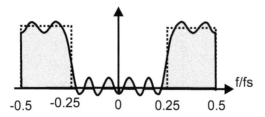

Figure 8.3   Spectral characteristics of half-band high-pass filter.

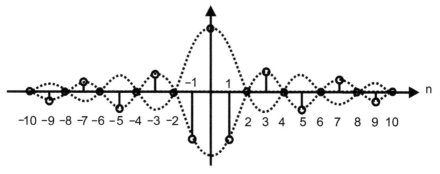

Figure 8.4   Impulse response of non-causal half-band high-pass filter.

sample at the symmetry point. The resultant impulse response is shown in Figure 8.4.

We note that the sum of the impulse responses of the low-pass half-band filter and high-pass half-band filter is a single unit-valued sample at its point of symmetry. The canceling of all but one of the weights in the summation of the two impulse responses is shown in Equation (8.6) for the non-causal form of the filter and in Equation (8.7) for the causal form.

$$h_{LP(n)} + h_{HP}(n) = \delta(n): \quad -N \leq n \leq +N \tag{8.6}$$

$$h_{LP(n)} + h_{HP}(n) = \delta(n-N): \quad 0 \leq n \leq +2N. \tag{8.7}$$

Equations (8.6) and (8.7) tell us that the low-pass and high-pass half-band filters are complimentary filters and satisfy the relationships shown in equations the following equations:

$$h_{HP}(n) = \delta(n) - h_{LP}(n): \quad -N \leq n \leq +N \tag{8.8}$$

$$h_{HP}(n) = \delta(n-N) - h_{LP}(n): \quad 0 \leq n \leq +2N \tag{8.9}$$

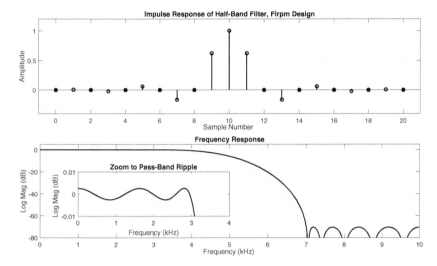

Figure 8.6 Impulse response and frequency response of half-band low-pass filter designed with firpm algorithm.

script that uses a call to the *firpm* algorithm to satisfy the specifications for the example presented in the previous section is shown here.

```
h2=firpm(20,[0 5-2 5+2 10]/10,[1 1 0 0],[1 1]);
```

The impulse response and frequency response of this filter are shown in Figure 8.6. The spectrum is seen to satisfy the filter specifications as well as exhibit the equal ripple side lobes in the pass band and stop band. Note that the side lobe levels are lower than those obtained in the window design shown in Figure 8.5.

8.4.1 Half-Band Firpm Algorithm Design Trick

In the previous section, we obtained the zero-valued samples of the half-band filter impulse response by restricting the band edge specifications to be symmetric about the quarter sample rate. A half-band design trick can be used to have the *firpm* algorithm design only, the odd-indexed coefficients of the desired half-band filter. We finish the design by inserting the even indexed samples composed of the zero-valued samples and the center tap sample. The odd indexed samples form a one-band filter with a transition bandwidth at the half sample rate. We ask the *firpm* algorithm to design this filter and then

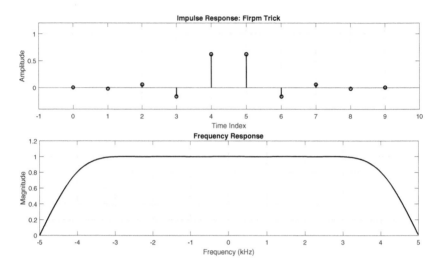

Figure 8.7   Impulse response and frequency response of trick one-band filter designed with Remez algorithm.

modify it. A MATLAB script that calls the *firpm* algorithm to form the design trick impulse response matching the example presented in the previous two sections is shown here. The impulse response and frequency response of the h3a, the *firpm*-designed filter, are shown in Figure 8.7. The spectrum is seen to be the pass band of the desired filter with half its transition band.

```
h3a=firpm(9, [0 5-2 5 5]/5, [1 1 0 0]);
h3 = zeros(1, 21);
h3(2:2:21)=0.5*h3a;
h3(10)=0.5;
```

The impulse response and frequency response of h3, the corrected filter response, are shown in Figure 8.8. The correction scales the filter by 0.5 and inserts the zero-valued samples at the even-indexed positions with a single 0.5-valued sample at the filter center position. The spectrum is seen to now contain both the pass band and stop band of the filter. The zero packing is seen to double the filter sample rate and offer two copies of the spectrum with the copy at the half sample rate exhibiting 180° phase shift. The scaling by 0.5 sets the gain in the two spectral regions to nominal values of +0.5 and –0.5. The inserted 0.5 at the filter midpoint raises the spectrum by amplitude

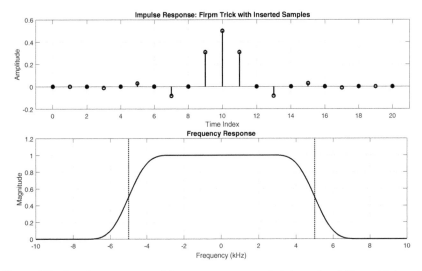

Figure 8.8 Impulse response and frequency response of trick one-band filter with inserted zero samples and symmetry point sample.

0.5. The effect of the raised spectrum is that in the spectral interval centered at 0 frequency the added 0.5 raises the 0.5 gain to the pass-band gain with a nominal value of 1.0, while in the spectral interval centered at the half sample rate, the added 0.5 raises the –0.5 gain to the stop-band gain, a nominal value of 0.0.

8.5 Hilbert Transform Band-Pass Filter

The half-band filters described and designed in the previous sections have been low-pass and high-pass filters. A trivial variation of this design leads to a half-band filter that performs the Hilbert transform. The Hilbert transform has a frequency response with a nominal unity gain over the positive frequencies and nominal zero gain over the negative frequencies. The frequency response of the Hilbert transform form of the half-band filter is shown in Figure 8.9.

The spectrum of the Hilbert transform filter is obtained by translating the low-pass half-band filter to the quarter sample rate. The form of this translation for a non-causal filter is shown in the following equation:

$$h_{\mathrm{HT}}(n) = h_{LP}(n)\,e^{jn\pi/2}: \quad -N \le n \le +N. \tag{8.12}$$

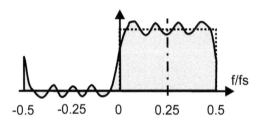

Figure 8.9   Spectral characteristics of Hilbert transform half-band filter.

The heterodyne indicated in Equation (8.12) could be visualized as a quadrature heterodyne with a cosine and a sine series as shown in in the following equation:

$$h_{\mathrm{HT}}(n) = h_{\mathrm{LP}}(n)\left\{\cos(n\pi/2) + j\sin(n\pi/2)\right\}: \quad -N \le n \le +N.$$
$$(8.13)$$

We note that the cosine component of the heterodyne is a sequence composed of unit amplitude samples with alternating signs on the even-indices and zero-valued samples on the odd-indices. The sequence is of the form $\{\ldots 1\ 0\ {-}1\ 0\ldots\}$. The half-band filter is zero-valued at the even-indices except for the sample at the origin, the symmetry point. Thus, the cosine heterodyne, the real part of the heterodyne, contains a single non-zero sample at the origin. The sine component of the heterodyne is a sequence composed of unit amplitude samples with alternating signs on the odd indices and zero-valued samples on the even indices. The sequence is of the form $\{\ldots 0\ 1\ 0\ {-}1\ldots\}$. The odd-indexed samples of the half-band filter contain the alternating sign side lobe samples of the $\mathrm{sinc}(n\,\pi/2)$ series. The samples of the sine heterodyne interact with the samples of the sinc sequence to remove the alternating signs of the sequence. The complex impulse response and the corresponding spectrum are shown in Figure 8.10.

8.5.1 Applying the Hilbert Transform Filter

The discrete Hilbert transform filter can be used to generate a complex signal from a real input sequence by eliminating the spectral components residing in the negative frequency band. After suppressing the negative frequencies, thus reducing the bandwidth by a factor of 2, it is common to also reduce the sample rate by the same factor. Thus, the Hilbert transform filter can be used as a half-band multirate filter. As an aside, this differs from the conventional Weaver modulator that translates the center of the positive frequency interval

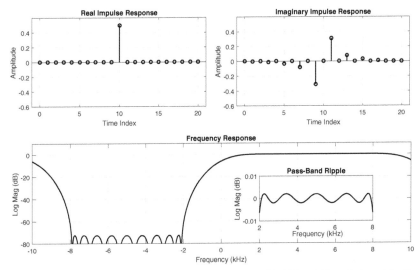

Figure 8.10 Impulse response and frequency response of half-band filter designed by firpm algorithm and converted to Hilbert transform by complex heterodyne to $f_s/4$.

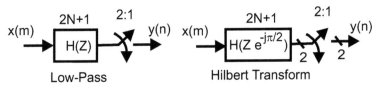

Figure 8.11 Hilbert transform filter as a spectrally translated version of low-pass half-band filter.

to baseband for processing by the half-band filter. The Hilbert transform filter translates the baseband half-band filter to the center of the positive frequency interval. This is the "which do we move, filter or signal spectrum?" question we discussed in Chapter 6. Moving the filter to the spectral center is equivalent to moving the spectral center to the filter. Figure 8.11 illustrates spectral translation that forms the Hilbert transform filter.

The complex heterodyne of the low-pass filter's impulse response results in a filter with complex impulse response. Such a filter can be formed as a two-path filter, one forming the real part and one forming the imaginary part of the impulse response. This form of the Hilbert transform filter is shown in Figure 8.12.

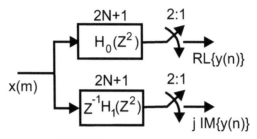

Figure 8.12   Two-path model of Hilbert transform filter with complex impulse response.

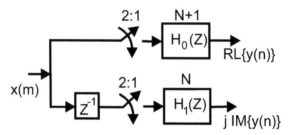

Figure 8.13   Noble identity applied to Hilbert transform filter.

Due to the zeros of the quarter sample rate cosine and sine, the impulse response of the upper path is seen to be zero at the even indices while the impulse response of the lower path is seen to be zero at the odd indices. The zeroes permit application of the noble identity in which we interchange the filter with the 2-to-1 resample to operate the filters at the reduced output sample rate. This interchange is demonstrated in Figure 8.13.

Finally, the interaction of the pair of 2-to-1 resampler switches and the input delay line can be replaced with a two-input commutator that performs the same function. This form of the resampling half-band Hilbert transform filter is shown in Figure 8.14.

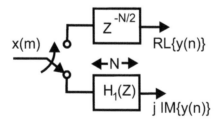

Figure 8.14   Two-path commutator driven Hilbert transform filter.

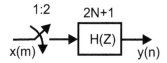

Figure 8.15 One-to-two up-sampling process with half-band filter.

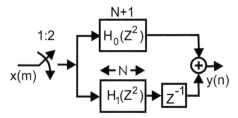

Figure 8.16 One-to-two up-sampling with polyphase half-band filter.

8.6 Interpolating With Low-Pass Half-Band Filters

The half-band low-pass filter can be used to up-sample a time series by a factor of 2. The initial form of 1-to-2 up-sampling process, based on zero insertion to raise the input sample rate, is shown in Figure 8.15.

The half-band filter $H(Z)$ can be partitioned into a pair of polyphase filters as shown in Equations (8.14) and (8.15) and illustrated in Figure 8.16.

$$H(Z) = \sum_{n=0}^{2N} h(n)Z^{-n} \tag{8.14}$$

$$H(Z) = \sum_{n=0}^{N} h(2n)Z^{-2n} + Z^{-1}\sum_{n=0}^{N} h(2n+1)Z^{-2n}. \tag{8.15}$$

The order of the resampling and the filtering performed in the filter can be reversed leading to the form shown in Figure 8.17.

Finally, as shown in Figure 8.18, we can replace the pair of 1-to-2 up-sampling switches and sample delay with a two-tap commutator that performs the equivalent scheduling of path outputs to the output sample stream.

We note that the prototype filter is designed with $2N + 1$ taps and then partitioned into two paths with one path containing the zero-valued samples and implemented as a delay-only path with the other path containing the remaining N non-zero samples. The filter requires N arithmetic operations (ops) to generate two output samples in response to each input sample. When we distribute the N ops per input over the two outputs, we find that the filter

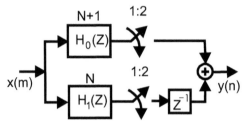

Figure 8.17   One-to-two up-sampling at output of polyphase half-band filter.

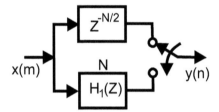

Figure 8.18   One-to-two up-sampling with commutated two-path half-band filter.

workload is $N/2$ ops per output. The number of multiplies per output can be reduced by another factor of 2 by taking advantage of the even symmetry of the coefficient set in the lower path filter.

The question we now address is: what is the workload per output for a half-band resampling filter operating at a particular set of performance specifications? We can examine Figure 8.19 to see the performance parameters that affect the filter length and workload per output point.

The length of the filter shown in Figure 8.19 can be estimated from the harris approximation presented in Chapter 3 and reproduced here as Equation (8.16).

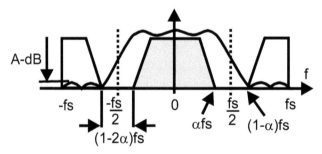

Figure 8.19   Spectral characteristics of half-band filter suppressing spectral replicate.

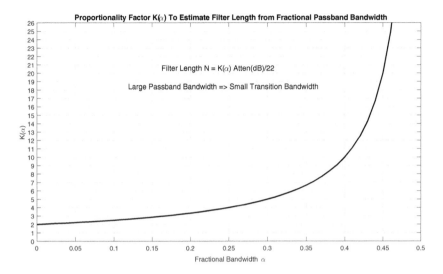

Figure 8.20 Filter length parameter $K(\alpha)$ as function of fractional bandwidth α.

$$N \cong \frac{f_{\text{SMPL}}}{\Delta f} \cdot \frac{\text{Atten(dB)}}{22}. \tag{8.16}$$

Substituting the transition bandwidth and sample rate shown in Figure 8.19 into Equation (8.16), we obtain the following equation:

$$N \simeq \frac{2f_S}{(1-2\alpha)f_S} \cdot \frac{\text{Atten(dB)}}{22}$$
$$= \frac{2}{(1-2\alpha)} \cdot \frac{\text{Atten(dB)}}{22} = K_1 \cdot \frac{\text{Atten(dB)}}{22}. \tag{8.17}$$

The relationship between filter length and fractional bandwidth is shown in Figure 8.20. We can clearly see that as the fractional bandwidth of the filter increases, which causes a decrease of transition bandwidth, the filter length increases. For a specific example, consider a half-band filter with fractional bandwidth of 0.45 with 60 dB attenuation. From Figure 8.20, we can estimate the half-band filter length N_{Len} as 20 60/22 or 54.54 taps which gets converted into 57 taps in order to locate the center tap in the upper path of a two-path partition.

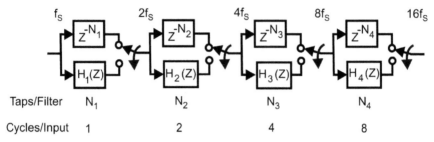

Taps/Filter	N_1		N_2		N_3		N_4
Cycles/Input	1		2		4		8

Figure 8.21 Cascade of four half-band filters to raise the sample rate by 16 in steps of increase by 2 per stage.

8.7 Dyadic Half-Band Filters

We now consider the use of a cascade of half-band filters to obtain a sample rate increase of any power of 2 such as increase by 8 or by 16. Suppose, for instance, we want to increase the sample rate of an input sequence by a factor 8. We have two primary options available to us. We can use an eight-path polyphase filter to accomplish this task, or we can use a cascade of three half-band filters. We first examine the workload for the sequence of half-band filters and then compare this workload to the M-path filter. The sequence of half-band filters operates at successively higher sample rates but with transfer functions that have successively wider transition bandwidths. There is a processing advantage to the cascade when the reduction in processing due to the wider transition bandwidth in successive filter stages compensates for operating the consecutive filter stages at successively higher sample rates.

Figure 8.21 is a block diagram showing four stages of half-band filters that raise the sample rate by a factor of 16. Shown here is the length of each polyphase filter stage as well as an index indication on how often each stage is used each time a data sample is delivered to the input port.

The frequency response of successive filters in the cascade is shown in Figure 8.22. Here we see that the successive filters operate at twice the sample rate of the previous stage and have transition bandwidths successively wider than the previous stage.

As shown in Equation (8.16), the length of successive filters is proportional to the ratio of their sample rates to their transition bandwidths. A sequence of proportionality factors for the first four filters is shown in Table 8.1. Also indicated is the workload performed by each filter in response to an input sample delivered to the input of the cascade.

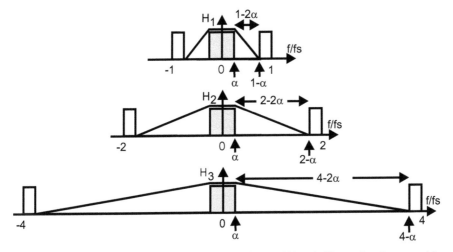

Figure 8.22 Frequency plan of three successive half-band filters showing transition bandwidth and sample rate.

Table 8.1   Length of four successive half-band filters and number of arithmetic operations performed by each filter per input sample.

	First stage	Second stage	Third stage	Fourth stage
$\dfrac{f_s}{\Delta f}$	$\dfrac{2}{1-2\alpha}$	$\dfrac{4}{2-2\alpha}$	$\dfrac{8}{4-2\alpha}$	$\dfrac{16}{8-2\alpha}$
Ops/input $K_1(\alpha)$	$\dfrac{2}{1-2\alpha}$	$2\dfrac{4}{2-2\alpha}$	$4\dfrac{8}{4-2\alpha}$	$8\dfrac{16}{8-2\alpha}$

Table 8.2 lists the workload proportionality factors for the sequence of cascade stages containing one to four stages of half-band filtering. The number of output samples has been normalized so that the proportionality factors in this list have units of ops/output sample.

Table 8.3 presents essentially the same information as Table 8.2 but has factored the normalization factor into the terms in each sum. This form of presenting the proportionality factor emphasizes the reduction in contribution of early stages in the cascade to each output of the cascade as the number of stages, and hence the number of output samples, increases. With a higher sample rate increase, the input workload can be amortized over a greater number of output samples. Note, for instance, that the contribution of the first stage to the workload per output sample is reduced by a factor of 2 each time another stage is added to the cascade.

Table 8.2   Workload proportionality factor per output sample for cascade of half-band filters.

	Workload proportionality factor per output sample
One stage $K_2(\alpha)$	$\dfrac{1}{2}\left[\dfrac{2}{1-2\alpha}\right]$
Two stages $K_2(\alpha)$	$\dfrac{1}{4}\left[\dfrac{2}{1-2\alpha}+2\dfrac{4}{2-2\alpha}\right]$
Three stages $K_2(\alpha)$	$\dfrac{1}{8}\left[\dfrac{2}{1-2\alpha}+2\dfrac{4}{2-2\alpha}+4\dfrac{8}{4-2\alpha}\right]$
Four stages $K_2(\alpha)$	$\dfrac{1}{16}\left[\dfrac{2}{1-2\alpha}+2\dfrac{4}{2-2\alpha}+4\dfrac{8}{4-2\alpha}+8\dfrac{16}{8-2\alpha}\right]$

Table 8.3   Alternate presentation of workload proportionality factor per output sample for cascade of half-band filters.

	Workload proportionality factor per output sample
One stage $K_2(\alpha)$	$\left[\dfrac{1}{1-2\alpha}\right]$
Two stages $K_2(\alpha)$	$\dfrac{1}{2}\left[\dfrac{2}{1-2\alpha}\right]+\left[\dfrac{1}{2-2\alpha}\right]$
Three stages $K_2(\alpha)$	$\dfrac{1}{4}\left[\dfrac{1}{1-2\alpha}\right]+\dfrac{1}{2}\left[\dfrac{1}{1-\alpha}\right]+\left[\dfrac{1}{1-\alpha/2}\right]$
Four stages $K_2(\alpha)$	$\dfrac{1}{8}\left[\dfrac{1}{1-2\alpha}\right]+\dfrac{1}{4}\left[\dfrac{1}{1-\alpha}\right]+\dfrac{1}{2}\left[\dfrac{1}{1-\alpha/2}\right]+\left[\dfrac{1}{1-\alpha/4}\right]$

Figure 8.23 presents curves of the expressions shown in Tables 8.2 and 8.3. We see that as the number of stages increases, the proportionality factor $K_1(\alpha)$ in ops per input sample increases, but when distributed over the number of output samples, the factor $K_2(\alpha)$ in ops per output sample is seen to decrease. We also see that the factor $K_2(\alpha)$ in ops per output sample asymptotically approaches 2 and in fact is close to 2 for a wide range of the fractional bandwidth and number of stages.

For comparison, we can use an M-path polyphase filter to change the sample rate by a factor of M. We can recast Equations (8.16) and (8.17) for the M-path filter to obtain the following equation:

$$
\begin{aligned}
N &\cong \frac{M\,f_S}{(1-2\alpha)\,f_S}\frac{\text{Atten}(dB)}{22} \\
&= \frac{M}{(1-2\alpha)}\frac{\text{Atten}(dB)}{22} = K_3(\alpha)\frac{M\,f_S}{(1-2\alpha)\,f_S}\frac{\text{Atten}(dB)}{22}.
\end{aligned}
\tag{8.18}
$$

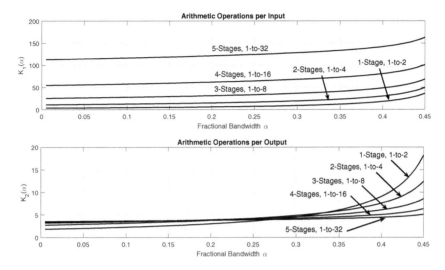

Figure 8.23 Proportionality factor $K_1(\alpha)$ in ops/input and $K_2(\alpha)$ in ops/output for one to five cascade stages of half-band filters.

As shown in Equation (8.19), we can determine the length of each path of the M-path filter by distributing the N weights over the M paths. If we assume that the top path, path-0, of the M-path filter contains only delays, then only $(M-1)$ of the paths contributes to the workload and removing one of the M paths from the workload estimate reduces the average workload. This scaled workload is shown in Equation (8.20). Figure 8.24 presents graphical representations of Equations (8.18) and (8.20).

$$\frac{N}{M} = \frac{1}{(1-2\alpha)}\frac{A(dB)}{22} \tag{8.19}$$

$$\frac{N}{M}\left(1-\frac{1}{M}\right) = \frac{1}{(1-2\alpha)}\left(1-\frac{1}{M}\right)\frac{A(dB)}{22}$$
$$= K_4(\alpha)\frac{A(dB)}{22} \tag{8.20}$$

We see in Figure 8.24 that the workload proportionality factor $K_4(\alpha)$ in ops per output point in the M-path filter always exceeds the factor $K_4(\alpha)$ of a two-path filter. By contrast, the related proportionality factor $K_2(\alpha)$ for the multiple half-band stages shown in Figure 8.23 is almost everywhere below the factor $K_2(\alpha)$ of the single half-band or two path filter. We thus conclude

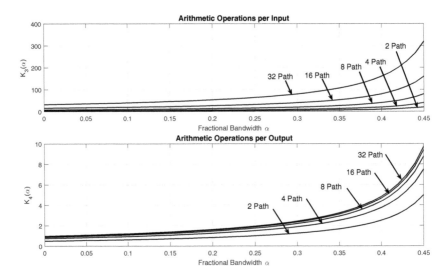

Figure 8.24   Proportionality factor $K_3(\alpha)$ in ops/input and $K_4(\alpha)$ in ops/output in M-path filter for values of $M = 2, 4, 8, 16,$ and 32.

that the cascade half-band filter is less costly in ops/output point than the M-path filter for the same interpolation factor. The ratio of workloads for the two options, cascade half-band filters and the M-path filter, approaches 1-to-10 for large resampling ratios.

References

Crochiere, Ronald and Lawrence Rabiner. *Multirate Signal Processing*, Prentice-Hall Inc., Englewood Cliff, NJ, 1983.

Fliege, Norbert. *Multirate Digital Signal Processing: Multirate Systems, Filter Banks, Wavelets*, West Sussex, John Wiley & Sons, Ltd, 1994.

Hentschel, Tim. *Sample Rate Conversion in Software Configurable Radios*, Norwood, MA, Artech House, Inc., 2002.

Jovanovic-Dolecek, Gordana. *Multirate Systems: Design and Applications*, London, Idea Group, 2002. Mitra, Sanjit and James Kaiser. *Handbook for Digital Signal Processing*, New York, John Wiley & Sons, 1993.

Mitra, Sanjit. *Digital Signal Processing: A Computer-Based Approach*, 2nd ed., New York, McGraw- Hill, 2001.

Vaidyanathan, P. P., and T.O. Ngyuen, "A Trick for the Design of FIR Half-band Filters", IEEE Trans. On Circuits and Systems - II, Vol. 34, Mar. 1987, pp 297-300.

Vaidyanathan, P. P., *Multirate Systems and Filter Banks*, Englewood Cliff, NJ, Prentice-Hall Inc., 1993.

Problems

8.1 Use the MATLAB *sinc* script file to form a 21-tap low-pass half-band filter.

Use subplots to stem the impulse response and to plot the magnitude spectrum and then the log magnitude spectrum of the filter. What is the DC gain of the filter? What is the appropriate scale factor to apply to the impulse response to set the DC gain to unity? Note the location of the impulse response zeros. After scaling, note the value of normalized frequency corresponding to amplitude 0.5.

8.2 Use the MATLAB *sinc* script file to form a 23-tap low-pass half-band filter.

Use subplots to stem the impulse response and to plot the magnitude spectrum and then the log magnitude spectrum of the filter. What is the DC gain of the filter? What is the appropriate scale factor to apply to the impulse response to set the DC gain to unity? Note the locations of the impulse response zeros. Compare the zero locations of this filter to the zero locations of the filter in problem 8.1. What can you say about the length of the filter to set the first sample to zero? This is a desired property for a polyphase partition! After scaling, note the value of normalized frequency corresponding to amplitude 0.5.

8.3 Use the MATLAB *sinc* script file to form a 22-tap low-pass half-band filter.

Use subplots to stem the impulse response and to plot the magnitude spectrum and then the log magnitude spectrum of the filter. What is the DC gain of the filter? What is the appropriate scale factor to apply to the impulse response to set the DC gain to unity? Note the locations of the impulse response zeros. Where did they go? What can you say about the length of the filter to have alternate values of the impulse response be zero? This is a desired property for a half-band filter and for a polyphase partition! After scaling, note the value of normalized frequency corresponding to amplitude 0.5.

8.4 Design a low-pass half-band finite impulse response (FIR) filter using a windowed sinc function to obtain a transition bandwidth of 0.1 and a stop-band attenuation of 60 dB. Use subplots to plot the impulse response and the log magnitude spectrum of the filter.

8.5 Design a low-pass half-band FIR filter using the Remez algorithm to obtain a transition bandwidth of 0.1 and a stop-band attenuation of 60 dB. Use subplots to plot the impulse response and the log magnitude spectrum of the filter.

8.6 Design a low-pass half-band FIR filter using the Remez algorithm trick to obtain a transition bandwidth of 0.1 and a stop-band attenuation of 60 dB. Use subplots to plot the impulse response and the log magnitude spectrum of the filter.

8.7 Design a low-pass half-band FIR filter using the Remez algorithm to obtain a transition bandwidth of 0.1 and a stop-band attenuation of 60 dB. Heterodyne the impulse response to center the band center at the quarter sample rate. Use subplots to plot the real and imaginary parts of the impulse response and the log magnitude spectrum of the filter.

8.8 Repeat the design of the Hilbert transform filter described in problem 8.7 and then apply to the filter 200 samples a real cosine of frequency 0.11. Plot the input and output time series (real and imaginary) and the windowed log magnitude spectrum of the input and output sequences.

8.9 Repeat the design of the Hilbert transform filter described in problem 8.7 and then apply to the filter 200 samples of a real cosine of frequency 0.11. Plot the absolute value of the input and output time series. Suggest how the Hilbert transform can be used to estimate signal amplitude for operation of an automatic gain control (AGC) circuit.

8.10 We are to design a 1-to-4 up-sampler as a cascade of two half-band filters. The input signal is the 200 samples of equal amplitude random phase sinusoids of normalized frequencies 0, 0.05, 0.11, 0.155, 0.21, and 0.26. Design two half-band filters that will reject spectral replicates of this signal in two steps. Apply a 1-to-2 zero-packed version of the input signal to the first half-band filter. Plot the time response of the first filter output and the input and output log-magnitude spectrum of the input and output signals. Zero-pack 1-to-2 the output of the first filter and deliver the zero packed sequence to the second half-band filter. Plot the time response of the second filter output and the input and output log-magnitude spectrum of the input and output signals.

9

Polyphase Channelizers

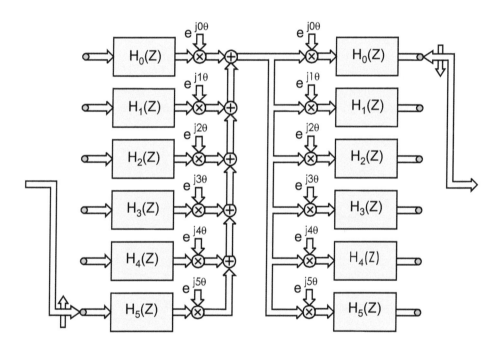

We have discussed and demonstrated in earlier chapters that multirate filters exhibit unique properties related to them being operated as LTV filters. One interesting property is the use of aliasing rather than heterodyning to move a spectral region to or from baseband while down-sampling or up-sampling, respectively. We learned that, when down-sampling, any multiple of the output sample rate could be aliased to baseband. Similarly, when up-sampling, the baseband signal can be aliased to any multiple of the input sample rate. We quickly conclude that since any of the multiple bands are available, perhaps all the bands are available. In fact, that is true. We now examine how the architecture of the polyphase filter structure can be applied to the task of simultaneously servicing multiple narrowband channels. The multiple channel version of the polyphase filter is often described as an analysis channelizer. We will first examine the use of the system as a multiple channel analysis filter bank and then its dual form as a multiple channel synthesis filter bank. Variants of the two techniques are then examined in the final section illustrating a number of specific applications of the process.

9.1 Analysis Channel Bank

In Chapter 6, we derived the polyphase partition of a band-pass finite impulse response (FIR) filter. We showed how the M-to-1 down-sampled input series aliases spectra centered at any multiple of the output sample rate to baseband. We then described how the complex heterodyne embedded in the band-pass filter is mapped through the polyphase partition to a set of frequency-dependent phase rotators on the separate paths of the polyphase partition. This structure, repeated here, is shown in Figure 9.1.

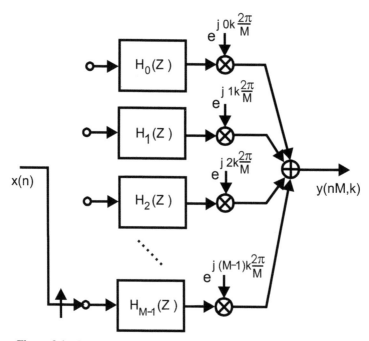

Figure 9.1   Polyphase partition of M-to-1 down-sampled band-pass filter.

The partition and signal flow shown in Figure 9.1 collectively perform the three operations of band-pass filtering, spectral translation, and resampling. The phase coherent summation performed at the final output accumulator has the form shown in Equation (9.1). Here, the variable $y_r(nM)$ represents the nMth time sample from the rth path of the M-path polyphase partition, while the variable $y(nM,k)$ represents the nMth time sample of the time series from the center frequency at the kth multiple of the output sample rate.

$$y(nM, k) = \sum_{r=0}^{M-1} y_r(nM) \, e^{j\frac{2\pi}{M}rk}. \tag{9.1}$$

We recognize that the summation in Equation (9.1) is the inverse discrete Fourier transform (IDFT) of the vector of M time samples formed at time nM from the array of M polyphase output paths. Since the summation can be performed for any index k, it can, in fact, be performed for all M indices. When the process is performed for a range of indices, it is described as a polyphase filter bank implementation of a channelized filter bank. In this mode, the phase rotators are applied by the IDFT for each output index. The block diagram of this form of the process is shown in Figure 9.2. Here we see that the input commutator can deliver a frequency division multiplexing (FDM) signal to the polyphase stages and that the output commutator extracts a time division multiplexed (TDM) frame from the IDFT output. This change from one multiplex form to another is why the system was original known as a transmultiplexer. We see that the output ports contain the time series of the aliased to baseband Nyquist bands originally centered at multiples of f_s/M; the down-sampled output rate of each baseband signal component. We call this realization of the channelizer a polyphase analysis channelizer.

We now address the interaction and coupling, or lack of coupling, between the parameters that define the polyphase analysis bank. We observe that the IDFT performs the task of separating the channels after the polyphase filter, so it is natural to conclude that the transform size M determines the number of channels spanning the input Nyquist span defined by the input sample rate. This is a correct assessment! We also note that the filter spectral characteristics are defined by the weights of the low-pass prototype filter residing in the polyphase partition and that all channel filters in the filter bank exhibit these same spectral characteristics.

In standard channelizer designs, the bandwidth of the prototype is specified in accordance with the end use of the channelizers. For instance, when the channelizer is used for spectral decomposition as in a spectrum

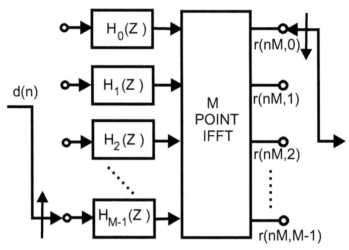

Figure 9.2   Polyphase analysis filter bank as a polyphase filter input and an IDFT output process.

analyzer, or for source coding, the channels are designed to exhibit a specified attenuation such as −3.0, −1.0, or −0.1 dB at their crossover frequency with a specified attenuation at their adjacent center frequency. Overlap of adjacent channel responses permits a narrowband input signal to straddle one or more output channel filters. This is a common practice in the spectral analysis of signals with arbitrary bandwidths and center frequencies. On the other hand, when a channelizer is used to separate adjacent communication channels, channels with known center frequencies and known non-overlapping bandwidths, the channelizers must maintain the separation of the channel outputs. Typical spectral responses for channel bandwidths corresponding to the two conditions just described are shown in Figure 9.3.

The polyphase filter normally operates with an input M-to-1 resampling commutator. This M-to-1 down-sampling aliases to baseband the spectral terms residing at multiples of the output sample rate. The IDFT following the polyphase filter defines the channel spacing, which is one-Mth of the input sample rate. Consequently, the standard polyphase channelizer has an output sample rate matching the channel spacing. For the case of the spectrum analyzer application, operating at this sample rate permits aliasing of the spectral band edge into the down-sampled pass band. When operated in this mode, the system is called a maximally decimated analysis filter bank. By proper design of the prototype filter, this aliasing can be reversed if the signal

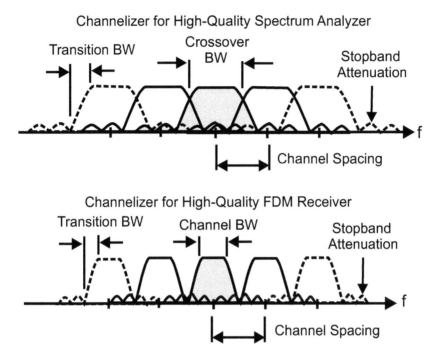

Figure 9.3 Spectral response of two analysis filter banks with same channel spacing: one for spectral analysis and one for FDM channel separation.

is synthesized from the down-sampled and aliased analysis process. A system so designed is known as a perfect reconstruction filter bank.

9.2 Arbitrary Output Sample Rates

For the case of the communication filter bank, an output sample rate matching the channel spacing can avoid adjacent channel crosstalk since the two-sided bandwidth of each channel is less than the channel spacing. An example of a signal that would require this mode of operation is the quadrature amplitude modulation (QAM) channels of a digital cable system. In North America, the channels were separated by 6 MHz centers and operate with square-root (SQRT) cosine tapered Nyquist shaped spectra with 18% or 12% excess bandwidth, at symbol rates of approximately 5.0 MHz. The minimum sample rate required of a cable channelizer to satisfy the Nyquist criterion would be 6.0 MHZ. (The European cable plants had channel spacing of 8.0 MHz and symbol rates of 7.0 MHz.). In support of the subsequent demodulation process, the sample rate would likely be selected to be a multiple of the symbol rate rather than a multiple of the channel spacing.

Systems that channelize and form samples of the Nyquist-shaped spectrum often present the sampled data to an interpolator to resample the time series from the channelizer sample rate of twice the channel spacing to the rate of two samples per symbol or twice symbol rate. For the cable TV plant example just cited, the 6 Ms/s, 5 M symbol signal would have to be resampled by 5/3 to obtain the desired 10 Ms/s. This is not a difficult task and it is done quite regularly in single channel receivers. This resampling may represent a significant computational burden if we are required to perform this interpolative resampling for every output channel and we may elect to embed the resampler in the polyphase analysis channelizer.

The conventional way we use the M-path polyphase filter bank is to deliver M input samples to the M paths and then compute outputs from each channel at the rate f_s/M. The thought may occur to us, "Is it possible to operate the polyphase filter bank in a manner that the output rate is higher than one-Mth of the input rate?" For instance, can we operate the bank so that we deliver $M/2$ inputs prior to computing an output sample rather than delivering M input samples before computing an output sample? Increasing the output sample rate of the polyphase channel bank by a factor of 2 makes subsequent interpolation tasks less expensive since the spectra of the output signals would already be oversampled by a factor of 2 with increased spectral separation. Operation in this mode would also permit channelization of overlapped channels without aliasing of the spectral transition bands. The alias-free partition is handy in applications requiring perfect reconstruction of the original time series from spectrally partitioned subchannels. For the record, a polyphase filter bank can be operated with an output sample rate of any rational ratio times the input sample rate. With minor modifications, the filter can be operated with totally arbitrary ratios between input and output sample rates. This is true for the sample rate reduction imbedded in a polyphase analysis filter bank at a receiver as well as for the sample rate increase embedded in a polyphase synthesis filter bank at a transmitter.

We first examine the task of doubling the output sample rate from the polyphase filter bank from f_s/M to $2 f_s/M$. It is easy to visualize this task if we revert to the original single-path band-pass prototype filter prior to its polyphase partition. In that filter, we repeatedly delivered the next M input samples to the filter and computed one output sample in response to the M input samples. Conceptually, we did this efficiently in two steps; we first shifted all input data samples in the filter registers M samples to the right to vacate the M left most input registers. We then deliver the next M input samples to the M vacated addresses and compute one output sample. Delivering M input samples and extract 1 output sample is how we simultaneously perform the two tasks of bandwidth reduction and M-to-1 down-sampling. To perform $M/2$-to-1 or M-to-2 down-sampling, we simply replace the M sample shift and M sample replacement with an $M/2$ sample shift and an $M/2$ sample replacement. This simple modification doubles the output sample rate.

The successive $M/2$ shifts in the one-dimensional filter can be observed in the two-dimensional polyphase version of the same filter. We accomplish this change from M inputs to $M/2$ inputs with the commutator delivering input data samples to the polyphase stages. We normally deliver M inputs to the M-path filter by delivering successive input samples starting at Port M-1 progressing up the stack to Port 0 and, by doing so, deliver M inputs per output for an M-to-1 down-sampling operation. To obtain the desired $(M/2)$-to-1 down-sampling, we deliver $M/2$ successive input samples starting at Port $(M/2)$-1 progressing up the stack to Port 0. The $M/2$ addresses to which the new $M/2$ input samples are delivered are first vacated of their former contents, the $M/2$ previous input samples. All the samples in the two-dimensional filter undergo a serpentine shift of $M/2$ samples with the $M/2$ samples in the top half of the first column sliding into the $M/2$ bottom addresses of the first column, while the $M/2$ samples in the bottom half of the first column slide into the $M/2$ addresses in the top half of the second column and so on. This is precisely the address shifts performed through the prototype one-dimensional filter prior to the polyphase partition. In reality, we do not perform the serpentine shift, but rather perform an addressing manipulation that swaps the upper and lower halves of the filter memory banks. This is shown in Figure 9.4 where successive sequences of length 32 are delivered to a filter bank with 64 paths.

We continue this discussion with comments on the 64-stage example. After each 32-point data sequence is delivered to the partitioned 64-stage polyphase filter, the outputs of the 64 stages are computed and conditioned for delivery to the 64-point inverse fast Fourier transform (IFFT). The reduced length data shifting into the polyphase filter stages causes a frequency-dependent phase shift of the form shown in Equation (9.2). The time delay due to shifting is nT, where n is the number of samples and T is the interval between samples. The frequencies of interest are integer multiple k of $1/M$th of the sample rate $2\pi/T$. Substituting these terms in Equation (9.2) and canceling terms, we obtain the frequency-dependent phase shift shown in

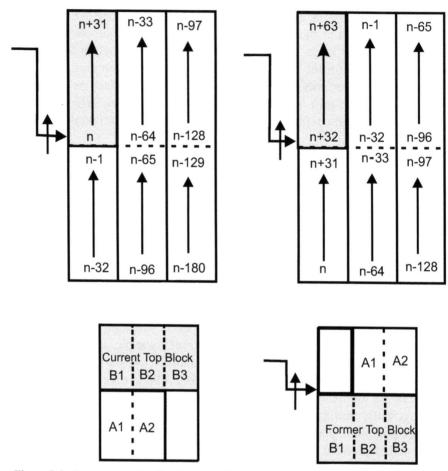

Figure 9.4   Data memory loading for successive 32-point sequences in a 64-stage polyphase filter.

Equation (9.3). From this relationship, we see that for time shifts n equal to multiples of M, as demonstrated in Equation (9.4), the phase shift is a multiple of 2π and contributes no offset to the spectra observed at the output of the IFFT. The M-sample time shift is the time shift applied to the data in the normal use of the polyphase filter. Now suppose that the time shift is $M/2$ time samples. When substituted in Equation (9.3), we find, as shown in Equation (9.5), a frequency-dependent phase shift of $k\pi$ from which we conclude that odd-indexed frequency terms experience a phase shift of π radians for each successive $N/2$ shift of input data.

$$\theta(\omega) = \Delta t \cdot \omega \tag{9.2}$$

$$\theta(\omega_k) = nT \cdot k \frac{1}{M} \frac{2\pi}{T} = \frac{nk}{M} 2\pi \tag{9.3}$$

$$\theta(\omega_k)|_{n=M} = \frac{nk}{M} 2\pi \bigg|_{n=M} = k \cdot 2\pi \tag{9.4}$$

$$\theta(\omega_k)|_{n=M} = \frac{nk}{M} 2\pi \bigg|_{n=M/2} = k \cdot \pi. \tag{9.5}$$

This π radian phase shift is due to the fact that the odd-indexed frequencies alias to the half sample rate when the input signal is down-sampled by $M/2$. Successive samples at the half sample rate differ by π radians. We can compensate for the alternating signs in successive output samples by applying the appropriate phase correction as we extract successive time samples from the odd-indexed frequency bins of the IFFT. The phase correction here is trivial, but for other down-sampling ratios, the residual phase correction would require a complex multiply at each transform output port. Alternatively, since time delay imposes a frequency-dependent phase shift, we can use time shifts to cancel these frequency-dependent phase shifts. We accomplish this by applying a circular time shift of $M/2$ samples to the vector of polyphase filter samples prior to their presentation to the IFFT. As in the case of the serpentine shift of the input data, the circular shift of the polyphase filter output data is implemented as an address manipulated data swap. This data swap occurs on alternate input cycles and a simple two-state machine determines for which input cycle of the output data swap is applied. This option is shown in Figure 9.5.

For any amount of input down-sampling, each input Nyquist zone will alias to some output frequency and the successive output samples have to be de-spun by the successive complex rotators defined by that aliased output frequency. A simple way to visualize the frequency-dependent phase correction of the M-path analysis filter bank when the input commutator delivers $M/2$ input samples is seen in Figure 9.6. In the upper subfigure, we see that successive input Nyquist zones are at multiples of the output sample rate f_s/M when the filter performs M-to-1 down-sampling to f_s/M. In the lower subfigure, we see that the even indexed input Nyquist zones are at multiples of the output sample rate $2f_s/M$ when the filter performs M-to-2 down-sampling and consequently alias to baseband. The odd multiples of f_s/M are located midway to the output sample rate $2f_s/M$ and hence are aliased to the half

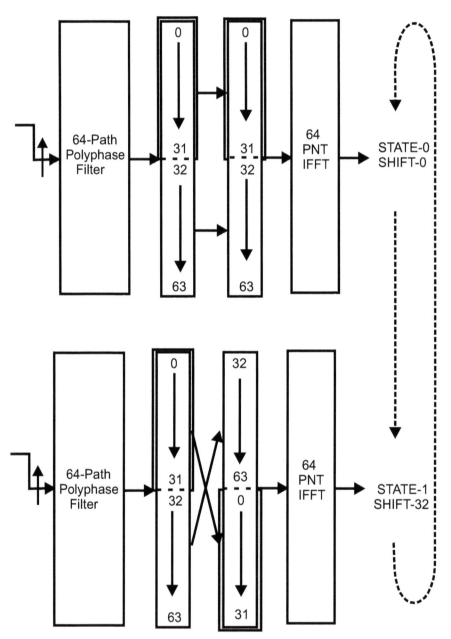

Figure 9.5   Cyclic shift of input data to FFT to absorb phase shift due to 32 sample time shift of data in the polyphase filter.

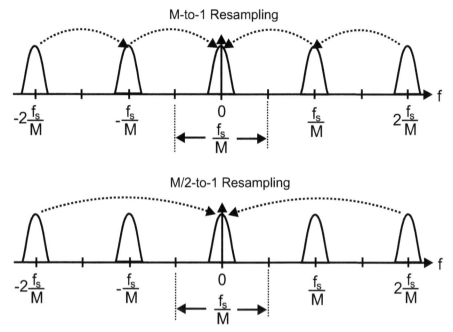

Figure 9.6   Alias output frequencies for successive Nyquist zones for *M*-to-1 and for *M*-to-2 down-sampling of *M*-path polyphase analysis filter bank.

sample rate. This explains the alternating signs on successive output samples from the odd indexed Nyquist zones.

It is easy to see that if we had performed an *M*-to-3 down-sampling to an output sample rate of $3f_s/M$, every third indexed input Nyquist zone would alias to baseband. Nyquist zone 1, located at f_s/M, is 1/3 of the way to the output sample rate. This bin and every bin index with residue 1 mod 3 aliases to 1/3 of the output rate and requires successive $2\pi/3$ correcting phase shift complex multiplies to shift the aliased offset spectrum to baseband. In a similar fashion, Nyquist zone indices with residue 2 mod 3, aliases to $-1/3$ of the output sample rate, and requires successive $-2\pi/3$ correcting phase shift complex multiplies to shift the aliased offset spectrum to baseband.

There are a number of ways to correct the offset alias terms formed by the increased output sample rate due to non-maximal down-sampling with the input commutator. One way, of course, is a set of frequency-dependent phase rotators applied to the output ports of the IDFT that separates the multiple aliases. The math will indicate the residual rotation of each index and the designer can respond with the correcting counter rotating heterodyne. In a second approach, we look at an *ad hoc* correction that counters the successive residual phase shifts with appropriate multiplier-free successive time shifts. This proves to be a satisfying solution till we examine the dual channelizer and realize the dual does not work there. We then develop a third method based on a series of *noble identity* transformations to form a multiple free, time shift, correction scheme that works in both the analysis and the synthesis channelizers.

We are now prepared to examine the process that permits resampling of the polyphase filter bank by any rational ratio. We first demonstrated the modification to the standard polyphase structure to support $M/2$ down-sampling. The modifications involved a serpentine shift of input memory and a circular shift of output memory that are both implemented by memory bank data swaps.

This technique is based on the observation that the commutator is the component in the polyphase filter bank that effects and controls the resampling, not the spacing between the adjacent channels. This is true even though it was the resampling process that first guided us to select the channel spacing so that we could access the aliasing to baseband. We now enlarge the set of options to permit aliasing to other frequencies and then be shifted to baseband. As we just demonstrated, there are two modifications to the polyphase-resampling filter required to obtain arbitrary resampling. These modifications would normally lead to an exercise in time varying residue index mapping of the two-dimensional input data array. If we limit the presentation to the index mapping process, we would develop little insight into the process and further would be bored to tears. Instead, we derive and illustrate the modifications by examining a specific example and observe the process developing.

We have just examined the case in which the embedded resampling ratio of a polyphase filter was trivially modified from M-to-1 to $M/2$-to-1. We now examine the case in which the embedded resampling ratio is modified to $(3/4\ M)$-to-1. The extension to any other rational ratio is a trivial extension of this process. To aid in understanding the process, we describe a specific resampling channelizer and modify the process to guide us to the desired solution.

The problem we examine is this: We have a signal containing 50 FDM channels separated by 192 kHz centers containing symbols modulated at 128 kHz by SQRT Nyquist filters with 50% excess bandwidth. Our task is to baseband channelize all 50 channels and output data samples from each channel at 256 ks/s, which is two samples per symbol.

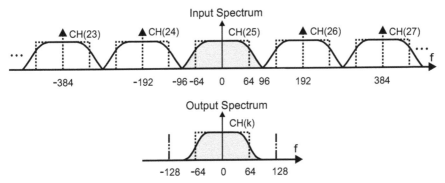

Figure 9.7 Input and output spectra of 50-channel channelizer and resampler.

The specifications of the process are listed next and the spectrum of the FDM input signal and of one of the 50 output signals is shown in Figure 9.7.

Number of Channels	50
Channels Spacing	192 kHz
Channel Symbol Rate	128 kHz
Shaping Filter	SQRT cosine taper
Roll-Off Factor α	50%
Output Sample Rate	256 kHz

We start by selecting a transform size N greater than the number of FDM channels to be processed. As indicated in Equation (9.6), the product of the transform size and the channel spacing defines the input sample rate of the data collection process. As shown in Equation (9.7), a restatement of the Nyquist sampling criterion, the excess bandwidth spanned by the extra channels in the transform, is allocated to the transition bandwidth of the analog anti-alias filter.

$$f_S = N \Delta f \tag{9.6}$$

$$f_S = \text{Two-Sided BW} + \text{Transition BW}$$

Two-Sided BW = # Data Channels . Channel Spacing

Transition BW = # Non-Data Channels + Channel Spacing $\qquad$ (9.7)

Transform Size = # Data Channels + # Non-Data Channels.

We note the engineering compromise to be made here: A larger transform size would reduce the cost of the analog anti-alias filter by permitting wider transition bandwidth, while a larger transform would increase the

Table 9.1   List of highly composite transform sizes considered for 50-output channels of M-path polyphase analysis channelizer.

Transform size N	Factors	Sample rate (MHz)	Pass-band edge in FFT bins	Stop-Band edge in FFT bins	Filter order 0.1 dB pass 60 dB stop
54	2,3,3,3	10.368	25.5	28.5	9
60	3,4,5	11.520	25.5	34.5	7
64	2,2,2,2,2,2	12.288	25.5	38.5	6
72	2,2,2,3,3	13.824	25.5	46.5	6
80	2,2,2,2,5	15.360	25.5	54.5	5

input sample rate, the number of polyphase stages in the filter bank, and the arithmetic processing burden required to implement the bank and the transform.

The size of the selected transform should be highly composite so that it can be implemented as an efficient IFFT. Table 9.1 lists transform sizes appropriate for the polyphase channelizer along with required data sample rate and order of anti-aliasing filter.

We select a 64-point fast Fourier transform (FFT) to span the 50 channels with the excess bandwidth allocated to the analog anti-alias filter. Thus, the sample rate for the collected spectra is 64 times the 192 kHz channel spacing or 12.288 MHz. These are complex samples formed from either a baseband block conversion or a digital down-conversion and resampling from a digital IF, often centered at the quarter sample rate. The desired output sample rate is 2 times 128 or 256 kHz. The ratio between the input and output sample rates is the resampling ratio, which is 12,288/256 or 48-to-1. Thus, our task is to use the 64-point discrete Fourier transform (DFT) to separate and deliver 50 of the possible 64 channels spanned by the sample rate, but to deliver one output sample for every 48 input samples.

Figure 9.8 is a block diagram of the original maximally decimated version of the 64-stage polyphase channelizer and the modified form of the same channelizer. The difference in the two systems resides in the block inserted between the 64-stage polyphase filter and the 64-point FFT. As indicated earlier, the inserted block performs no computation, but rather only performs a set of scheduled circular buffer shifts. Next we will develop and describe the operation of the circular buffer stage and state machine scheduler.

Our first task is to modify the input commutator to support the 48-to-1 down-sample rather than the standard 64-to-1 down-sample. This is an almost trivial task. We arrange for the modified resampling by keeping the 64-path filter but stripping 16-ports from the commutator. The commutator for the

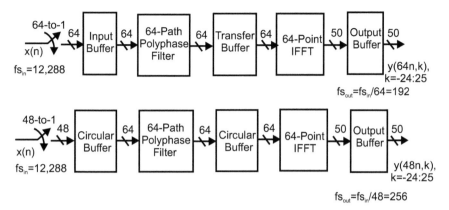

Figure 9.8   Maximally decimated filter bank structure and modified two-samples-per-symbol filter bank structure.

standard 64-point polyphase filter starts at Port 63 and delivers 64 successive inputs to Ports 64, 63, 62, and so on through 0. The modified commutator starts at Port 47 and delivers 48 successive inputs to Ports 47, 46, 45, and so on through 0. Input memory for the 64-path filter must be modified to support this shortened commutator input schedule. The mapping structure of the reindexing scheme is best seen in the original, one-dimensional prototype filter shown in Figure 9.8 and then transferred to the two-dimensional polyphase partition.

Figure 9.9 presents the memory content for a sequence of successive 48-point input data blocks presented to the 64-point partitioned prototype filter. In this figure, we have indicated the interval of 64-tap boundaries that become the columns of the two-dimensional array as well as the boundaries of successive 48-point input blocks that are presented to the input array. Successive input blocks start loading at Address 47 and work up to Address 0. The beginning and end of this interval are denoted by the tail and arrow, respectively, of the left most input interval in the filter array. As each new 48-point input array is delivered, the earlier arrays must shift to the right. These shifting array blocks cross the 64-point column boundaries and, hence, move to adjacent columns in the equivalent two-dimensional partition. This crossing can be visualized as a serpentine shift of data in the two-dimensional array or, equivalently, as a circular row buffer down shift of 48 rows in the polyphase memory with a simultaneous column buffer right shift of the input data column. The operation of this circular buffer is illustrated in Figure 9.10, which indicates the indices of input data for two input cycles. Here we see

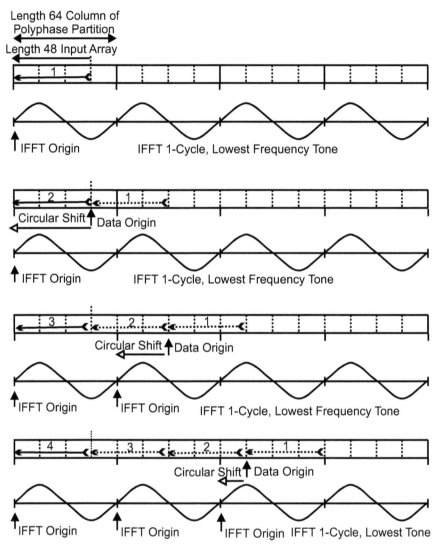

Figure 9.9 Memory contents for successive 48-point input data blocks into a 64-point prototype pre-polyphase partitioned filter and FFT.

that between two successive input cycles, the rows in the top one-fourth of memory translate to the bottom fourth, while the bottom three-fourths of rows translate up to one-fourth of memory. We also see that the columns in the bottom three-fourths shift to the right one column during the circular row translations. The next input array is loaded in the left most column of this group of addresses.

Returning to Figure 9.9, the one-dimensional prototype, we note that every new data block shifts the input data origin to the right by 48 samples. The vector $\hat{y}(r,48n)$ formed as the polyphase filter output from all 64 path filters is processed by the IFFT to form the vector $\hat{Y}(k,48n)$ of channelized (index k) output time series (index $48n$). On each successive call to the IFFT, the origin of the sinusoids in the IFFT is reset to the beginning of the input array. Since the origin of the input array shifts to the right on successive inputs while the origin of the IFFT simultaneously resets to the beginning of the input array, a processing offset exists between the origins of the polyphase filter and of the IFFT.

We align the origins, removing the offsets, by performing a circular shift of the vector $\hat{y}(r,48n)$ prior to passing it to the IFFT. Since the offset is periodic and is a known function of the input array index, the circular offset of the vector can be scheduled and controlled by a simple state machine. Figure 9.9 shows the location of the two origins for four successive 48-point input arrays and the amount of circular offset required to align the two prior to the IFFT. Note that the offset schedule repeats in four cycles, four being the number of input intervals of length 48 that is a multiple of 64. The cyclic shift schedule for the array $\hat{y}(r,48n)$ prior to the IFFT is shown in Figure 9.10. This non-maximal resampled polyphase analysis filter demonstrated the flexibility of the polyphase filter with the ability to independently control the number of channels and the spectral spacing with the transform length, the channel bandwidth with the polyphase filter weights, and the output sample rate with the commutator and required cyclic shifts of data buffer and filter outputs to drive all aliased bands to baseband.

9.2.1 Noble Identity Based Analysis Filter Bank

We now concentrate on analysis channelizer with a non-maximal sample rate change, M-to-2 in the same M-path analysis filter. We have already observed that the $M/2$ point data shifts in the M-path filter data register could be described as a serpentine shift through the successive columns. We modified the way successive input data vectors interacted with addresses of the input registers using block memory swaps to avoid shifting data through the actual serpentine shift. We also learned that increasing the output sample rate of the M-path filter caused different Nyquist zones of the input spectrum to alias to other locations besides baseband. The frequency offset locations of

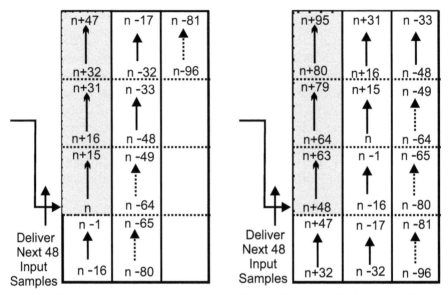

Figure 9.10   Memory contents for successive 48-point input data blocks into a 64-point polyphase filter.

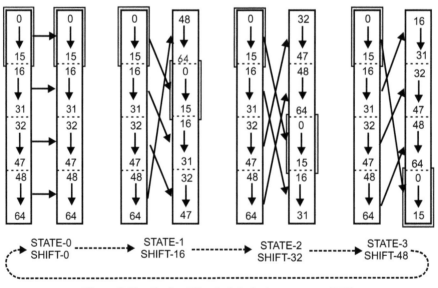

Figure 9.11   Cyclic shift schedule for input array to FFT.

the various alias bands were easily identified and could be heterodyned to baseband as the successive time samples were output from their IDFT output ports. We identified a technique that used selected circular time delays of the *M*-path filter output vector to assure delivery of all input Nyquist zones to baseband. We alluded to the fact that the *ad hoc* response to the serpentine shift did not have a structure that mapped well into the yet to be examined dual channelizer, the synthesis filter bank. We now derive another approach to the channelizer design based on successive applications of the noble identity that has a structure applicable to both the analysis and synthesis channelizers. Following this section, we proceed to examine the synthesis channelizer and examine interesting ways the pair of channelizer can be combined to access valuable processing tasks.

We start our journey with the structure shown in Figure 9.12. This is an *M*-path band centered polyphase filter followed by an *M*/2-to-1 down-sampler.

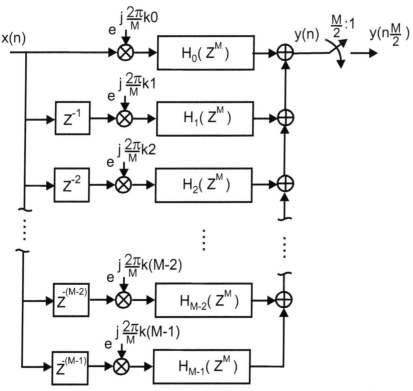

Figure 9.12   *M*/2-to-1 down-sampler following filter polynomials in Z^M.

Earlier designs had an output port M-to-1 down-sampler which we moved through the M-path filters formed by polynomials in Z^M. When we slid the resampler through the filters in each arm, they became polynomials in Z^1 fed by M-to-1 resamplers. We now pull the $M/2$-to-1 resamplers through the filters and find that, as shown in Figure 9.13, they become polynomials in Z^2 fed by $M/2$-to-1 resamplers. At the same time, we also interchanged the order of complex rotators and path filters. We note that the bottom half of the delays on the M-path arms are equal to or greater than $M/2$; so the $M/2$ resampler can

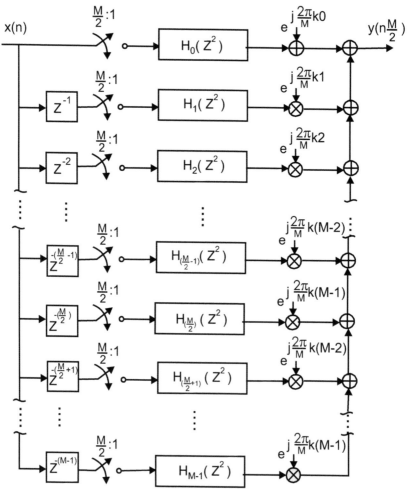

Figure 9.13   *$M/2$-to-1 down-samplers preceding filter polynomials in Z^2.*

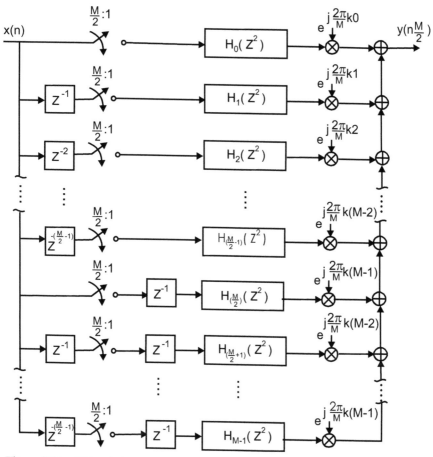

Figure 9.14 *M/2-to-1 down-samplers moved through delays in bottom half of delay chain.*

slide through those delays to obtain a reduced length input delay, the *M/2*-to-1 resampler, and a single output delay as shown in Figure 9.14.

In Figure 9.15, we replace the multiple *M/2*-to-1 resamplers and the sequential delays on the *M/2* paths with the standard commutator equivalent process. We note that we have a pair of commutators, one servicing the upper *M/2* paths and one the lower *M/2* paths. We are approaching the end of this derivation. As the upper port of the input commutator delivers *M/2* inputs to the top half of the *M*-path filter, the lower port of the input commutator delivers the same *M/2* input sample to a delay line which will deliver the same input sequence to the lower half of the *M*-path filter to be used when

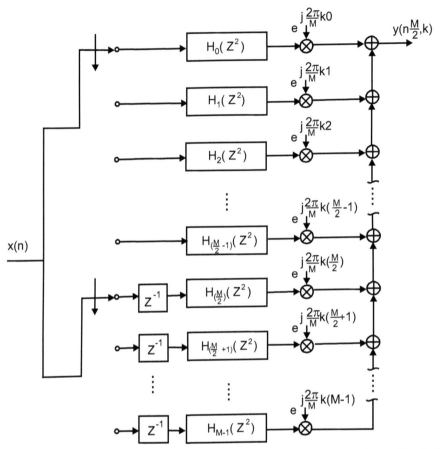

Figure 9.15 Replace input delay chains and $M/2$-to-1 down-samplers with dual commutators.

the next $M/2$ input samples are delivered to the top half. This matches the $M/2$ data shifts we described as the serpentine shift of input data.

What this little derivation did, is deliver the same data to both upper and lower paths, with appropriate delays, to align the data segments with the successive set of filter weights as the data segments appear to serpentine shift through the filter without actually performing the serpentine shift.

This seems like a waste of memory, storing two copies of the same data; and it is. We address that objection shortly. The real value of this architecture will be apparent in the upcoming section where we examine the dual channelizer, the synthesis up-sampling channelizer.

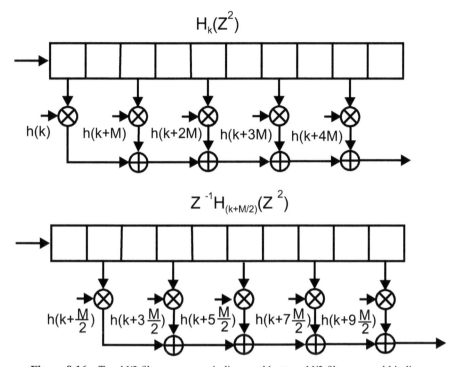

Figure 9.16 Top *M/2* filters use even indices and bottom *M/2* filters use odd indices.

We now address the filters on the lower *M/2* paths of the polyphase filter, the filter preceded by the single delay blocks. The filters on the paths are polynomials in Z^2, which means they have delay element between successive coefficient taps. Rather than using an external delay in each of the *M/2* lower paths, we can shift the taps to the odd indexed delays. The upper *M/2* paths leave their tap connections on the even indexed delays. This is shown in Figure 9.16. We call attention to a detail that effects implementation; the filter with *N* coefficients per path now requires *2N* addresses to accommodate the polynomials in Z^2. This is seen in Figure 9.16 in which the five-tap filters shown uses 10 delay registers.

The final step in the polyphase *M/2*-to-1 down-sample analysis channelizer is the matter we addressed earlier, the input registers holding two copies of the same *M/2* input sequence with a 1-sample delay between the two sequences. We fix that here. The two filter structures shown in Figure 9.16 contain the current *M/2* samples in the even indices of the top paths and the previous *M/2* samples in odd indices of the bottom paths. Both input registers

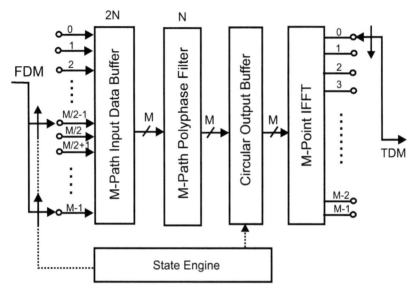

Figure 9.17   *M*-path, *M*/2-to-1 down-sample polyphase analysis filter architecture formed by noble identity transformations.

are fed by identical commutators; see Figure 9.15 with the bottom delays moved into the path filters as shown in Figure 9.16. Since they contain the same samples, the upper and lower filters can access the same $M/2$ deep single copy of samples, with the upper half of filters accessing the even indices and the lower half of filters accessing the odd indices. Thus, the depth and width of the data register of the M-path analysis filter is not M by $2N$, but rather $M/2$ by $2N$. We will see shortly that the data register of the M-path synthesis filter will stay M by $2N$ and why that is so. What we have derived here is the $M/2$-to-1 down-sample M-path analysis structure shown in Figure 9.17 and it is a more compact memory form shown in Figure 9.18 obtained by interleaving the upper and lower half data buffers.

The $M/2$-to-1 down-sampling analysis filter bank architectures shown in Figures 9.17 and 9.18 have four distinct segments. They are the two-dimensional input data buffer fed by $M/2$ port input commutator, the M-path polyphase partitioned prototype low pass filter, the circular buffer using circular time delay to drive all output aliases to baseband, and the M-point IFFT which unwraps the M-fold aliases and defines the output Nyquist interval locations and spacing of the input signal's spectrum. The architecture also contains a simple two-state state machine that controls the circular output buffer and the $M/2$ port commutator.

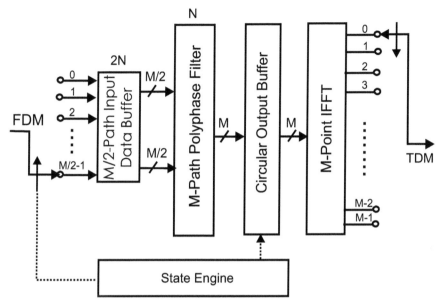

Figure 9.18 *M*-path, *M*/2-to-1 down-sample polyphase analysis filter architecture formed by interleaving top and bottom half data buffers.

The segment of MATLAB script shown in Script 9.1 illustrates the compactness of the synthesis channelizer. The actual channelizer loop is performed by 13 lines of script. Two lines collect a vector *v1*, the next six input data samples, and inserts it in the filter register *reg*. The inner FOR loop ($k = 1{:}6$) computes the 12 polyphase output samples $v2(k)$ and $v2(k + 6)$. They are computed as inner products between the alternate samples of row *k* in the *reg* and rows *k* and $k + 6$ of the polyphase coefficients *hh1*. Following the formation of the polyphase filter output vector *v2*, the vector is ended around time delayed on alternate output clock indices and presented to the scaled IFFT to form the output time vector *v3* which is stored in successive columns in the output time array *v4*. This version of the channelizer has the smallest amount of interaction between the input data register and the length *M*/2 serpentine shift we described in earlier *ad hoc* architecture. This architecture easily accommodates the *M*-to-12/9 down-sampling analysis channelizer we presented earlier but may not do as well for an *M*-to-12/10 down-sampling analysis channelizer.

MATLAB Script 9.1. 12-path, 12-to-2 down-sampled analysis filter bank, even and odd indexed data column for six-top and six-bottom filter coefficients.

```
% Analysis Channelizer 12-to-2,

% Use Even and Odd Indexed data columns for 6-top and 6-bottom coefficients

% h1=remez(118,[0 4 8 60]/60,{'myfrf',[1 1 0 0]},[1 80]) computed elsewhere

  hh1=reshape([0 12*h1],12,10);   % map 1-D filter coefficients to 2-D array

  reg=zeros(6,20);      % 2-D filter register

  v1=zeros(1,6)';       % input sample vector

  v2=zeros(1,12)';      % filter output vector

  v3=zeros(1,12)';      % IFFT output vector

  v4=zeros(12,21);      % 21 sample time series from 12 output ports


  x0=zeros(1,124);      % Input data samples for impulse response

  x0(6)=1;              % Impulse at 1 of 12 locations. aligned with center
  tap

  m1=1;                 % output clock

  flg=0;                % State of state machine

  for n=1:6:124

      v1(1:6)=fliplr(x0(n:n+5)).';    % fetch next 6 sample input vector

      reg=[v1 reg(:,1:19)];           % input next 6 sample input vector

    for k=1:6

        v2(k)=reg(k,1:2:20)*hh1(k,:)';     % compute 6-upper filter outputs

        v2(k+6)=reg(k,2:2:20)*hh1(k+6,:)'; % compute 6-lower filter outputs

    end

  if flg==0        % No phase rotation, filter output without swap

      flg=1;       % Reset flag

  else             % Prepare to swap top and bottom halves of filter output

      flg=0;       % Reset flag

      v2=[v2(7:12);v2(1:6)];  % swap here

  end

  v3=fftshift(6*ifft(v2));    % IFFT Separate Aliases, Shift DC to bin 7

  v4(:,m1)=v3;                % output vector of time series samples

  m1=m1+1;                    % increment output clock

  end
```

The segment of MATLAB script shown in Script 9.2 differs from the Script 9.1 in the dimensions of the input data register, 12 by 10 here in Script 9.2 and 6 by 20 in Script 9.1. Here in Script 9.2, the serpentine shift is performed at the second line of the outer FOR loop by the block data swap of the upper and lower halves of the input data register. The input data vector is inserted in the upper half of the input data register following the block swap. There is no serpentine shift in Script 9.1. In 9.1, the data samples reside in the even and odd register addresses and two lines of code access the data samples for the upper 6 and lower 6 filter output terms. In Script 9.2, the data samples are not interleaved and one line of code accesses the data samples for all 12 filter output terms. Otherwise, the code in Script 9.1 and Script 9.2 are identical.

MATLAB Script 9.2. 12-path, 12-to-2 down-sampled analysis filter bank swap top and bottom halves of input data register.

```
% Analysis Channelizer 12-to-2

% Swap top and bottom reg halves input reg

% h1=remez(118,[0 4 8 60]/60,{'myfrf',[1 1 0 0]},[1 80]) computed elsewhere

  hh1=reshape([0 12*h1],12,10);  % map 1-D filter coefficients to 2-D array

  reg=zeros(12,10);    % 2-D filter register

  v1=zeros(1,6)';      % input sample vector

  v2=zeros(1,12)';     % filter output vector

  v3=zeros(1,12)';     % IFFT output vector

  v4=zeros(12,21);     % 21 sample time series from 12 output ports


  x0=zeros(1,124);     % Input data samples for impulse response

  x0(6)=1;             % Impulse at 1 of 12 locations. aligned with center
tap

  m1=1;                % output clock

  flg=0;               % State of state machine
for n=1:6:124-5

     v1=fliplr(x0(n:n+5)).';           % fetch next 6 sample input vector

     reg=[reg(7:12,:);reg(1:6,:)];     % reg swap halves

     reg(1:6,:)=[v1 reg(1:6,1:9)];     % input next 6 sample input vector
```

```
    for k=1:12

        v2(k)=reg(k,:)*hh1(k,:)';     % compute 12 filter outputs

    end

if flg==0        %No phase rotation, filter output without swap

    flg=1;       % Reset flag

else             % Prepare to swap top and bottom halves of filter output

    flg=0;       % Reset flag

    v2=[v2(7:12);v2(1:6)];   % swap here

end

v3=fftshift(6*ifft(v2));    % IFFT Separate Aliases, Shift DC to bin 7

v4(:,m1)=v3;                % output vector of time series samples

m1=m1+1;                    % increment output clock

end
```

9.3 Noble Identity Based Synthesis Filter Bank

Till this point in discussing filter banks, we emphasized non-maximal down-sampling analysis filter banks. In these analysis filter banks, down-sampling causes multiple narrow bandwidth spectral Nyquist zones to alias to baseband. Following this reasoning, we consider reversing this process and use up-sampling to cause multiple narrow bandwidth baseband signals to alias up to multiple spectral Nyquist zones. In the first case, we decompose a wide bandwidth signal into multiple narrow bandwidth baseband signals by aliasing them to baseband by down-sampling the composite time series. This is the analysis process. In the reverse case, we compose a wide bandwidth signal from multiple narrow bandwidth baseband signals by aliasing them from baseband by up-sampling and combining the multiple time series. This is the synthesis process.

In an earlier chapter, we commented that a linear time invariant (LTI) filter block diagram has a dual graph. The dual graph block diagram is formed by replacing summing junctions by nodes, replacing nodes by summing junctions, and reversing the directions of signal flow on the graph. The dual graph has precisely the same transfer function as the original graph. This relationship is not true for a linear time varying (LTV) filter and is complicated by the fact that an LTV filter does not have a transfer function.

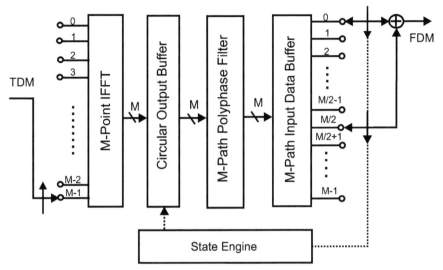

Figure 9.19 *M*-path, 1-to-*M*/2 up-sample polyphase synthesis filter architecture formed by dual of noble identity transformations.

So rather than discussing the lack of transfer functions, we can discuss the process performed by the LTV filter. What we find is that the process performed by the dual graph of an LTV filter is the opposite of the process performed by the original graph. Thus, if an original graph does down-sampling, the dual graph will do up-sampling. If the original graph performs down-sampling to alias multiple Nyquist zones to baseband, then the dual graph performs up-sampling to alias baseband signals to multiple Nyquist zones. Thus, we have immediate access to a non-maximum up-sampling polyphase synthesis filter bank by forming the dual of a non-maximum down-sampling polyphase analysis filter bank. Such an analysis filter bank is shown in Figure 9.17. Its dual, the synthesis filter bank, is shown here in Figure 9.19.

References

Bellanger, M. and J. Daguet, "TDM-FDM Transmultiplxer: Digital Polyphase and FFT", IEEE Trans. Communications, Vol. COM-22, Sept 1974, pp.1199-1204.

Crochiere, Ronald and Lawrence Rabiner, *Multirate Signal Processing*, Englewood Cliff, NJ, Prentice-Hall Inc., 1983.

Elliot, Doug, Editor. *Handbook of Digital Signal Processing: Engineering Applications*, ,Chapter 8, "Time Domain Signal Processing with the DFT", pp 639-666, Academic Press, 1987.

Fliege, Norbert. *Multirate Digital Signal Processing: Multirate Systems, Filter Banks, Wavelets*, West Sussex, John Wiley & Sons, Ltd, 1994.

Fliege, Norbert J. "Polyphase FFT Filter Bank for QAM Data Transmission", Proc. IEEE ISCAS'90, pp.654-657, 1990.

harris, fred j. "On the Relationship Between Multirate Polyphase FIR Filters and Windowed, Overlapped FFT Processing", Twenty-third Annual Asilomar Conference on Signals and Computers, 1989.

Jovanovic-Dolecek, Gordana. *Multirate Systems: Design and Applications*, London, Idea Group, 2002. Koilpillai, R. D., T.Q. Nguyen, and P.P. Vaidyanathan, "Some Results in the Theory of Crosstalk Free Transmultiplexer", IEEE Trans. SP. Vol.39, pp. 2174-2183, October 1991.

Mitra, Sanjit. *Digital Signal Processing: A Computer-Based Approach*, 2nd ed., New York, McGraw- Hill, 2001.

Mitra, Sanjit and James Kaiser. *Handbook for Digital Signal Processing*, New York, John Wiley & Sons, 1993.

Renfors, Markku, and T. Saramaki, "Recursive N-th Band Digital Filters, Parts I and II", IEEE Trans. CAS, Vol. 34, pp. 24-51, January 1987.

Scheuermann, H., and H. Göckler, "A Comprehensive Survey of Digital Transmultiplexing Methods", Proc. IEEE, Vol. 69, No. 11, November 1981, pp-1419-1450.

Vaidyanathan, P. P. *Multirate Systems and Filter Banks*, Englewood Cliff, NJ, Prentice-Hall Inc., 1993.

Vaidyanathan, P. P. "Multirate Digital Filters, Filter Banks, Polyphase Networks and Applications: A Tutorial", Proc. IEEE, Vol. 78, pp 56-93, January 1990.

Problems

9.1 Design a prototype low pass filter for a 10-to-1 polyphase partition with the following specifications:

Sample Rate	100 kHz		
Pass Band	0–4 kHz	Pass Band Ripple	0.1 dB
Stop Band	6–50 kHz	Stop Band Attenuation	60 dB,
			−6 dB/Octave

What filter length satisfies the specifications?

What is the length of each path in a 10-path polyphase partition?

What is the frequency spacing obtained from the 10-path polyphase filter followed by a 10-point FFT?

Sketch the spectral response of three spectral bands at frequencies −10, 0, and 10 kHz when the polyphase filter is operated at 10 kHz output rate, 10 inputs per output.

Sketch the spectral response of three spectral bands at frequencies −10, 0, and 10 kHz when the polyphase filter is operated at 20 kHz output rate, 5 inputs per output.

9.2 Design a prototype low pass filter for a 10-to-1 polyphase partition with the following specifications:

Sample Rate	100 kHz		
Pass Band	0–4 kHz	Pass-Band Ripple	0.1 dB
Stop Band	5–50 kHz	Stop-Band Attenuation	60 dB, −6 dB/Octave

What filter length satisfies the specifications?

What is the length of each path in a 10-path polyphase partition?

What is the frequency spacing obtained from the 10-path polyphase filter followed by a 10-point FFT?

Sketch the spectral response of three spectral bands at frequency −10, 0, and 10 kHz when the polyphase filter is operated at 10 kHz output rate, 10 inputs per output.

Sketch the spectral response of three spectral bands at frequency −10, 0, and 10 kHz when the polyphase filter is operated at 20 kHz output rate, 5 inputs per output.

9.3 Design a prototype low-pass filter for a 10-to-1 polyphase partition with the following specifications:

Sample Rate	100 kHz		
Pass Band	0–4 kHz	Pass-Band Ripple	0.1 dB
Stop Band	6–50 kHz	Stop-Band Attenuation	60 dB, −6 dB/Octave

Partition the prototype into a 10-path polyphase filter with a 10-point FFT following the partition. Deliver 1000 samples of the input signal formed as a sum of equal amplitude, real sinusoids with random phase and of frequencies 0.5, 1.1, 1.6, 2.1, 2.6, 3.1, 3.6, and 4.1 kHz, and the same list of frequencies with a 20 kHz offset.

Perform the fully decimated, 10-to-1 down-sample, polyphase filtering and plot the time series and the spectra of each time series output from the 10 ports of the FFT following the polyphase partition. Comment on the signal levels in the channels adjacent to the occupied channels.

9.4 Design a prototype low-pass filter for a 10-to-1 polyphase partition with the following specifications:

Sample Rate	100 kHz		
Pass Band	0–4 kHz	Pass-Band Ripple	0.1 dB
Stop Band	5–50 kHz	Stop-Band Attenuation	60 dB,
			−6 dB/Octave

Partition the prototype into a 10-path polyphase filter with a 10-point FFT following the partition. Deliver 1000 samples of the input signal formed as a sum of equal amplitude, real sinusoids with random phase, and of frequencies 0.5, 1.1, 1.6, 2.1, 2.6, 3.1, 3.6, and 4.1 kHz, and the same list of frequencies with a 20 kHz offset.

Perform the fully decimated, 10-to-1 down-sample, polyphase filtering, and plot the time series and the spectra of each time series output from the 10 ports of the FFT following the polyphase partition. Comment on the signal levels in the channels adjacent to the occupied channels.

9.5 Repeat problem 9.4 except that the filter should now be operated as a 5-to-1 down-sample polyphase filter with output sample rate of 20 kHz rather than 10 kHz. Plot the time and spectra of each time series output from the 10 ports of the FFT following the polyphase partition. Comment on the signal levels in the channels adjacent to the occupied channels.

9.6 Design a prototype low-pass filter for a 16-to-1 polyphase partition with the following specifications:

Sample Rate	320 kHz		
Pass Band	0–5 kHz	Pass-Band Ripple	0.1 dB
Stop Band	10–160 kHz	Stop-Band Attenuation	60 dB,
			−6 dB/Octave

What filter length satisfies the specifications?
What is the length of each path in a 16-path polyphase partition?
What is the frequency spacing obtained from the 16-path polyphase filter followed by a 16-point FFT?

Sketch the spectral response of three spectral bands at frequencies –20, 0, and 20 kHz when the polyphase filter is operated at 20 kHz output rate, 16 inputs per output.

Sketch the spectral response of three spectral bands at frequencies –20, 0, and 20 kHz when the polyphase filter is operated at 40 kHz output rate, 8 inputs per output.

10

Cascade Channelizers

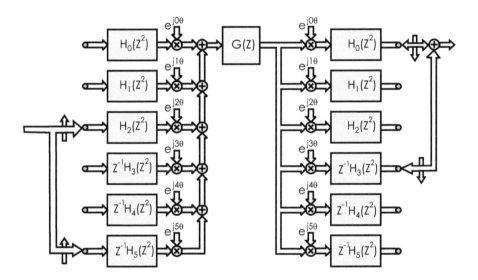

We have derived the 1-to-$M/2$ synthesis channelizer which we can easily imagine resides at the transmitter of a communication system where it combines a number of narrow bandwidth baseband signals to form a broad bandwidth composite signal containing multiple frequency division up-converted channels. We have derived the $M/2$-to-1 analysis channelizer which we can easily imagine resides at the receiver of the same communication system where it processes the received broad bandwidth composite signal to decompose it back into multiple narrow bandwidth baseband signals. Even though the two channelizers are duals, they perform inverse transformations at the transmitter and receiver. Cable TV and satellite links as well as the AM and FM radio bands are standard examples of this workload partition. I will put the signals together at my side and you disassemble them at your side. They do it quite well and we are pleased with their performance and cost.

What comes as a pleasant surprise is we may want both the analysis and synthesis channelizer to be coupled as a single entity at both sides of the communication link. The coupled pair will be shown to offer a wide range of very desirable capabilities at very low cost. We will find that the coupled channelizer pair may perform common processing tasks with order of magnitude cost reductions. We will also find unusual flexibility in the processing options offered by the coupled system, capabilities that will make us wonder how we managed prior to learning about them. Among the capabilities will be the ability to form wide bandwidth supper channels from multiple arbitrary bandwidth input channels translated to arbitrary center frequencies. Similarly, we will have the ability to decompose a super channel containing multiple bandwidth signals at arbitrary center frequencies. Another desirable ability is the reduced cost implementation of various high sample rate wide bandwidth filters which would be prohibitively expensive using traditional architectures. These useful and desirable capabilities are possible when design constraints allow the analysis–synthesis channelizer pair to faithfully reproduce the input signal from its channelized partitions. Our particular interest then is the design of non-maximally decimated perfect reconstruction filter banks.

10.1 Perfect Reconstruction Analysis–Synthesis Filter Banks

In the previous chapter, we developed the high-level block diagrams of the non-maximally decimated analysis filter bank and its dual, the non-maximally decimated synthesis filter bank. The diagrams were a bit busy, containing four processing blocks and a control block. In this chapter, we will use less detailed diagrams containing only two blocks, the M-path filter, and the M-point inverse fast Fourier transform (IFFT) with input ports and output ports. We will immediately know which filter bank is displayed because we know that the analysis filter input is the polyphase filter and its output is the inverse discrete Fourier transform (IDFT), while the synthesis filter input is the IFFT and its output is the polyphase filter. Another visual clue is that the analysis commutator delivers inputs to two inputs, a bottom port and a center port. The synthesis commutator accepts and adds two outputs, a top port and a center port. One way we can put them in cascade is by entering through analysis filter and leaving through the synthesis filter. In this configuration, the analysis filter's output IFFT connects to the synthesis

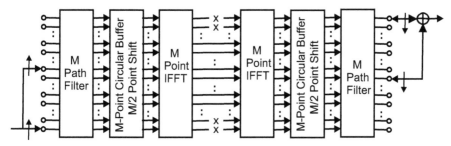

Figure 10.1 *M*-path analysis filter bank coupled to an *M*-path synthesis filter bank.

filter's input IFFT. This configuration is shown in Figure 10.1 and displays all the visual properties we just described.

We note that there are connection paths between the output IFFT vector of the analysis filter bank and the input IFFT vector of the synthesis filter bank. The time samples being passed from analysis to synthesis blocks are the low sample rate time series from each of the baseband spectral Nyquist zones of the input signal. We can modify the signal output by the synthesis filter by inserting gain terms along the path connections between the two IFFTs. We can trivially modify the spectral profile of the output time series by adjusting these gains. Be careful here! This looks similar to fast convolution, a process by which we move a signal to the frequency domain by a fast Fourier transform (FFT), modify its spectrum with a spectral mask, and return the signal, with the modified spectrum, to the time domain by an IFFT. We are not doing that here. We are scaling the narrow bandwidth time series formed by the analysis channelizer and then reconstructing the output time series by the synthesis channelizer. Every signal touched along the way is a time signal! We call this process *filtering in the channel domain*!

The channel gain terms for the example presented in Figure 10.1 represent a binary mask. If the path between the transforms is connected, the path gain is 1, and if the path is interrupted by an *x*, the path gain is 0 (we actually have to insert zeros for those paths). This proves to be an interesting way to vary the bandwidth of an output signal without varying filter weights. In fact, the task, *"Change the bandwidth but don't change any coefficients!"* led to this design.

The question now is, what conditions must be satisfied by the channel filters in the analysis and in the synthesis filter banks to be perfect reconstruction? We know the answer (see Section 4.2)! The combined impulse response of the cascade filters must be a Nyquist filter. For our application, the Nyquist filter has a two sided 1/2 amplitude (-6 dB) bandwidth equal to the channel spacing which is f_s/M. Adjacent channels, with spectral centers separated by f_s/M cross their -6 dB gain levels. The sum of the adjacent channels transition bandwidths is exactly 1.0 if the transition spectrum is odd symmetric about the band edge and the filter spectrum has no in-band ripple. The transition bandwidth is always odd symmetric if the sinc impulse response is windowed by an even symmetric window. It is too bad that the standard Nyquist filter has in-band ripple. In the communication community, the filters on each side of the communication link are a cosine tapered square-root (SQRT) Nyquist pulse. What we do is perform half the spectral shaping at the modulator and half the spectral shaping at the demodulator. The cascade of the two filters is approximately a Nyquist pulse. The approximation is limited by the SQRT-Nyquist filter's high level of pass-band ripple. The spectral shaping is split between the two ends of the link to permit bandwidth limiting at the modulator and to suppress channel noise at the demodulator. Our two filters in the analysis and in the synthesis banks are directly coupled without additive noise. We have no need to split the shaping. Thus, we can use a high-quality Nyquist filter in the analysis channelizer and a wider pass-band low-pass filter in the synthesis channelizer. Using the non-maximally decimated $M/2$-to-1 resampling satisfies the Nyquist criterion for the analysis filter and leaves some excess bandwidth for the wider pass band of the synthesis filter. The pass band of the synthesis filter must include the pass band and the two transition bandwidths of the analysis filter. Altering the transition bandwidth of the analysis filter would break the odd symmetry about the band edge and prevents it from being a perfect reconstruction filter.

We now examine the filter specifications and their design process for the two filter banks. Figure 10.2 contains three subplots. The top subplot illustrates the shape and spacing of the analysis filter Nyquist spectra. Adjacent channels have half amplitude bandwidths of ± 10 kHz with symmetric transition bandwidths. The transition bandwidth, maximized to minimize filter length, extends from 5 to 15 kHz. Adjacent Nyquist spectra centers are separated by 20 kHz. The center subplot shows a single half-band Nyquist spectrum with sample rate of 40 kHz formed by the analysis $M/2$-to-1 down-sample process. This is the spectral response of each baseband channel seen at each output port of the analysis channelizer with un-aliased transition bandwidths due to the increased output sample rate.

Note that the half-band Nyquist filter pass-band and stop-band widths are equal. The bottom subplot shows two spectra. The first is the periodic replica spectra of the half-band Nyquist filters with replicas at the 40 kHz sample

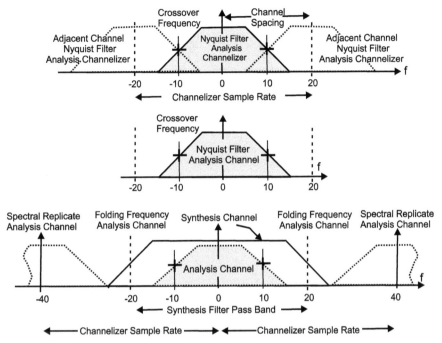

Figure 10.2 Top subplot: analysis Nyquist spectra spaced at 20 kHz centers with half-amplitude crossings at 10 kHz. Center subplot: single Nyquist spectrum with transition bandwidth of 5–15 kHz sampled at 40 kHz. Bottom subplot: periodic Nyquist spectra with replica spectra at multiples of 40 kHz sample rate and overlay synthesis spectrum transition bandwidth of 15–25 kHz.

rate. The second is an overlay of the wider bandwidth synthesis filter. This filter performs the anti-alias function in the synthesis filter bank. It shows the pass-band bandwidth spanning the full two sided bandwidth of the analysis filter spectrum in the baseband centered Nyquist zone. Its pass band protects the fidelity of the Nyquist filter pass band and transition bandwidths. Its stop band rejects all spectral replicates in other output Nyquist bands formed by the synthesis filter bank. Note that the synthesis filter has the same 10 kHz wide transition bandwidth as that of the analysis filter. If both filters are designed with the same pass-band and stop-band ripple levels, the two filters will have equal lengths.

Figure 10.3 presents the spectra of finite impulse response (FIR) filters designed to satisfy the design parameters of the analysis Nyquist filter and of the wide band synthesis filter. In the upper and center subplots, we see the transition bandwidth crossover between adjacent Nyquist zone bands

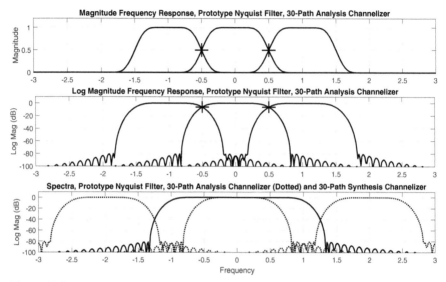

Figure 10.3   Top subplot: three adjacent Nyquist magnitude spectra of synthesis filter bank. Center subplot: three adjacent Nyquist log magnitude spectra of synthesis filter bank. Bottom subplot: single half-band Nyquist spectra with adjacent spectral replicas and overlay spectrum of synthesis filter.

in linear and log magnitude coordinates. Markers indicate the location at normalized frequencies of ±0.5 at crossover level of 0.5 and -6.02 dB. Both analysis filter and synthesis filter have the same transition bandwidth of 0.25 and stop-band attenuation level of -80 dB. A clever compromise in the design of the synthesis filter is the crossover gain of the synthesis filter transition and the replica Nyquist spectra at normalized frequency 1.25. The crossover level exceeds the desired -80 dB. The coupling of the attenuation levels of the filter and the spectra being suppressed permitted us to widen the synthesis filter's transition bandwidth beyond 1.25 and purchase additional out-of-band attenuation or a reduced length filter meeting the design specification.

We now have to actually design the filters. We have already discussed that the analysis filter should be a Nyquist filter. Recalling that the Nyquist filter must exhibit odd symmetry about its half amplitude band edge, we use a windowed sinc as our starting point. An example of the design for the analysis filter for a 30-path channelizer is shown in MATLAB Script 10.1. In preparing for this example, the sinc samples started at -4 and incremented in steps of $1/30$ to $+4$. The $+4$ and -4 are the filter widths in number of zero crossings of the sin$x$/x sequence, and the count 1/30 tells us the ratio of sample rate to bandwidth. This means the sample rate is 30 times the -6db bandwidth which is required for the 30-path channelizer. While we do not have to start on zero crossings, we like to have the number of samples in the filter to be a multiple of M, the number of paths. We do this so that when we map to the polyphase partition, all the paths will have the same number of coefficients. The symmetric argument $(-4:1/30:+4)$ would deliver an odd number of taps due to the sample at index 0, specifically $8 \times 30 + 1$ for this example. We respond to this problem by discarding the two end points as shown in the Script 10.1. This number, while odd, is also not a multiple of 30, a matter we fix by appending a leading zero to the coefficient list. We complete the design by applying a Kaiser window to the sinc weights. The knob yet to be adjusted is the β term (the time BW product) of the window. Increasing the β term has the effect of reducing the filter's spectral side lobe levels. Reducing spectral side lobes is purchased with increased transition bandwidth. The designer applies a window to the sinc series with a particular β selected from Figure 3.8 (in chapter 3), approximately 8 for -80 dB stop-band levels. We then examine its spectrum to verify side lobe levels and to determine the stop-band frequency, the frequency at which the filter achieves the required stop-band levels. If the edge is less than the design specified frequency, the filter is longer than necessary, and if it exceeds the required stop-band frequency, the filter requires additional taps. This process converges quite quickly to the correct length and β value to meet specified performance. A pleasant property of the windowed filter design is that the pass-band ripple and stop-band ripple are the same levels: one indicating deviation from unity gain and one deviation from zero gain. The windowed spectrum falls about 60 dB per decade, a useful property for the following reason. The reconstruction error of the cascade has contributions from both pass-band and aliased stop-band ripple. Constant or slowly falling stop-band side lobe levels raises the level of pass-band spectral artifacts, an easy measure of reconstruction error. We will verify performance of this design shortly.

MATLAB Script 10.1 Prototype Nyquist Filter for Polyphase Analysis Filter

```
h1=sinc(-(4-1/30):1/30:(4-1/30)).*kaiser(239,8)'; %windowed sinc
hh1=reshape([0 h1],30,8); % append leading zero and map to 2D array
```

We now consider the design of the synthesis filter. Prior to realizing that the aliased stop-band ripple contributes to the composite pass-band ripple, we used a Remez algorithm based design modified to have stop-band ripple fall-off at 20 dB per decade. We then realized that a faster decay rate would be advantageous and replaced the Remez design with a modified windowed sinc design. Both options are shown in Script 10.2. Here we see the modification to the sinc argument, the scale factor 2.015 multiplying the argument of the sinc. The effect of the scale factor is to reduce the effective sample rate which reduces the number of samples in the main lobe (or side lobe). This changes the ratio of sample rate to bandwidth and widens the filter's bandwidth relative to the sample rate without changing the number of coefficients.

MATLAB Script 10.2 Prototype Nyquist Filter for Polyphase Synthesis Filter

```
% h2=firpm(238,[0 0.703 1.28 15]/15,{'myfrf',[1 1 0 0]},[1 1]);
% Original Remez Design with 1/f stop band, replaced by scaled sinc
h2=sinc((-(4-1/30):1/30:(4-1/30))*2.015).*kaiser(239,8)';% modified
h2=h2/sum(h2);
hh2=reshape([0 h2],30,8);% append leading zero and map to 2D array
```

10.2 Cascade Analysis and Synthesis Channelizers

Now that we know how to design the filters, we can couple them in the form shown in Figure 10.1. After coupling them, we should test the pair under different conditions to learn what their capabilities are and demonstrate effective ways to use them. The first test condition should verify signal fidelity as it passes through the many stages of the analysis channelizer down-sampling polyphase filter, circular buffer, and IFFT and the synthesis channelizer up-sampling IFFT, circular buffer, and polyphase filter. We have found that the most dramatic test is the system impulse response. One minor problem is that the system has $M/2$ different impulse responses. We should examine a couple of them to develop some level of confidence in the system operation. An advantage of the impulse response test is that it offers some measure of system robustness to, what may think, is insignificant filter design and implementation changes. We will identify one important operating condition that must be strictly adhered to.

Figure 10.4 shows the result of the impulse response test of the 30-channel cascade analysis and synthesis filter banks with all channel outputs of the

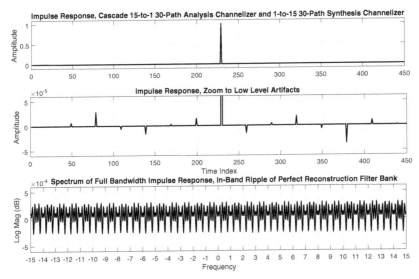

Figure 10.4 Top subplot: impulse response of full bandwidth cascade analysis and synthesis filter banks. Center subplot: zoom to low-level details of impulse response. Bottom subplot: spectrum ripple in-band ripple levels, perfect reconstruction errors.

analysis bank presented as inputs to the synthesis bank. Here, the binary mask option to pass and discard selected channels has enabled all input channels to contribute to the synthesized output signal. We applied a single impulse to input port 3, one of the 15 input ports, followed by trailing zeros to move the impulse through the processing chain. We expect the channelizer pair to reproduce the impulse at its output with the causal delay of the cascade filters. The top subplot is the time response, an amplitude 1.0 impulse located at index 228. The center subplot is a zoom to the low-level artifacts of the impulse response. We see pre- and post-echoes located at multiples of 30 samples from the primary impulse location. The levels of the echoes are all smaller than $3.6.10^{-5}$ or about 40 parts per million (ppm). These errors are more than 90 dB below the impulse amplitude and correspond to the 2 least significant bits of a 16-bit analog-to-digital converter (ADC). The bottom subplot shows the log magnitude spectrum of the impulse response. It has an average level of 0 dB and a periodic ripple with spectral periods of 1, the bandwidth of the analysis channelizer baseband components. The amplitude of the ripple is less than 3.10^{-4} dB or about 35 ppm which is also about -90 dB below 0 dB, the standard spectral reference level. Note that the error levels are slightly below two filter side lobe levels. We verified that if we redesign the filter pair for 100 dB stop-band ripple, the time domain artifacts and the spectral ripple amplitude of the impulse response test will also drop another order of magnitude.

This is a good place to discuss an unexpected result. Note in the filter design process that we designed odd length filters and then appended a leading zero when we performed the mapping to the 2D arrays with the reshape command as shown: hh1 = reshape([0 h1],30,8). We might think we could append a trailing zero as shown: hh1a = reshape([h1 0], 30, 8). We did this and found the pre- and post-echo levels increased by three orders of magnitude from $3.6.10^{-5}$ to $5.0.10^{-2}$. Similarly, when we designed the filters to have an even number of taps so that there is no need to append a 0 when applying the reshape operation, we again found that pre- and post-echo levels increased again by three orders of magnitude. Figure 10.5 shows the spectral ripple levels obtained by the cascade analysis and synthesis filter with 239 taps with an appended leading zero and the analysis and synthesis filters with 240 taps. We observe that the increase levels of spectral ripple occur in the transition bandwidths between channel Nyquist zones. We believe that this is due to a frequency-dependent phase mismatch between adjacent Nyquist zone filters which we would be able to cancel with appropriate time delays in the processing chain. We chose not to pursue this because we have perfectly well-behaved solutions with odd length filters with an appended leading zero which obtain the very low levels of reconstruction artifacts!

10.2.1 Cascade Channelizers with Channel Masks

In the previous section, we examined the impulse response of the coupled M-path, $M/2$-to-1, and 1-to-$M/2$ resampled analysis and synthesis filter banks when all channels of the analysis band were processed by the synthesis bank. We did not use the binary mask option between the two channelizers that are shown in Figure 10.1. In this section, we use the binary mask to select which baseband channels will contribute to the synthesized output time series. Such a mask would allow us to form variable bandwidth super channels from the multiple baseband analysis channelizer. If the variable bandwidth filter happens to form super channels which have significantly reduced bandwidth, the synthesis channelizer can permit sample rate reductions commensurate with the bandwidth reductions. The mask can also shift spectral contributions from the baseband signal set to arbitrary spectral spans that differ from the frequency centers at which it resided on the way into the analysis. Thus,

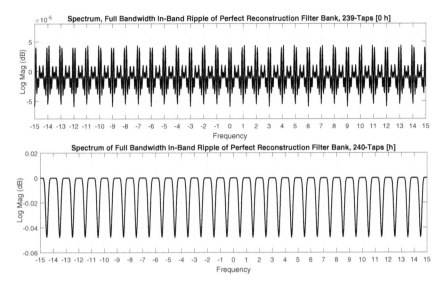

Figure 10.5 Top subplot: spectrum ripple in-band ripple levels, for 239-tap filter with appended leading zero. Bottom subplot: spectrum ripple in-band ripple levels, for 240-tap filter.

the mask and pair of channelizers offer the option to reduce bandwidth, shift frequency centers, and change sample rate in the same process with no additional arithmetic operations beyond those of the channelizer. We still have to examine the workload for the channelizer to learn if it makes sense to even use the channelizer pair.

Figure 10.6 shows the frequency response of three different filters synthesized by the cascade channelizer with binary channel selection mask. The selection mask formed three filters: a wide bandwidth, a narrow bandwidth, and a Hilbert transform. Overlaid on the filter spectra response are the spectra of the analysis channelizer channel responses which were combined by the synthesis filter. The overlays only show alternate channels to avoid making the figures too busy. The three synthesized filters have the same signal processing workload and only differ in the binary mask between the coupled analysis and synthesis channel banks.

Figure 10.7 presents four subplots showing spectra at the input and output of the coupled channelizer of Figure 10.1. The top subplot presents the spectrum of a comb of sinusoids input signal presented to the coupled channelizer of Figure 10.1. The tone frequencies are equally spaced with

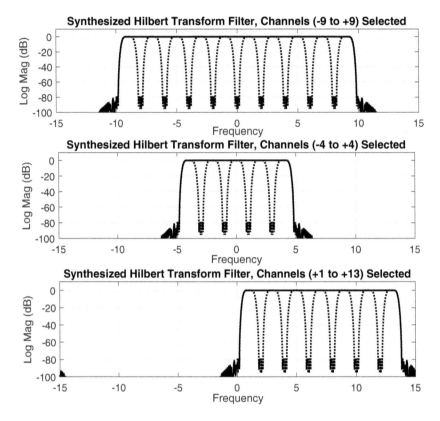

Figure 10.6 Top subplot: spectrum of synthesized wide bandwidth filter from analysis channels (−9 to +9). Center subplot: spectrum of synthesized narrow bandwidth filter from analysis channels (−4 to 4). Bottom subplot: spectrum of synthesized Hilbert transform filter from channels (+1 to +13).

successive tones having reduced amplitudes for ease of distinguishing spectral intervals. The second subplot presents the spectrum of the time series formed at the output of the channelizers with the wide bandwidth mask. The spectral response of the synthesized filter is also shown in the same subplot. The third subplot similarly presents the spectrum of the time series formed at the output of the channelizers with the narrow bandwidth mask. The spectral response of the synthesized filter is also shown in the same subplot. The bottom subplot presents the spectrum of the time series formed at the output of the channelizers with the Hilbert transform mask. The spectral response of the synthesized filter is also shown in the same subplot. All the filters have the same transition bandwidth: the transition bandwidth of the prototype filter in the analysis channelizer. If the filters were to be implemented by conventional FIR filters, they would all be of the same length.

MATLAB Script 3. Analysis, Binary Mask, and Synthesis Channelizer Band-Pass Filter

```
x0=[zeros(1,2) 1 zeros(1,500)];
x2=zeros(1,500);
hh1=reshape([0 h1],30,8);        hh2=reshape([0 h2],30,8);
m2=1;                    % output clock, synthesis channelizer

v1=zeros(1,30)';         % Input vector, Analysis
v2=zeros(1,30)';         % Output of Polyphase filter
v3=zeros(1,30)';         % output of IFFT
reg1=zeros(30,16);       % polyphase filter 2-D register

u3=zeros(1,30)';         % Input vector, Synthesis
u2=zeros(1,30)';         % output of IFFT
u1=zeros(1,15)';         % Output of Polyphase filter
reg2=zeros(30,16);       % polyphase filter 2-D register
flg1=0;                  %state flag for analysis circular buffer
flg2=0;                  %state flag for synthesis circular buffer

for n=1:15:length(x0)-14
 v1(1:15)=flipud(x0(n:n+14).');  % select next 15 input samples
 v1(16:30)=v1(1:15);             % copy top input to bottom input
 reg1=[v1 reg1(:,1:15)];         % input vector in filter register
  for k=1:15
   v2(k)=reg1(k,1:2:16)*hh1(k,:)';         % Top 15, odd index
   v2(k+15)=reg1(k+15,2:2:16)*hh1(k+15,:)';% bottom 15, even index
  end
  if flg1==0                     % No circular buffer shift
     flg1=1;                     % toggle flag
  else
     flg1=0;                     % toggle flag
     v2=[v2(16:30);v2(1:15)];    % circular buffer shift
  end
   v3=30*ifft(v2);               % Output IFFT vector time samples
   u3=v3.*[ones(1,5)';zeros(1,21)';ones(1,4)']; %synthesis mask
   u2=15*ifft(u3);               % Output IFFT vector time samples
  if flg2==0                     % No circular buffer shift
     flg2=1;                     % toggle flag
  else
     flg2=0;                     % toggle flag
     u2=[u2(16:30);u2(1:15)];    % circular buffer shift
  end
  reg2=[u2 reg2(:,1:15)];        % input vector in filter register
   for k=1:15
    u_top=reg2(k,1:2:16)*hh2(k,:)';        % Top 15, odd index
    u_bot=reg2(k+15,2:2:16)*hh2(k+15,:)';% Bottom 15,even index
    u1(k)=u_top+u_bot;           % sum top and bottom
   end
  x2(m2:m2+14)=u1.';             % Output 15 synthesized time samples
  m2=m2+15;                      % Increment output clock
     end
```

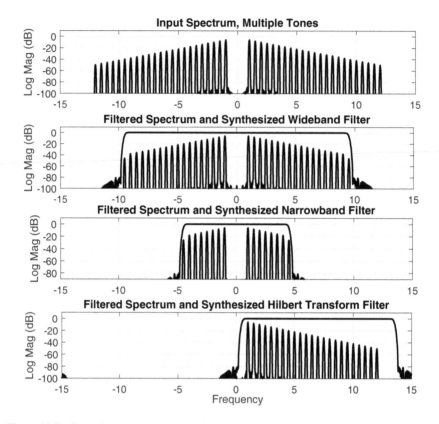

Figure 10.7   Top subplot: spectrum of multiple sinusoid input signal to coupled channelizers. Second subplot: spectrum of synthesized wide bandwidth filter response. Third subplot: spectrum of synthesized narrow bandwidth filter response. Bottom subplot: spectrum of synthesized Hilbert transform filter response.

This is because filter length is a function of sample rate, transition bandwidth, and pass-band and stop-band ripple levels. Since these are all the same for the three filters, they would be of the same length. The important difference here is that the FIR filter implementations would require different sets of coefficients, while the channelizer implementation requires only a change in the binary mask. We will shortly continue examining interesting variations of the channelizer structure. This is a good point to compare the relative computational workloads of the direct FIR filter and coupled channelizer realizations of the filters to see if it is worth our time. The full

MATLAB script to implement the channel masked cascade analysis and synthesis band-pass filter is shown in MATLAB Script 10.3.

10.2.2 Compare Cascade Channelizers to Direct Implementation FIR

We start this comparison with the specifications of a convenient FIR filter with the following specifications. The estimated length of an FIR filter that meets these specifications is 600 taps.

> Filter Specifications for Implementation Comparison
> Sample Rate   60 kHz
> Pass Band   0-to-15.25 kHz
> Stop Band   15.75-to-30 kHz
> Pass-Band Ripple 10^{-3} dB
> Stop-Band Attenuation −80 dB

We have already commented that any FIR filter with the same sample rate, transition bandwidth and pass-band ripple and stop-band attenuation will have approximately the same number of taps. With this perspective, we ask, "can we build a more efficient filter to meet the specification with 600 taps?" We know we can build a more efficient filter if we have a large ratio of sample rate to bandwidth. For that case, we implement an M-path down-sampling version of the filter which reduces the workload per input sample by a factor of M and then implement M path up-sample filter which returns the output rate to the original input rate. We did this in Chapter 5. The problem may be that we cannot reduce the sample rate when the filter bandwidth is a large fraction of the sample rate. If we did reduce the sample rate, the signal would experience multiple aliases from Nyquist zones which are multiples of the reduced sample rates. It works out that is a manageable problem. We have become very good at separating multiple aliases and then reassembling them with polyphase filter bank. So that is the plan! We can visualize the processing scenario shown in Figure 10.8.

The top two subplots show the available spectral bandwidth and the shape of the desired filter. The filter is seen to have a wide bandwidth with a narrow transition bandwidth. This is the first indication that the filter will contain many coefficients; in fact, 600 when we include the stop-band attenuation and in-band ripple levels. The third subplot shows us our channel spectrum with the same transition bandwidth designed with the number of taps equal to the same 600. The bandwidth of the filter is f_s/M, where M is selected from a list

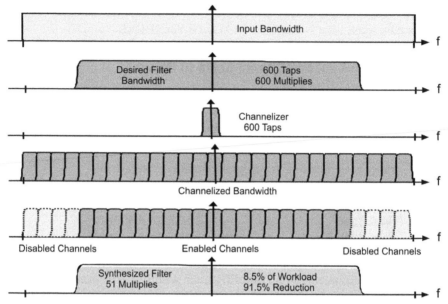

Figure 10.8   Spectra of desired, same length channel filter, spectral span of polyphase Nyquist zones, mask enabled Nyquist channels, and synthesized filter bandwidth from selected Nyquist channel bands.

of convenient size IFFT algorithms. We chose M to be 60, a transform length which is a product of small primes: $60 = 3 \times 4 \times 5$. A transform with these composite lengths, can be implemented by the very efficient, prime length, Good-Thomas (G-T) transform, a transform that does not have a trig table or the twiddle factors we associate with the Cooley–Tukey (C-T) algorithm. The short prime length transforms ($N = 3, 4, 5$) can be implemented by the Winograd algorithms which also do not use a trig table or twiddle factors. The Winograd transforms have a workload on the order of $2N$ real multiplies. We can always fall back on mixed radix C-T algorithms because the transform contributes a small fraction of the workload. The fourth subplot shows how the 60-channel bandwidths span the available spectral bandwidth. The fifth subplot shows the enabled and disabled channels identified by the binary mask between the analysis channelizer and the synthesis channelizer. The final subplot shows the synthesized bandwidth assembled from the selected baseband channels by the synthesis channelizer. Of course, it is not an accident that the desired bandwidth can be assembled from a subset of the 60 available channels. We mentioned that this is a convenient set of specification

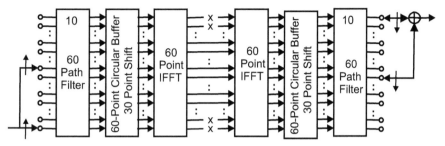

Figure 10.9 60-Path analysis and synthesis channelizer implementation of 600-tap FIR filter. 60-Path filters designed with 600 coefficients to obtain the same filter performance.

for the target specification with the filter bandwidth being an integer multiple of f_s/M. Oddly enough, this is an artificial requirement which will be relaxed shortly when we discuss channelizer enhancements. Figure 10.9 shows the 60-path polyphase filter realization of channelizer realization of the target 600 tap filter. The figures show the number of taps or coefficients (10) per path in both filters and the computational workload (144 real multiplies) of the 60-point G-T IFFTs. We can compute the workload of the channelizer pair. It goes as follows: for every 30 inputs, the system performs 600 multiplies per filter and 144 multiplies per IFFT. The total workload is $2 \times (600 + 144) = 1488$ multiplies. This workload is distributed over 30 inputs for a workload per input of 1288/30 = 49.6 multiplies/input. The workload for the FIR filter is 600 multiplies per input and the ratio of the two workloads is 49.6/600 = 0.083. The workload of the polyphase filter is 8.3% of the direct implementation of the FIR filter. That is, a 91.7% workload reduction. Do we have your attention yet? This result often comes with a surprise. How can a system that contains two 600-tap filters and two 60-point IFFTs, the entire block diagram shown in Figure 10.9, outperform a simple tapped delay FIR filter by more than an order of magnitude? The remarkable performance improvement comes from reducing the input sample rate and operating the process at the reduced internal rate of $f_s/30$! We learn again; it is not how many operations do we perform per output, but rather how many operations do we perform per input!

Figure 10.10 shows the time and spectral response of the equivalent filter formed by the coupled 60-path analysis and synthesis filter with 31 enabled channels. The top subplot shows the filter impulse response. The peak appears at sample index 571 for an input impulse located at sample 0. The peak will shift with changes in which, of the 30 input positions, the

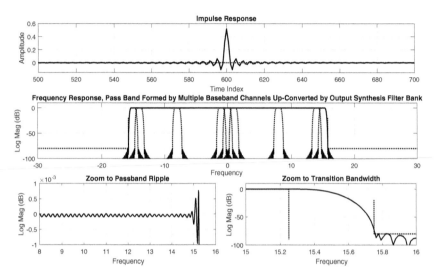

Figure 10.10   Time and frequency responses of 60-path coupled channelizers. Top subplot: impulse response. Middle subplot: frequency response with overlay of analysis filter spectra input to synthesizer. Bottom-left subplot: zoom to in-band spectral ripple. Bottom-right subplot: zoom to filter transition bandwidth.

impulse enters the analysis filter. With the impulse applied at position 30, the peak impulse position shifted sample index 600. A delayed output for a delayed input is expected for a linear time-invariant (LTI) filter. It also is the response of the LTV even though the input impulse sample enters the system through different path filters of the polyphase analysis filter. The delay of the synthesized filter is about twice the delay of the direct FIR filter because the input signal has passed through two filters. The second subplot shows the spectral response of the dual channelizers with a number of offset spectral responses of the analysis channelizer's baseband frequency responses. Note the fast spectral decay rate after the stop-band edge. The bottom two subplots are zooms to the in-band ripple levels and the transition bandwidth. Except at the band edge, the in-band ripple levels are an order of magnitude below the specified 10^{-3} dB level.

Figure 10.11 shows the time and spectral response of the 599-tap directly implemented FIR filter. The top subplot shows the filter impulse response. The peak appears at the filter midpoint at sample index 299. Note that this delay is about half the delay of the synthesized filter. This is the coin we spend when we replace the direct implantation with the channelized

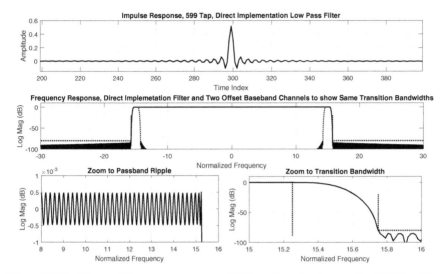

Figure 10.11 Time and frequency responses of 599-tap direct implementation FIR filter. Top subplot: impulse response. Middle subplot: frequency response with overlay of analysis filter spectra input to synthesizer. Bottom-left subplot: zoom to in-band spectral ripple. Bottom-right subplot: zoom to filter transition bandwidth.

implementation, additional delay! The second subplot shows the spectral response of the dual channelizers with two offset spectral responses of the analysis channelizer's baseband frequency responses. Note the slow spectral decay rate after the stop-band edge. This filter was designed with a 1/f stop-band decay rate which we see here. The bottom two subplots are zooms to the in-band ripple levels and the transition bandwidth. We note that the in-band ripple is uniform with a ripple level of 10^{-3} dB. This is a larger level that the synthesized filter in-band ripple.

10.3 Enhanced Capabilities of Coupled Channelizers

In Section 10.2.1, we demonstrated that the binary mask between the two channelizers could form variable bandwidth filters by enabling a selected subset of the baseband channel outputs from the analysis channelizer to form a selected bandwidth super channel by the synthesis channelizer. The cascade polyphase filter banks can synthesize broad bandwidth filters from any even or odd integer multiple of the baseband channel bandwidth. The example in the previous section formed a filter from 31 of 60 available channels.

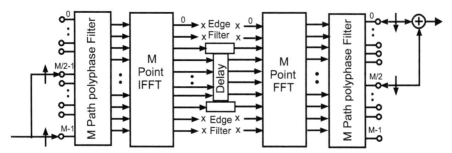

Figure 10.12   Efficient two-tier channelizer with spectral modification of edge channels to enable arbitrary variable bandwidth filter synthesizer.

Now suppose we want additional flexibility in bandwidth selection. Say we want a filter with bandwidth spanned by a non-integer multiple of channel bandwidths such as 31.5 channels instead of 31 channels. We present two methods to accommodate the requirement for finer or arbitrary increments of synthesized bandwidths. In the first method, we alter the bandwidth of the two edge channels with inner tier filters as shown in Figure 10.12. The two inner tier filters have complex conjugate weights that form offset bandwidth reducing filters between the output of the analysis filter and the input to the synthesis filter as shown in Figure 10.13. These filters operate at the reduced sample rate, $2f_s/M$, and thus have a small number of coefficients with workload distributed over $M/2$ input samples. The workload to implement the inner filter is a very small fraction of the channelizer work, typically increasing the workload of the cascade channelizers by only a few percent. The edge filters introduce delay to the two edge paths. The non-modified filter paths between the edge paths must be delayed in a buffer to assure phase alignment of all inputs to the synthesis filter bank. The effect of the end channel bandwidth reduction is shown in Figure 10.14 for three examples of the process. Note that each sample delay of the inner tier filter is responsible for $M/2$ sample delays at the output of the synthesizer channelizer output. We examine other options that reduce bandwidth without increasing system delay.

We now discuss a second option. Rather than altering the bandwidth of only the edge filters and incurring addition signal delay, we can alter the bandwidth, but not center frequency, of all the filters in the analysis filter bank. We increase (or decrease) the bandwidth of the even indexed channel filters by an amount β while decreasing (or increasing) the bandwidth of the odd indexed channel filters by the same amount. The channelizer so formed

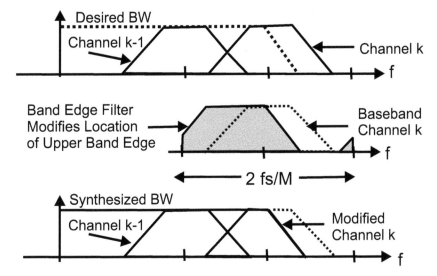

Figure 10.13 Spectra of channelizer channel responses, inner tier filter applied to end channel, and modified end channel response to obtain the desired synthesized filter with reduced bandwidth.

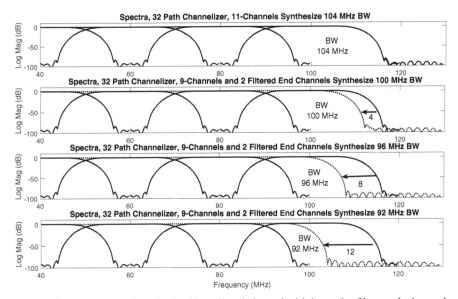

Figure 10.14 Spectra of synthesized broadband channel with inner tier filters reducing end channel bandwidth to reduce synthesized bandwidth.

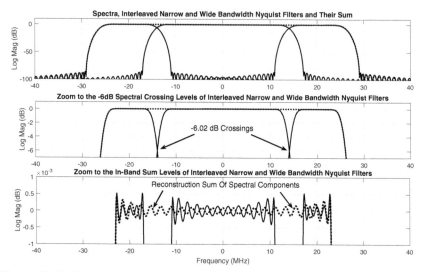

Figure 10.15 Perfect reconstruction spectra of adjacent channel widths of modified channelizer with interleaved complementary bandwidth channels.

will have two sets of interleaved complementary bandwidth channels as shown in Figure 10.15. The complementary channel bandwidths are formed in two analysis filter banks and then interleaved and binary masked in a single synthesis filter bank as shown in Figure 10.16. The effect of the variable complimentary channel bandwidths is shown in Figure 10.17 for three examples. Since we only require alternate bins from the two length M analysis filters in Figure 10.15, we can alter the analysis channelizers to be two $M/2$ length filters and interleave the two $M/2$ point output vectors to implement the M length analysis IFFT with the reassembled interleaved channels.

10.3.1 IFFT Centered Channelizer Enhancements

One of the channelizer filter options shown in Figures 10.6 and 10.7 was a Hilbert transform filter with pass-band spanning the positive frequency channels. Since the Hilbert transform filter reduces the signal bandwidth, we can modify the channelizer to reduce the sample rate during the synthesizing process. We do this by halving the synthesis channelizer to be half length of the analysis channelizer as shown in Figure 10.18. If we want the filter to have symmetry about its midpoint, we must select an odd number of channels from

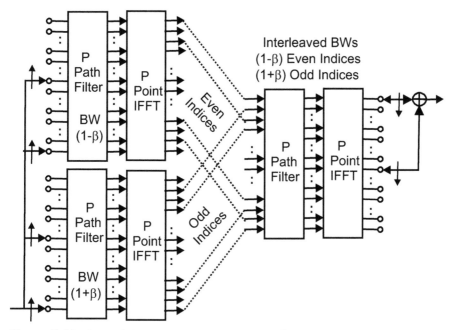

Figure 10.16 Synthesizing interleaved narrow and wide channel channelizer from dual analysis channelizers.

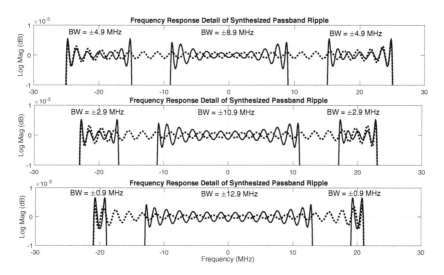

Figure 10.17 Spectra of synthesized broadband channel with interleaved alternating wide and narrow bandwidth channel filters.

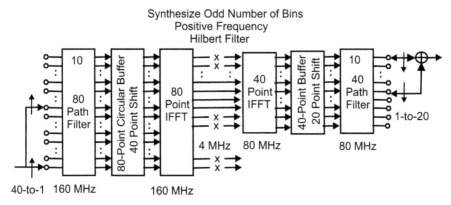

Figure 10.18   Block diagram of channelizer-based variable bandwidth filter extracting positive frequencies and reducing sample rate with half-length synthesis filter.

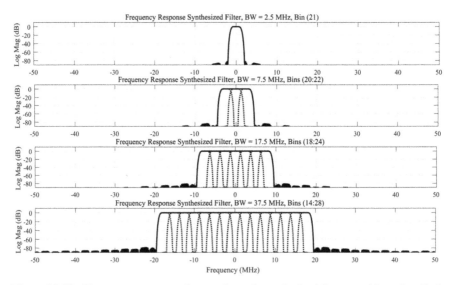

Figure 10.19   Frequency response of super channels synthesized from an odd number (1, 3, 7, and 15) of baseband channels: even stacking of BW intervals.

the analysis channelizer. If we did this at baseband, we would select the dc channel and a matching number of left and right channels. Possible spectral shapes of the synthesized channels are seen in Figure 10.19.

Figure 10.20 shows a variant of the channelizer shown in Figure 10.18, in that rather than having the center frequencies be aligned with those of the IFFT, the center frequencies are offset midway between these frequencies.

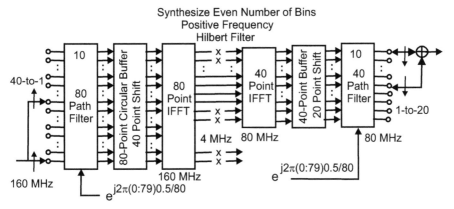

Figure 10.20 Block diagram of channelizer-based variable bandwidth filter: BW adjusted by binary mask between analysis and synthesis filter banks.

The half-cycle offset relative to the integer frequencies of the IFFT are obtained by a complex heterodyne of the 80-path and 40-path filter weights. This heterodyne is seen in Figure 10.20 at the bottom of the filter blocks. The frequency offset also alters the state machines controlling the circular buffers between the 80-path filter and 80-point IFFT and between the 40-point IFFT and the 40-path filter. The circular shifts here become quarter-length shifts instead of half-length shifts. The half-cycle offset means that DC resides at the 6-dB crossover frequencies of the subchannels in the merged super channel instead of at the center of the subchannel in the merged super channel. Figure 10.21 shows four examples of the super channels formed by an even number of subchannels taken from the positive frequency span of the 80-path analysis channelizer.

The primary difference between the even and odd stack indexed filter banks resides in the complex heterodyne of the filter weights in the analysis and synthesis filters. Although this is performed offline, their use in the 80-path and 40-path filters doubles the workload at the input and output ports of the filter. The second problem is that it is not a trivial matter to switch between the two modes of even number and odd number of subchannels in the merged super channel. We have to reload the coefficients in the 80-path and 40-path filters as well as switch state machine for the circular buffers.

We now present a modified design of a cascade channelizer that will support both even and odd stacked super channel implementation with real coefficients and without changes in coefficients or state machine. One channelizer uses frequencies starting at 0 and counting by 2, while others start

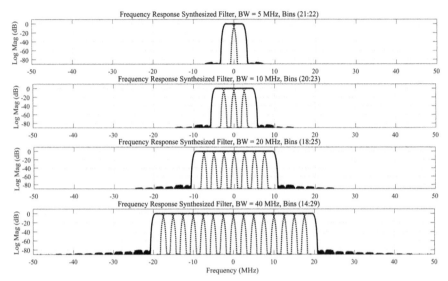

Figure 10.21   Frequency response of super channels synthesized from an even number (2, 4, 8, and 16) of baseband channels: Odd stacking of BW intervals.

at 1 and also count by 2. Both have channel bandwidths of 2. If we double the size of the IFFT, we obtain twice as many channel centers now separated by 1. This will support both the even and odd stack channelizer. One starts at index 0 and skips channels to access the even indices, and the other starts at index 1 and also skips channel centers to access the odd indexed channels.

Without developing the details of the altered channelizer, we simply describe the difference. When the filter centers are separated by 2-MHz centers, we require 80 channels to span the 160 MHz sample rate wide Nyquist interval. In the new design, when the filter centers are separated by 1 MHz centers, we require 160 channels to span the same 160 MHz sample rate wide Nyquist interval. This means the number of paths required for the 1 MHz channel spacing is 160. We implement a 160-path analysis channelizer with 5 taps per path from our 799-tap prototype filter. The channel filter bandwidth output to their stop band is 3.0 MHz for which we had selected the analysis filter's output sample rate to be 4.0 MHz. We obtain the 4.0 MHz sample rate from the 160 MHz input sample rate by operating the analysis filter with a 40-to-1 down-sample schedule. All the options we have described did this. Finally, the 40-to-1 down-sampling in the 160-path filter requires a 160-point circular buffer with 40-point shifts to alias all aliased channels to baseband. The channelizer will have a four-state state machine to control the

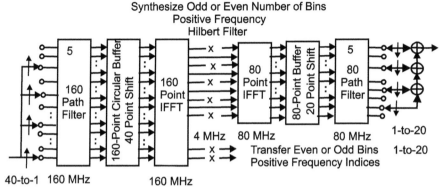

Figure 10.22 Block diagram of channelizer-based variable bandwidth filter: BW adjusted by even or odd indexed binary mask between analysis and synthesis filter banks.

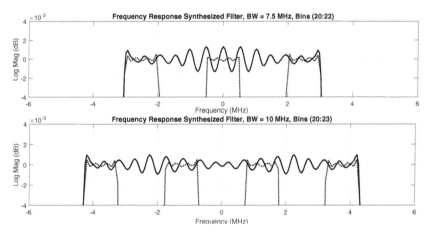

Figure 10.23 Spectrum, zoom to in-band ripple of super channels synthesized from three alternate even stacked and four alternate odd stacked subchannels.

shifts in both the analysis and synthesis channelizers. The polyphase filter workload is not changed in this alternate design. What does change is the IFFT size. Figure 10.22 shows the block diagram of the 160-path odd and even stacking channelizer. Figure 10.23 presents the zooms to the in-band ripple of super channels synthesized from three even stacked and four odd stacked subchannels. Overlaid on the super channel spectral responses are the subchannel spectral responses. The center frequencies of the subchannels are clearly seen to match the even and odd frequency indices. We see the ripple levels are less than 0.001 dB, which is, in fact, 1 part in 10,000, the same level as the 80 dB stop-band ripple. The in-band ripple level is formed from three contributor sources: the in-band ripple of the analysis filter and the in-band ripple and stop-band ripple levels of the synthesis filter. It is quite impressive that the subchannel responses cross 6 dB below the displayed region and still sum to unity (0 dB) within the ripple levels of the analysis and synthesis filters.

10.4 Multiple Bandwidths Arbitrary Frequency Center Channelizers

We have presented a number of clever ways to apply the M-path, coupled analysis, and synthesis channelizers to signal processing tasks. We have only skimmed the long list of applications, and, given the opportunity, we would be pleased to present many more remarkable sample applications. Unfortunately, page limitations keep us from doing so here. Thus, we present the last of our interesting applications and refer the reader to a long list of papers to help fill the gap. One of the more novel variations of the analysis channelizer is obtained by appending small processing engines to its output ports. These engines implement small synthesis channelizers, complex heterodynes, and, occasionally, an arbitrary interpolator. These processors combine the time series from selected subsets of the analysis filter's baseband output ports to synthesize super channel signals spanning variable bandwidths from arbitrary center frequencies. Perfect reconstruction filters in the analysis–synthesis cascade guarantees seamless assembly of channel segments forming these super channels.

The composite super channel spectrum will be offset from baseband by the same offset that the spectral block was offset from a channel center, at some integer multiple of fs/M. This offset is always less than $f_s/(2M)$. The residual offset frequency of the super channel signal is removed by the complex heterodyne following the synthesis block. Note that the heterodyne occurs at the reduced output sample rate of the synthesizer filter bank rather than at the high input rate of the analysis filter bank. The frequency offset occurs in two steps, the first being the aliasing caused by the input down-sampling, and the second being the low data rate heterodyne. If desired, an arbitrary interpolator can change the output sample rate from a multiple of f_s/M, the synthesis channelizer's output rate, to an output sample rate twice the composite signal's symbol rate. Figure 10.24 shows the channelizer configuration just described.

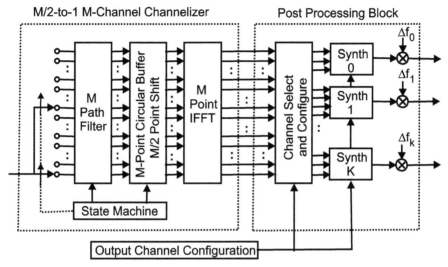

Figure 10.24 Analysis filter bank with multiple arbitrary bandwidth output signals down-converted to baseband from arbitrary input center frequencies.

The dual graph of Figure 10.24 is presented in Figure 10.25. This enhanced synthesis filter is formed by appending small processing engines to its input ports. These engines implement, if required, an arbitrary interpolator, a complex heterodyne, and an analysis channelizer. These processors partition each input time series into subchannels matched to the bandwidth and sample rate of the input ports of the synthesis filter's baseband input ports. The synthesized signal is a composite output time series containing multiple signals of variable bandwidths at arbitrary center frequencies. Perfect reconstruction filters in the analysis–synthesis cascade guarantees seamless assembly of the output wide bandwidth signal.

In past presentations, we would demonstrate the performance of an enhanced *M*-path analysis filter bank of the type shown in Figure 10.24. We would then comment that there was no need to examine the dual process, the *M*-path synthesis filter bank, because they were simple dual processors. While that is true, we learned that students would like to see the synthesis filter bank because there are many presentations of the analysis bank and few of the synthesis bank. Thus, we now discuss and present an example of the enhanced synthesizer filter bank.

The synthesis channelizer we simulated is a 30-path filter performing 1-to15 up-sampling. The output sample rate is 600 MHz which means the

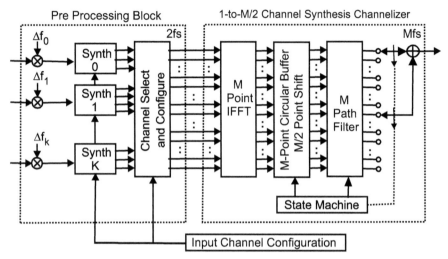

Figure 10.25 Synthesis filter bank with wide bandwidth output signal containing multiple up-converted input signals with arbitrary bandwidths and center frequencies.

up-converted filter centers are 20 MHz apart and the input sample rate to the channelizer is 40 MHz. The channel filter −6dB bandwidth is 10 MHz with stop-band frequency edges of 7.5 MHz on either side of center frequencies. We deliver signals with four different bandwidths and sample rates to the synthesizer. The spectra and sample rates are indicated in Figure 10.26. The important thing to notice is that the channelizer wants to see input signals at 40 MHz sample rate. Two of the signals do have that rate and the other two signal rates are 120 and 160 MHz.

These rates are multiples of 40 MHz, but they are not 40 MHz. This is the reason we have the input processing blocks, the small analysis channelizers. One analysis channelizer processes the signal with the 120 MHz sample rate. This process is a six-path polyphase filter that performs 3-to-1 down-sampling to form six baseband channels at the desired 40 MHz rate. The spectral partitioning performed by the six-path analysis channelizer is shown in Figure 10.27. The top subplot shows the spectrum of the 40 MHz symbol rate sampled at 120 MHz. Overlaid on the spectrum are the spectra of the channel responses of the analysis channelizer. The input spectrum is partitioned into six baseband channels, each sampled at 40 MHz. The six lower subplots of Figure 10.27 show the spectra of the time series output by the six analyzer channels. Note that five of the channels, −2 to +2, contain the spectral segments of the original input signal. The time series

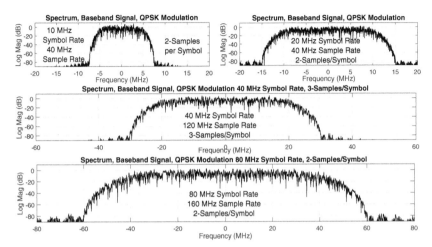

Figure 10.26 Spectra of four QPSK input signals presented to 30-channel synthesis channelizer with symbol rates of 10, 20, 40, and 80 MHz at sample rates of 40, 40, 120, and 160 MHz, respectively.

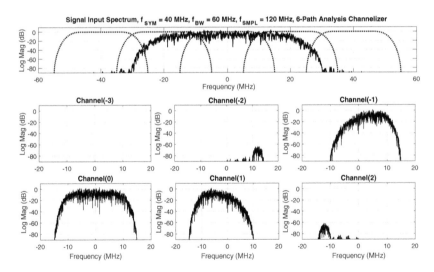

Figure 10.27 Top subplot: input spectrum of 40 MHz symbol rate input signal with spectral overlays of six-path channelizer filter responses. Six lower subplots show spectra of baseband channel output time series, each sampled at 40 MHz.

from these five channels are presented to five contiguous input ports of the output synthesizer at the channelizer indices corresponding to the spectral location (center frequency) of the reassembled spectrum.

The other analysis channelizer processes the signal with the 160 MHz sample rate. This process is an eight-path polyphase filter that performs 4-to-1 down-sampling to form eight baseband channels at the desired 40 MHz rate. We have two versions of the 80 MHz symbol rate time series. The center frequency of one of these wideband signals is placed at the center of one of the 20 MHz band centers of the 30-path synthesis filter bank. The other 80 MHz symbol rate time series is frequency shifted by the input heterodyne to the center frequency at the crossover frequency of the channelizer, a 10 MHz offset from channelizer filter's center frequency. This frequency shift is performed on the baseband signal prior to being processed by its signal conditioning eight-path analysis filter bank. Any frequency offset applied to the baseband signal is observed as a frequency offset relative to the center frequency of the synthesis channelizer. We, tongue in cheek, liken this to a worm hole: we shift the spectrum at baseband at a low input sample rate and its aliased image shifts the same amount from a selected channel center frequency at the high output ample rate. The spectral partitioning performed by the eight-path analysis channelizer in its band centered version and its frequency shifted version is shown in Figures 10.28(a) and (b). The top subplots show the spectrum of the 80 MHz symbol rate sampled at 160 MHz. In Figure 10.28(a), the spectrum is centered at 0 frequency, and in Figure 10.28(b), the spectrum is frequency shifted to be centered at -10 MHz. Overlaid on these spectra are the spectra of the channel responses of the eight-path analysis channelizer. The input spectrum is partitioned into eight baseband channels, each sampled at 40 MHz. The eight lower subplots of Figure 10.28 show the spectra of the time series output by the eight analyzer channels. Note that seven of the channels, -3 to $+3$, contain the spectral segments of the original input signal for the non-offset version of the signal, but six of the channels, -3 to $+2$, contain the spectral segments of the original input signals for the offset version of the signal. In either case, the time series from all seven channels are presented to seven contiguous input ports of the output synthesizer at the channelizer indices corresponding to the spectral location (center frequency) of the reassembled spectrum.

The 30 lower subplots of Figure 10.29 present baseband spectra presented to the 30 inputs ports of the 30-path synthesis channelizer. The sample rate of each signal at these ports is 40 MHz, the sample rate designed for the synthesizer. The five low bandwidth signals presented to the channelizer were

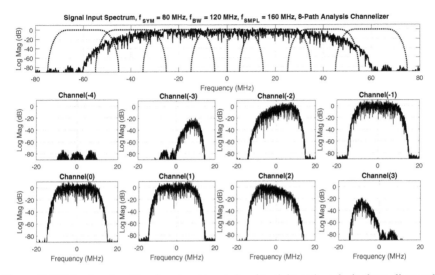

Figure 10.28(a) Spectra of band centered input signal to eight-path analysis channelizer and spectra of baseband output channels sampled at 40 MHz.

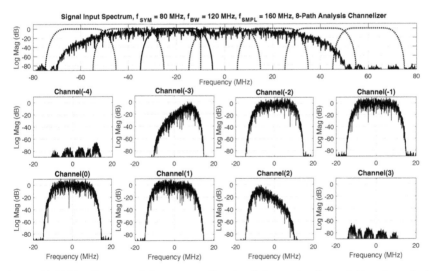

Figure 10.28(b) Spectra of frequency offset band input signal to eight-path analysis channelizer and spectra of baseband output channels sampled at 40 MHz.

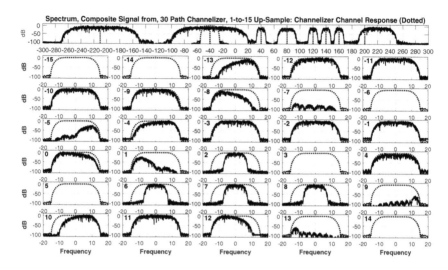

Figure 10.29   Top subplot: spectrum of synthesized super channel formed from multiple input signals with different bandwidths. Thirty lower subplots: spectrum of baseband signal presented to 30-path synthesizer channelizer by pre-processing analysis channelizers.

sampled at 40 MHz and required no signal conditioning on the way into the synthesizer. These signals were placed in the channel ports 2, 4, 6, 7, and 8 corresponding to center frequencies 40, 80, 120,140, and 160 MHz as can be seen in the top subplot. Three signals delivered to the synthesizer had bandwidths wider than the synthesizer channel widths and required the spectral decomposition and sample rate reduction performed by their input analysis channelizers. The wideband spectral segments spanned seven of the synthesizer's channels and one decomposed signal was delivered to channel ports -13 to -7 and one delivered to channel ports -5 to $+1$. Note that one of the wideband spectra is centered at -210 MHz, the midpoint between the channel center frequencies at -200 and -220 MHz. This is the spectrum of the signal that was frequency shifted -10 MHz on the way into the eight-path analysis channelizer. The second wideband signal is centered at -40 MHz, the center frequency of the input port -2.

The process illustrated by this 30-path synthesizer example is quite flexible and is able to accommodate a wide range of signal bandwidths and center frequencies. The only resource we did not use in this example was the use of an arbitrary interpolator. Suppose the wideband 80 MHz symbol signal had been presented to us at a 200 MHz sample rate. Since 200 is an even multiple of 40, we would have changed its eight-path analysis filter to a 10 path and performed 5-to-1 down-sampling. What if the input sample rate was 180 MHz? We could use a modified nine-path channelizer and perform a 4.5-to-1 down-sample (Can one do that? Yes we can; we go up by 2 and then down by 9!) We probably would choose another option. The most versatile option is the arbitrary interpolator. The interpolator can convert the 190 MHz input sample rate to a 200 MHz sample rate which is a rate we can easily accommodate.

10.5 Channelizers with Even and Odd Indexed Bin Centers

Having gotten this far into this textbook, the reader is comfortable with the perspective that the M-path polyphase analysis channelizer is formed by three distinct subsystems that perform essential operations in the channelization process. These subsystems are the M-path polyphase filter that performs the spectral shaping of the channels, the discrete Fourier transform that performs the phase coherent summation required to separate the aliased channel time series, and the input commutator that performs the P-to-1 down-sampling which is responsible for aliasing of the M multiple spectral regions. The fact that we can control each of these coupled functions somewhat independently is the reason the channelizer is so versatile and has so many embodiments. What we discuss and present in this section is a number of minor variations of the channelizer that lead to interesting and desirable capabilities. This performance benefit occurs with reductions in signal processing work load.

The M-path analysis channelizer center frequencies coincide with the M sampled data frequencies of the M-point discrete Fourier transform (DFT), the frequencies with integer number of samples per length of M samples. These are the M multiples of f_s/M, the frequencies that alias to DC when their sinusoids are down-sampled M-to-1. The spacing between center frequencies is also f_s/M as is the output sample rate when maximally decimated. The first channelizer variation we examine is shown in Figure 10.30. Here we present two channelizers with equally spaced center frequencies, say 2 MHz, but with different center frequency locations. In the upper subplot, the center frequencies reside on the even integer frequencies while in the lower subplot, the center frequency reside on the odd integer frequencies. The filters have the same bandwidth and have the same sample rate in their non-maximally decimated implementations.

The standard response to the problem that a signal and a filter do not reside at the same center frequency is to either move the signal to the filter

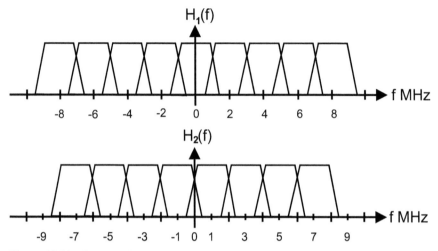

Figure 10.30   Spectra of two channelizers with different center frequencies but with same channel shape and same frequency spacing. Upper subplot centers match DFT center frequencies and are centered on the even integers on this figure. Lower subplot centers are offset by half their spacing and are centered on the odd integers in this figure.

(by the Armstrong heterodyne) or move the filter to the signal (using the equivalency theorem). These two options are shown in Figure 10.31. In both cases, a complex heterodyne at the input rate is required to perform the spectral alignment. The question may come to mind: "Is there a better option?" and the answer is yes! This requires some thinking outside of the box! It starts by considering a DFT for an odd number of points. Such a DFT can be implemented by a Good–Thomas algorithm or by a conventional mixed radix Cooley–Tukey algorithm. Figure 10.32 shows the root locations of $Z^{15} - 1$ which corresponds to the center frequencies of a 15-point DFT. On the left subplot, the zero frequency location of an unaltered input sequence is indicated on the circle. This, of course, coincides with index 0 of the 15-point DFT. On the right subplot, the zero frequency location of the input sequence following a heterodyne to the half sample rate by alternating signs is indicated at the half sample rate on the circle. The DC term is seen to reside midway between indices 7 and 8 of the 15-point DFT. This means that the indices 7 and 8 correspond to the two frequencies below and above DC by half the channel spacing. In this process, we do not have to apply complex heterodyne to the input series or to the filter weights to access the half-bandwidth offset frequency channelizer responses. The interaction of the odd length DFT and the alternating sign input heterodyne places the offset input frequency centers

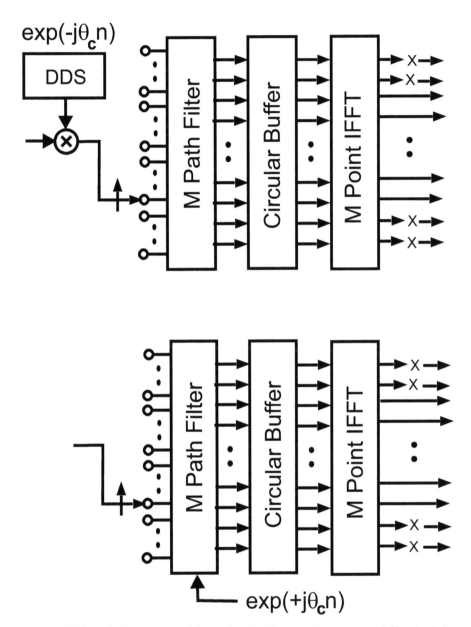

Figure 10.31 Aligning spectra of input signal with spectral responses of filter bank by complex heterodyne of input signal in upper subplot or by complex heterodyne of filter coefficient weights in lower subplot.

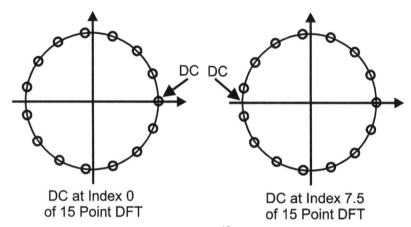

Figure 10.32   Two unit circles with roots of $(Z^{15} - 1)$, the frequencies corresponding to a 15-point DFT. The left subplot indicates the location of DC or zero frequency of an unaltered input sequence presented to the DFT. The right subplot indicates the location of DC or zero frequency heterodyned to the half sample rate by an alternating sign heterodyne of the input sequence.

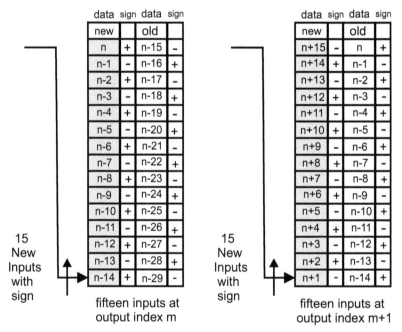

Figure 10.33   Polyphase filter input sample indices and sign of input heterodyne for two successive 15-point data samples in 15-path polyphase filter. Note the sign reversals of the two new input vectors.

at the DFT bin centers. We are not quite finished with this *thinking outside the box* example. We now examine how the alternating sign input data interacts with the filter coefficients. Figure 10.33 shows the input data index and the data signs for two successive inputs of 15 new input samples to 15-point polyphase filter operating in its maximally decimated form.

Figure 10.34 shows the input data index and the data signs for two successive inputs of 10 new input samples to 15-point polyphase filter operating in its non-maximally decimated 10-to-1 down-sampling form. We note that there are no sign changes in successive 10-sample input vectors in the non-maximally decimated version of the 15-path filter. This is because the length of the successive input vectors is 10 which is a multiple of the two sample periods of the sign changes of the input heterodyne. Here it comes! Because the signs do not change on successive inputs, we can associate the

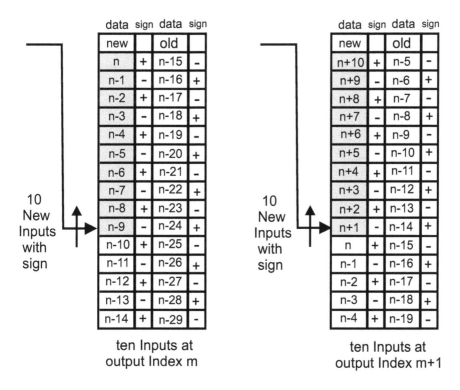

Figure 10.34 Polyphase filter input sample indices and sign of input heterodyne for two successive 10-point data sample sequences in 15-path polyphase filter. Note that there are no sign reversals of the two new input vectors.

signs with the filter weights. That is, rather than heterodyning the input samples to the half sample rate at the input sample rate, we heterodyne the filter weights as an offline operation. This is an interesting version of the equivalency theorem embedded in the polyphase filter. Figure 10.35 shows the application of the equivalency theorem to the non-maximally

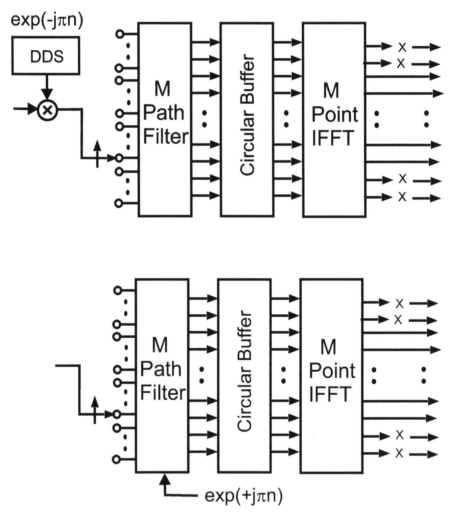

Figure 10.35   Aligning spectra of input signal with spectral responses of odd length, non-maximally decimated filter bank by alternating sign heterodyne of input signal in upper subplot, or by alternating sign heterodyne of filter coefficient weights in lower subplot.

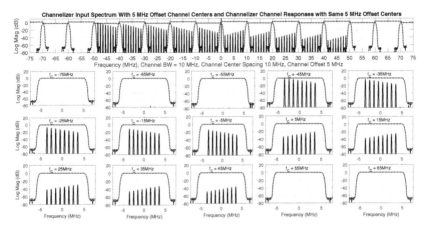

Figure 10.36 Spectra of input signal and channel centers of 15-path polyphase channelizer performing 10-to-1 down-sampling with alignment of channelizer spectra with half-channel bandwidth offset performed by embedding alternating sign heterodyne in filter weights. Lower 15 subplots show spectra obtained at each baseband channel output port.

decimated filter bank formed by an odd length polyphase filter. Interestingly, there is no online signal processing required to obtain the odd-indexed filter centers in this version of the M-path filter. Figure 10.36 shows the input and output spectrum formed by the 15-path polyphase filter with alternating sign heterodyne embedded in filter weights – very nice results.

Our last comments on this equivalency theorem application follow. If you have need of an even length transform, you would lose the half sample rate being located midway between DFT frequency indices. We can still use the spectral location between DFT indices at the quarter sample rate. As an example, Figure 10.37 shows DC at index 0 of an 18-point DFT without the heterodyne and midway between indices 4 and 5 of the 18-point DFT as a result of an input heterodyne by $\exp(j\,n\,\pi/2)$. To be able to embed the phase shifts in the polyphase filter, the down-sample rate P must be a multiple of 4 to keep the phase changes stationary in the filter on successive inputs of length P. We demonstrated successful operation of this modified process with an 18-path filter and 18-point FFT performing 12-to-1 down-sampling. There is, of course, a re-indexing required to locate the shifted frequency centers at the offset DFT output indices.

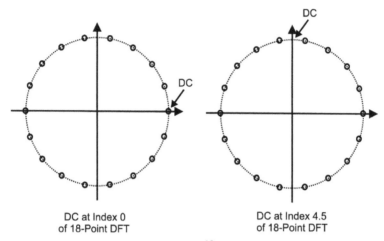

DC at Index 0
of 18-Point DFT

DC at Index 4.5
of 18-Point DFT

Figure 10.37   Two unit circles with roots of $(Z^{18} - 1)$, the frequencies corresponding to an 18-point DFT. The left subplot indicates the location of DC or zero frequency of an unaltered input sequence presented to the DFT. The right subplot indicates the location of DC or zero frequency heterodyned to the quarter sample rate by $\exp(j\ n\ \pi/2)$ heterodyne of the input sequence.

References

fred harris, Elettra Venosa, Chris Dick, *Implement Even and Odd Stacked Frequency Bins in Cascade Non-Maximally Decimated Analysis and Synthesis Filter Banks, DSP-2018,* Shanghai, China, 19-21-November 2018.

fred harris, Elettra Venosa, Xiaofei Chen, Chris Dick, *Cascade Non-Maximally Decimated Filter Banks Form Efficient Variable Bandwidth Filters for Wideband Digital Transceivers,* DSP-2015 Conference, Singapore, 21-24 July 2015.

fred harris, Elettra Venosa, Xiaofei Chen, Chris Dick, *Interleaving Different Bandwidth Narrowband Channels in Perfect Reconstruction Cascade Polyphase Filter Banks for Efficient Flexible Variable Bandwidth Filters in Wideband Digital Transceivers*, DSP-2015 Conference, Singapore, 21-24 July 2015.

Sudhi Sudharman, Athul D. Rajan and T. S. Bindiya, *Design of a Power-Efficient Low-Complexity Reconfigurable Non-maximally Decimated Filter Bank for High-Resolution Wideband Channels*, Circuirs, Systems, and Signal Processing, Springer Science+Business Media, Nov. 2018 fred harris, Behrokh Farzad, ElettraVenosa, Xiaofei Chen, *Multi-Resolution PR NMDFBs for Programmable Variable Bandwidth Filter in Wideband Digital Transceivers,* International Conference on Digital Signal Processing, Hong Kong, 20-23 August 2014

fred harris, Elettra Venosa, Xiaofei Chen, and Bhaskar Rao, *Polyphase Analysis Filter Bank Down-Converts Unequal Channel Bandwidths with Arbitrary Center Frequencies,* Springer Journal Analog Integrated Circuits and Signal Processing, June 2012, Vol. 71, Issue 3, pp 481-494.

Mehmood Ur-Rehman Awan, Yannick Le Moullec, Peter Koch, fred harris, *Hardware Architectures for Analysis of Polyphase Filter banks Performing Embedded Resampling for Software Defined Radio Front Ends*, Special Issue on Digital Front End and RF Processing for ZTE Communications: An International Journal, March 2012, Vol. 10 No. 1, pp. 54-62.

Mehmood Ur-Rehman Awan, Peter Koch, fred harris *Time and Power Optimization in FPGA Based Architectures for Polyphase Channelizers*, 45-th Annual Asilomar Conference on Signals, Systems, and Computers, Pacific Grove, CA, 6-9 November 2011.

Mehmood Ur-Rehman Awan, Yannick Le Moullec, Peter Koch, fred harris, *Polyphase Filter Banks for Embedded Sample Rate Changes in Digital Radio Front-Ends*, Special Issue on Digital Front End and RF Processing for ZTE Communications: An International Journal, December 2011, Vol. 9, No. 4, pp. 3-9.

fred harris, Chris Dick, Xiaofei Chen, and Elettra Venosa *Wideband 160 Channel Polyphase Filter Bank Cable TV Channelizer,* IET Signal Processing, Special Issue on Multirate Signal Processing, Vol. 5, Issue 3, June 2011, pp. 325-332.

fred harris, Elettra Venosa, Xiaofei Chen *Polyphase Analysis Filter Bank Down-Converts Unequal Channel Bandwidths with Arbitrary Center Frequencies, Design I,* Software Defined Radio Conference (SDR'10), Washington DC, 30 Nov. – 3 Dec. 2010.

fred harris, Elettra Venosa, Xiaofei Chen, *Polyphase Synthesis Filter Bank Up-Converts Unequal Channel Bandwidths with Arbitrary Center Frequencies, Design II*, Software Defined Radio Conference (SDR'10), Washington DC, 30 Nov. – 3 Dec. 2010.

fred harris, *Polyphase Filter Banks for Unequal Channel Bandwidths and Arbitrary Center Frequencies,* Software Defined Radio Conference (SDR'10), Washington DC, 30 Nov. – 3 Dec. 2010.

fred harris and Chris dick, *Polyphase Channelizer Performs Dual Sample Rate Change for Matched Filter Shaping and Channel Frequency Spacing,*

43-rd Annual Asilomar Conference on Signals, Systems, and Computers, Pacific Grove, CA, 1-4 November 2009

fred harris, Chris Dick, and Michael Rice, *Digital Receivers and Transmitters Using Polyphase Filter Banks for Wireless Communications,* (with Chris Dick and Michael Rice), Special Issue of Microwave Theory and Techniques, MTT, Vol. 51, No.4, April 2003, pp. 1395-1412.

fred harris and Chris dick, *Performing Simultaneous Arbitrary Spectral Translation and Sample Rate change in Polyphase Interpolating or Decimating Filters in Transmitters and Receivers,* Software Defined Radio Technical Conference, SDR'02, , San Diego, CA 11-12 November 2002.

Problems

10.1 Using the design examples shown in MATLAB Script 10.1 guide, design the prototype polyphase filter for a 16-path filter analysis channelizer. For ease of design, the sample rate is 16 MHz, with the two sided half amplitude bandwidth of 1.0 MHz, and peak stop-band side lobe level is -80 dB. Stop-band frequency is 0.75. This filter has more taps per path but uses the same β window as shown in MATLAB Script 10.1. On two subplots, plot the linear magnitude and log magnitude spectra of three adjacent overlapped channels. Verify the frequency and levels of their crossovers on both subplots.

10.2 Form the prototype filter of problem 10.1 and then form a synthesized impulse response as the sum of three frequency offset versions of that time response. The three versions are h1, h1*exp $(j*2*\text{pi}*(-N/2:+N/2)/16)$, and h1*exp$(j*2*\text{pi}*(-N/2:+N/2)/16)$. On three subplots, plot the synthesized impulse response, the log magnitude synthesized frequency response, and a zoom to the in-band ripple of the log magnitude synthesized frequency response.

10.3 Using the design examples shown in MATLAB Script 10.1 guide, design the prototype polyphase filter for a 16-path filter analysis channelizer. For ease of design, the sample rate is 16 MHz, with the two sided half amplitude bandwidth of 1.0 MHz, and peak stop-band side lobe level is -60 dB. Stop-band frequency is 0.75. This filter has the same number of taps per path but uses a smaller β window as shown in MATLAB Script 10.1. On two subplots, plot the linear magnitude and log magnitude spectra of three adjacent overlapped channels. Verify the frequency and levels of their crossovers on both subplots.

10.4 Form the prototype filter of problem 10.3 and then form a synthesized impulse response as the sum of three frequency offset versions of that time response. The three versions are h1, h1*exp $(j*2*pi*(-N/2:+N/2)/16)$, and h1*exp$(j*2*pi*(-N/2:+N/2)/16)$. On three subplots, plot the synthesized impulse response, the log magnitude synthesized frequency response, and a zoom to the in-band ripple of the log magnitude synthesized frequency response.

10.5 Form the prototype filter of problem 10.1 and then write the MATLAB script for a 16-path analysis filter bank that performs an 8-to-1 down-sample of an input time series. Test the analysis channelizer by applying a time series containing a single unit amplitude impulse. On 16 subplots of one figure, plot the 16 different impulse responses from the 16-output ports of the filter bank. On 16 subplots of a second figure, plot the 16 different log magnitude frequency responses from the 16-output ports of the filter bank.

10.6 Form the prototype filter of problem 10.1 and then write the MATLAB script for a 16-path analysis filter bank that performs an 8-to-1 down-sample of an input time series. Test the analysis channelizer by applying a time series containing 2400 samples of a composite time series formed as the sum of four complex sinusoids at frequencies [+1.1/16, +2.8/16, −4.0/16, and −6.8/16] with amplitudes [1.0, 0.25, 1.0, and 0.125] each with a random phase angle. On 16 subplots of one figure, plot the 16 different time series from the 16-output ports of the filter bank. On 16 subplots of a second figure, plot the 16 different log magnitude spectra from the 256-point windowed time series at the16-output ports of the filter bank.

10.7 Using the design examples shown in MATLAB Script 10.2 guide, design the prototype polyphase filter for a 16-path filter synthesis channelizer. For ease of design, the sample rate is 16 MHz with the two sided half amplitude bandwidth of 1.0 MHz, and peak stop-band side lobe level is −60 dB. Two sided pass band is −0.75 to 0.75 MHz and stop band spans 1.25 to 8 MHz. This filter has the same number of taps per path as the analysis filter of problem 10.3. On two subplots, plot the linear magnitude and log magnitude spectra of the synthesis channel filter overlaid on the corresponding synthesis channel filter. Verify that the synthesis pass-band is as wide as the span of the analysis pass-band and transition bandwidth.

10.8 Write the MATLAB script to synthesize a variable bandwidth filter by the cascade of a non-maximally decimated 16-path analysis filter bank and a 16-path synthesis filter bank. The analysis bank performs 8-to-1 down-sampling and the synthesis bank performs 1-to-8 up-sampling. A binary mask between the output of the analysis bank and the input to the synthesis bank changes the bandwidth of the synthesized filter. The analysis and synthesis filters satisfy the specifications of problem 10.7. To demonstrate the performance of the design, form the composite impulse response of the cascade when the binary mask passes all 16 outputs between the analysis and synthesis banks. On three subplots, plot the impulse response, a zoom to the low-level time domain side lobes to show reconstruction errors, and a zoom to the in-band ripple of the filter's frequency response.

10.9 Repeat problem 10.8. Write the MATLAB script to synthesize a variable bandwidth filter by the cascade of a non-maximally decimated 16-path analysis filter bank and a 16-path synthesis filter bank. The analysis bank performs 8-to-1 down-sampling and the synthesis bank performs 1-to-8 up-sampling. We will verify performance of the cascade with a different binary mask than the one used in problem 10.8. Here the mask will have 11 of the analysis channel's time series to the synthesis channel's corresponding input ports. The enabled channel bins are −5 to +5. On three subplots, plot the synthesized impulse response, the synthesized frequency response errors, and a zoom to the in-band ripple of the filter's frequency response.

10.10 Design a nine-path polyphase analysis filter bank that performs 6-to-1 down-sampling and outputs five even indexed output channels. Input sample rate is 180 kHz, channel spacing is 20 kHz, and output sample rate is 30 kHz. Transition bandwidth is 2 kHz, stop-band attenuation is 60 dB and in-band ripple is 0.1 dB. On one figure, show the impulse response of each output port. On a second figure show the frequency response of each output port. After forming the impulse response and frequency response of the nine channels, test the channelizer with a 2600-point sequence containing multiple sinusoids in the five even indexed channel center frequencies [−40, −20, 0, +20, +40] kHz. On one subplot, show the windowed input spectrum to the channelizer, and on nine additional subplots, show the windowed output spectra from each channel output port.

10.11 This problem is the frequency offset version of problem 10.10. Design a nine-path polyphase analysis filter bank that performs 6-to-1 downsampling and outputs six odd indexed output channels. Input sample rate is 180 kHz, channel spacing is 20 kHz, and output sample rate is 30 kHz. Transition bandwidth is 2 kHz, stop-band attenuation is 60 dB and in-band ripple is 0.1 dB. On one figure, show the impulse response of each output port. On a second figure, show the frequency response of each output port. After forming the impulse response and frequency response of the nine channels, test the channelizer with a 2600-point sequence containing multiple sinusoids in the six odd indexed offset channel center frequencies [−50, −30, 0, −10, +10, +30, +50] kHz. On one subplot, show the windowed input spectrum to the channelizer, and on nine additional subplots, show the windowed output spectra from each channel output port.

11

Recursive Polyphase Filters

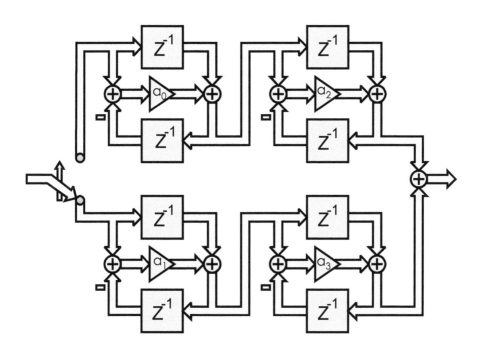

When we first start a description of the resampling process, we often use the model of a bandwidth-reducing filter followed by down-sampling to a reduced sample rate commensurate with the reduced bandwidth. We argue that it is foolish to compute output samples that are destined to be discarded by the down sampler. This leads us to systems that compute only the required output samples. Such a filter does not compute an output sample for each input sample. Somewhere in the back of our mind, we form this vague concept that the resampling operation can only be performed in non-recursive filters. After all, a non-recursive filter does not require a prior output to compute the next output, and, thus, we are permitted to skip output samples. A recursive filter on the other hand does require prior outputs to form the next output. This suggests that the recursive filter must compute each output for its own benefit even if the down sampler has no need for it in the output data stream.

Recursive filters can in fact be resampled and most of the structures we have developed for non-recursive filters are applicable to recursive structures. Having direct access to the impulse response of the non-recursive prototype filter facilitates the partition of the filter to form the stages of its polyphase version. While recursive filters can be similarly partitioned, not having direct access to the filter impulse response requires us to perform the partition with a different set of tools. The most effective approach to the partition of a recursive filter is to change the design procedure so that the partition is built into its structure. In this chapter, we describe this embedding process that leads to efficient resampling recursive filters

Nearly all of the filtering characteristics of a multirate partition are related to the phase versus frequency characteristics of the elementary building blocks forming the filter. It is the phase behavior that permits signals flowing through different paths to combine constructively or destructively and thus establish pass-band and stop-band regions. We know that spectral phase is preserved when a spectral region is translated by a heterodyne and it is pleasing to know this property is also valid when the spectral region is translated by aliasing. This permits the constructive or destructive combining to occur prior to, or following, an aliasing process. Since phase is an important key to understanding the behavior of multirate systems, we examine recursive filters that are characterized as all-pass structures, filters that pass all input frequencies only affecting their phase response.

11.1 All-Pass Recursive Filters

A very rich class of multirate filters can be formed from a set of elemental recursive all-pass subfilters. The structure of these filters is very different from traditional recursive digital filters. The filter structure can be introduced in a number of ways, and a particularly enlightening form is as a tapped delay line, reminiscent of an M-point finite impulse response (FIR) filter, whose taps are all-pass filters formed by polynomials in Z^M. This structure is illustrated in Figure 11.1 and then again in Figure 11.2 with the kth delay in

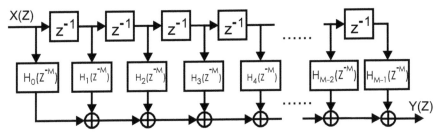

Figure 11.1 *M*-path polyphase all-pass filter structure.

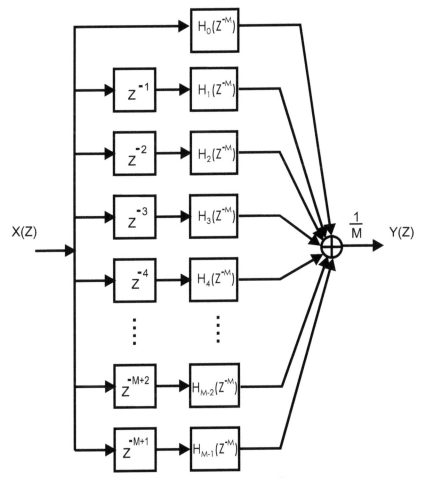

Figure 11.2 *M*-path polyphase filter with delays Z^{-K} allocated to the *K*th path.

the tapped delay line explicitly assigned to the kth path. The transfer function describing these structures is shown in Equation (11.1).

$$H(Z) = \sum_{m=0}^{M-1} H_m(Z^M) Z^{-m}. \tag{11.1}$$

We recognize that the transfer function is the same form of a polyphase partition of a FIR filter. The difference in this structure is that the path transfer functions $H_m(Z^M)$ are all-pass recursive filters rather than the non-recursive approximations to an all-pass filter obtained by partitioning a FIR filter. The most common application of the M-path polyphase filter is the case of a two-path filter ($M = 2$) and we will spend a fair amount of time examining this form. Before we do so, we will review the properties of recursive polyphase filters and examine useful architectures.

11.1.1 Properties of All-Pass Filters

All-pass filters have sinusoidal steady-state gains of unity for all input frequencies. This is stated concisely in the following equation:

$$|H(Z)|_{Z=\exp(j\theta)} = |H(\theta)| = 1. \tag{11.2}$$

The simplest non-trivial all-pass network is a single delay line, represented by a single pole at the origin, as shown in Equation (11.3). This single pole located at the origin contributes unity gain and a phase shift that varies linearly with input frequency as seen in Figure 11.3. We note that an all-pass network is characterized entirely by its phase shift.

$$H(Z) = \frac{1}{Z} \tag{11.3}$$

The next non-trivial all-pass network is the first-order filter in Z shown in Equation (11.4). This form is a first-order version of the general relationship presented in Equation (11.5). A characteristic of an all-pass network is that the sequence of coefficients that describe the numerator in descending powers of Z is the index-reversed sequence that describes the denominator in descending powers of Z. The numerator is the reciprocal polynomial of the denominator polynomial. Reciprocal polynomials have reciprocal roots; thus, all-pass networks have reciprocal poles and zeros. The pole-zero diagram for this simple first-order all-pass filter is shown in Figure 11.4. The phase response of this filter is the difference of the numerator phase and the

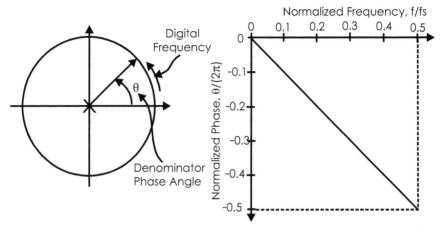

Figure 11.3 Pole diagram and phase as function of frequency for single delay Z^{-1}.

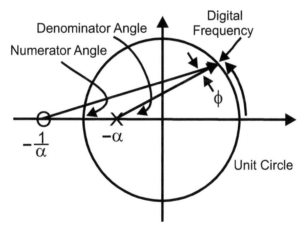

Figure 11.4 Pole-zero diagram of first-order all-pass networks $[(1 + \alpha Z)/(Z + \alpha)]$: showing phase angles that form output phase angle.

denominator phase, which, by simple trigonometric consideration, is the included angle on the unit circle as indicated in the figure as $-\phi$. The phase response corresponding to different values of the argument α is shown in Figure 11.5. From this figure, we see that the phase angle must be zero at central angle $\theta = 0$ and must be $-\pi$ at central angle π.

$$H(Z) = \frac{1 + \alpha Z}{Z + \alpha} \tag{11.4}$$

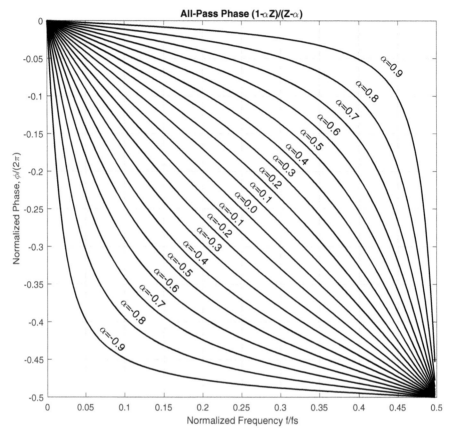

Figure 11.5   Phase response of first order in Z all-pass network as function of pole position α ($\alpha = 0.9, 0.8, \ldots, 0, \ldots, -0.8, -0.9$).

$$
\begin{aligned}
H(Z) &= \frac{N_M(Z)}{D_M(Z)} = \frac{Z^M D_M(Z^{-1})}{D_M(Z)} \\
&= \frac{1 + a_1 Z^1 + a_2 Z^2 + \cdots + a_{M-1} Z^{M-1} + a_M Z^M}{Z^M + a_1 Z^{M-1} + a_2 Z^{M-2} + \cdots + a_{M-1} Z^1 + a_M}.
\end{aligned}
\tag{11.5}
$$

Note that for the first-order all-pass filter, the transfer function for $\alpha = 0$ defaults to the pure delay, and, as expected, its phase function is linear with frequency. The phase function is a function of the pole position α and this network can be thought of, and will be used as, the generalized delay element. We observe that the phase function is anchored at its end points ($0°$ at zero frequency and $180°$ at the half sample rate) and that it warps with

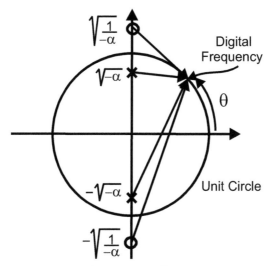

Figure 11.6 Pole-zero diagram of first order in Z^2 all-pass networks $[(1 + \alpha Z^2)/(Z^2 + \alpha)]$.

variation in α. It bows upward (less phase) for positive α and bows downward (more phase) for negative α. The bowing phase function permits us to use the generalized delay to obtain a specified phase angle at any frequency. For instance, we note that when $\alpha = 0$, the frequency for which we realize $90°$ phase shift is 0.25 (the quarter sample rate). We can determine a value of α for which the $90°$ phase shift is obtained at any normalized frequency such as at normalized frequency 0.45 ($\alpha = 0.8$) or at normalized frequency 0.05 ($\alpha = -0.73$).

An important variant of this first-order filter is obtained by converting it into a first-order polynomial in Z^2. This is equivalent to zero-packing its impulse response, which has the effect of replicating its spectrum as we traverse the unit circle. We see two copies of the spectrum or phase response as we pass the two poles of the transfer function rather than the single pole. The form of this transfer function is shown in the following equation:

$$H(Z) = \frac{1 + \alpha Z^2}{Z^2 + \alpha} \qquad (11.6)$$

The pole-zero diagram of this transfer function is shown in Figure 11.6 where we note that all roots reside on the imaginary axis.

The phase response of the first-order polynomial in Z^2 is seen in Figure 11.7. Note that the first half of the phase response matches the response seen in Figure 11.5. The second half is the phase-continuous

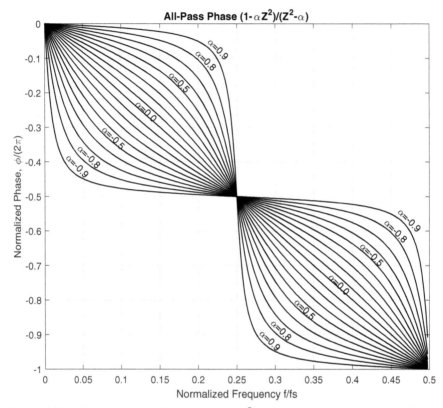

Figure 11.7   Phase response of first-order (in Z^2) all-pass network as function of pole position α ($\alpha = 0.9, 0.8, \ldots, 0, \ldots, -0.8, -0.9$).

extension we would have seen in Figure 11.5 had we finished traversing the unit circle. The phase response of the first-order filter in Z^2 is seen to be a *lazy S* which for positive α ramps slowly to the right, falls rapidly, and again continues to ramp slowly to the right. We will be using this lazy-*S* phase behavior shortly.

A closed-form expression for the phase function is obtained by evaluating the transfer function on the unit circle. This is done in the following derivation. Equation (11.7) presents the transfer function for the first-order filter in Z^M. Initially, we will be interested in this form of the all-pass filter for the two cases, $M = 1$ and $M = 2$.

$$H(z) = \frac{1 + \alpha Z^M}{Z^M + \alpha} \tag{11.7}$$

Equation (11.8) starts the process of evaluating the network on the unit circle by replacing the indeterminate Z with $e^{j\theta}$.

$$H(\theta) = \frac{1 + \alpha e^{jM\theta}}{e^{jM\theta} + \alpha}. \tag{11.8}$$

In Equation (11.9), we factor from numerator and denominator the square root of the exponential term $e^{jM\theta/2}$.

$$H(\theta) = \frac{e^{-jM\theta/2} + \alpha e^{jM\theta/2}}{e^{jM\theta/2} + \alpha e^{-jM\theta/2}}. \tag{11.9}$$

In Equation (11.10), we replace the complex exponential $e^{j(-)}$ with $\cos(-)$ $+ j\sin(-)$.

$$H(\theta) = \frac{\cos(M\theta/2) - j\sin(M\theta/2) + \alpha\cos(M\theta/2) + j\alpha\sin(M\theta/2)}{\cos(M\theta/2) + j\sin(M\theta/2) + \alpha\cos(M\theta/2) - j\alpha\sin(M\theta/2)}. \tag{11.10}$$

We next gather up the real and imaginary components of the numerator and denominator and divide the numerator and denominator by the common real part. These steps are shown in Equations (11.11)–(11.13).

$$H(\theta) = \frac{(1+\alpha)\cos(M\theta/2) - j(1-\alpha)\sin(M\theta/2)}{(1+\alpha)\cos(M\theta/2) + j(1-\alpha)\sin(M\theta/2)} \tag{11.11}$$

$$H(\theta) = \frac{1 - j\dfrac{(1-\alpha)\sin(M\theta/2)}{(1+\alpha)\cos(M\theta/2)}}{1 + j\dfrac{(1-\alpha)\sin(M\theta/2)}{(1+\alpha)\cos(M\theta/2)}}. \tag{11.12}$$

$$H(\theta) = \frac{1 - j\dfrac{(1-\alpha)}{(1+\alpha)}\tan\left(M\dfrac{\theta}{2}\right)}{1 + j\dfrac{(1-\alpha)}{(1+\alpha)}\tan\left(M\dfrac{\theta}{2}\right)}. \tag{11.13}$$

The components of the numerator and denominator of Equation (11.13) can be visualized with the aid of Figure 11.8. The numerator and denominator of Equation (11.13) share the same real part and differ only by the sign of their imaginary parts. The magnitude of the numerator and denominator of Equation (11.13) are seen to be the hypotenuse of the two right triangles with base of unit length. Thus, the magnitude of the transfer function (the

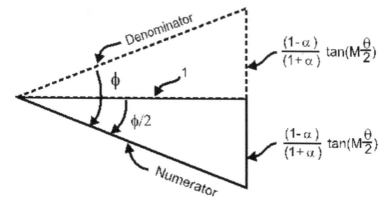

Figure 11.8   Visualization of components in Equation (11.13).

ratio of the numerator and denominator magnitudes) must be unity, and the phase angle of the transfer function is minus twice the angle formed by the numerator right triangle. The phase of the first-order low-pass network is shown in Equation (11.14). Note that if $\alpha = 0$, the phase 2ϕ reverts to a linear phase proportional to frequency θ, with the proportionality factor M, originally the degree of the first-order polynomial, now identifying the number of linear delay elements.

$$\phi = -2\text{ATAN}\left[\frac{1-\alpha}{1+\alpha}\text{TAN}\left(M\frac{\theta}{2}\right)\right]. \tag{11.14}$$

11.1.2 Implementing First-Order All-Pass Networks

The first-order all-pass network can be implemented in a number of architectures. We now review a few of them and identify those with desirable finite-precision arithmetic performance or with a reduced number of arithmetic operations. The obvious implementations of the all-pass transfer function, shown in Equation (11.15), is the factored form shown to represent the filter in a canonic form with one feedback and one feed forward path from the common delay element. The structure of this filter is presented in Figure 11.9.

$$H(Z) = \frac{1+\alpha Z}{Z+\alpha} = \frac{Z^{-1}+\alpha}{1+\alpha Z^{-1}} = \frac{1}{1+\alpha Z^{-1}}[\alpha+Z^{-1}]. \tag{11.15}$$

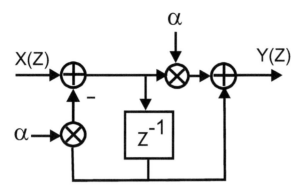

Figure 11.9 Canonic implementation of first-order all-pass filter.

This form of the all-pass filter does not have two different coefficients for the roots α and $1/\alpha$. This avoids the common problem in finite precision implementations related to the fact that $Q(\alpha)$ and $Q(1/\alpha)$, where $Q(\ )$ represents the quantization process, are not reciprocals so that the filter is not all-pass. For example, with $\alpha =1/3$, a 12-bit quantized version of $Q(\alpha) = 0.3330078$ while $Q(1/\alpha) = 3$, with product $Q(\alpha)\,Q(1/\alpha)$ equal to 0.9990234. The primary objection to this form of the filter is that it requires two multiplications by the same coefficient α.

An alternate representation of the all-pass filter reverses the order of generating the pole-zero pair. In the previous version, we implement the pole at the input to the filter and the zero at its output. In the second option, we implement the zero at the input to the filter and the pole at its output. The appropriate factoring of the transfer function is shown in Equation (11.16) and the form of the reordered implementation is shown in Figure 11.10.

$$H(Z) = \frac{1+\alpha Z}{Z+\alpha} = \frac{Z^{-1}+\alpha}{1+\alpha Z^{-1}} = [Z^{-1}+\alpha]\frac{1}{1+\alpha Z^{-1}} \qquad (11.16)$$

We note that the feed forward and feedback sections of the transfer function have the same valued coefficient and that the outputs of their products are summed to form the output $Y(Z)$. The two inputs to the multipliers can first be combined in a pre-sum and a single multiply can be used to apply the coefficient to both the feed forward and the feedback paths. This structure is seen in Figure 11.11 in the general form, as a first-order filter in Z^M and as a processing block indicating a single coefficient α.

A major attraction of this form of the all-pass filter structure is that since the same coefficient is applied to the numerator and denominator, the system

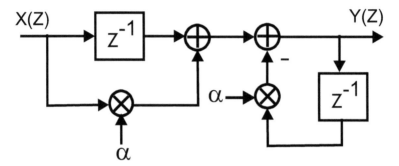

Figure 11.10   Reordered all-pass filter structure.

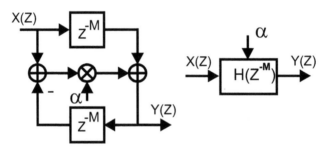

Figure 11.11   Single coefficient all-pass filter structure.

maintains all-pass properties for any degree of coefficient quantization (but not quantized arithmetic). The primary feature of this structure is that a single multiplication forms both the numerator and denominator. One multiply here replaces two in the previous version of the filter. An interesting property of this form is that the feed forward register, containing delayed versions of the input, and the feedback register, containing delayed versions of the unity gain output signal, do not require wider bit fields to accommodate processing gain word growth. The feedback register, the multiplier, and adders do however require extension into lower order bits to preserve significance in the difference formed at the input adder. This form has another advantage when all-pass stages are cascaded. Since the feedback register of one stage contains the same data as the feed forward register of the next stage, the two registers can be folded into a single register serving both feedback and feed forward. This will be demonstrated in the next section.

One final advantage of this structure is that we have easy access to a delayed version of the output of the all-pass filter. This delayed output is available from the feedback delay, and when $M = 1$, the delayed sample

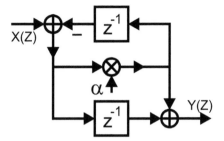

Figure 11.12 Single coefficient all-pass filter dual structure.

is available at the output of the feedback register. We will have need for a delayed version of the all-pass output when we implement the low-pass to band-pass transformation. We discuss this option in detail in a later section when we translate the frequency response of a prototype low-pass filter to a different center frequency. This transformation replaces a delay (a linear phase shift) with a delay plus a generalized delay (a non-linear phase shift). The transformation is shown in the following equation:

$$\frac{1}{Z} \Rightarrow -\frac{1}{Z}\frac{1-cZ}{Z-c}, \quad c = \cos\left(2\pi\frac{f_C}{f_S}\right). \tag{11.17}$$

The dual graph of the structure presented in Figure 11.11 is also a single coefficient version of the all-pass filter. Recall that in the dual structure, a summing junction replaces each node, a node replaces each summing junction, and the direction of the signal flow is reversed. The dual version of this graph is shown in Figure 11.12.

The dual graph structure is not as interesting as the original and is included only for completeness. This form of the filter does not exhibit any attractive efficiency options comparable to the register folding and sharing that is possible for the non-dual form when successive stages are operated in cascade. While on the topic of completeness, another common, single multiply all-pass structure is shown in Figure 11.13. This structure requires a single delay element but uses three adds instead of the two adds required in the structure presented in Figure 11.11.

The all-pass filter structures exhibit similar range of frequency-dependent sensitivity to finite arithmetic noise. The magnitude responses of the filters deviate slightly from unity gain when we truncate finite arithmetic multiplication and addition. Figure 11.14 compares the average spectral variance from the all-pass unity gain of two forms of the filters implemented

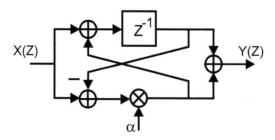

Figure 11.13   Alternate single coefficient all-pass structure.

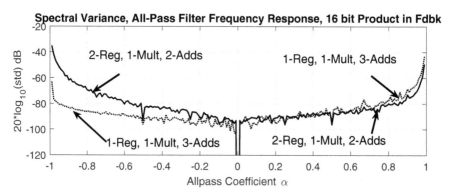

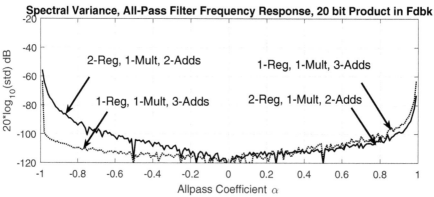

Figure 11.14   Spectral variance for two forms of all-pass filter for 16-bit and 20-bit coefficients as function of coefficient α.

with 16-bit and with 20-bit coefficients and feedback register widths, respectively. Note the asymmetry in variance between positive and negative values of the coefficient α and the increased variance as the roots approach the unit circle ($\alpha \to 1$) and become high Q roots. Since nearly all applications of the all-pass filter use positive values of α, the asymmetry does not influence the selection of our architecture. For positive values of α, the filters show similar sensitivity to filter feedback register finite bit width. The advantages of the register sharing in successive stages with the two-register form of the filter, which we will show shortly, warrants its use in most applications.

11.2 Two-Path All-Pass Recursive Filters

We now have access to elemental all-pass structures required to form the filters we will be using as building blocks for our final designs. Rather than examining other low-level building block filters, we change our approach and use a specific filter structure to view the composite filter from a top-down view and develop a perspective of how the all-pass structure leads to this different class of filters. Once we have developed the structure and perspective for a particularly simple class of filters, we will expand the class by invoking a simple set of frequency transformations. The structure we eventually master, formed with the all-pass filters, is a two-path polyphase filter structure. This is essentially a tapped delay line, of length 2, with each tap formed as a cascade of filters in powers of Z^2. This structure is presented in Figure 11.15.

The two-path structure can be redrawn as shown in Figure 11.16. The delay in the lower path can be placed on either side of the all-pass network. As observed earlier, this delay already resides in the feedback path of the all-pass filter and may be extracted from the filter as opposed to being inserted prior to the filter. When the filter participates in a multirate configuration, a commutator port replaces the delay, and the delay is originally positioned on the side of the filter operating at the higher of the two rates.

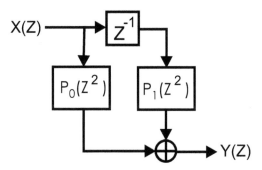

Figure 11.15   Two-path polyphase filter.

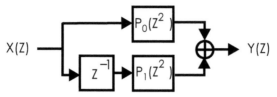

Figure 11.16 Redrawn two-path polyphase filter.

11.2.1 Two-Path Half-Band Filters: Non-Uniform Phase

This filter structure shown in Figure 11.16 offers a surprisingly rich class of filter responses that can be formed by the all-pass building blocks. By restricting our initial dialogue to two-path filters, we can visit familiar filter characteristics and demonstrate the wide versatility of the two-path structure while comparing relative efficiencies of the all-pass and classical structures. The two-path structure can implement half-band low-pass and high-pass filters, as well as Hilbert-transform filters that exhibit minimum or non-minimum phase response. The two-path filter can implement standard recursive filters such as the Butterworth and Elliptic filters. A MATLAB routine, *tony_des2*, that computes the coefficients of the two-path filter and a number of its variants is included in the software accompanying this textbook. Also, the half-band filters can be configured to embed a 1-to-2 up-sampling or a 2-to-1 down-sampling operation within the filtering process. The prototype half-band filters have their 3-dB band edge at the quarter sample rate. All-pass filters can be used to apply frequency transformations to the two-path prototype to form arbitrary-bandwidth low-pass and high-pass complementary filters, and arbitrary center frequency pass-band and stop-band complementary filters. These same all-pass frequency transformations can be applied to any standard filter structure to obtain desired parameter adjustable spectral transformations. Zero-packing the time response of the two-path filter, another trivial all-pass transformation, causes spectral scaling and replication. The zero-packed structures used in cascade with other filters in iterative filter designs achieve composite spectral responses exhibiting very narrow transition bandwidths with low-order filters.

The specific form of the prototype two-path filter is shown in Figure 11.17. The number of poles (or order of the polynomials) in the two paths differs by precisely one, with the one extra pole located at the origin and conveyed by the single delay in the lower leg. Since the denominator terms in the all-pass filters are of the form $(Z^2 + \alpha_k)$, for positive α_k, all poles of this filter are restricted to the imaginary axis. The structure forms

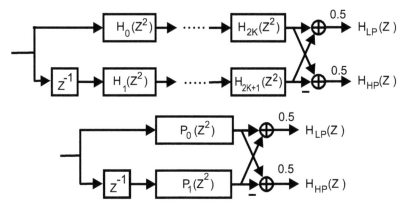

Figure 11.17 Two-path all-pass filter.

complementary low-pass and high-pass filters, from the scaled sum and difference respectively, of the outputs from the two paths. An important idea related to polyphase filter partitions is that the scaled sum and difference are in fact a two-point transform of the pair of outputs and represent a phase-rotated coherent sum of the path outputs. The scale factor of this sum and difference is 1/2.

The cascade of first-order polynomials in Z^2 in each of the legs of the two-path filter can be implemented by a set of identical stages with different coefficients. This model, shown in Figure 11.18, reflects our standard way of implementing a recursive filter as a sequence of first order in Z^2 filter stages. The particular form we present here can also be recast in a more efficient coupled architecture. This architecture takes advantage of the earlier observation that the feedback registers of one stage contain the same data as the feed forward registers of the next stage and consequently can be shared rather than replicated. The architecture that shares the feedback and feed forward registers is shown in Figure 11.19.

The transfer function of the two-path filter shown in Figure 11.17 is shown in the following equation:

$$H(Z) = P_0(Z^2) \pm Z^{-1}P_1(Z^2)$$

$$P_i(Z^2) = \prod_{k=0}^{K_1} H_{i,k}(Z^2), \quad i = 0, 1 \qquad (11.18)$$

$$H_{i,k}(Z^2) = \frac{1 + \alpha_{i,k}Z^2}{Z^2 + \alpha_{i,k}}.$$

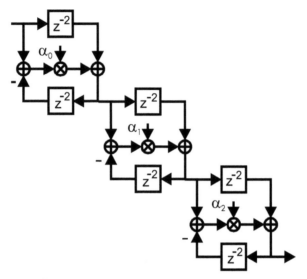

Figure 11.18   Direct implementation of cascade first-order filters in Z^2.

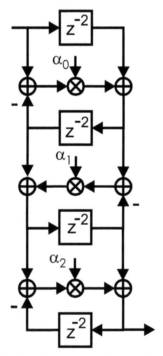

Figure 11.19   Folded implementation of cascade first-order filters in Z^2.

In particular, we can examine the trivial case of a single all-pass filter in path-0 and a delay only in path-1. The transfer function for this case is shown in the following equation:

$$H(Z) = \frac{1 + \alpha_0 Z^2}{Z^2 + \alpha_0} \pm \frac{1}{Z} = \frac{\alpha_0 Z^3 \pm Z^2 + Z \pm \alpha_0}{Z(Z^2 + \alpha_0)}$$
$$= \frac{\alpha_0 (Z \pm 1)[Z^2 \mp (\frac{1}{\alpha_0} - 1)Z + 1]}{Z(Z^2 + \alpha_0)}.$$

$$(11.19)$$

We note a number of interesting properties of this transfer function, applicable to all the two-path prototype filters. The denominator roots are on the imaginary axis restricted to the interval ± 1 to assure stability. The numerator is a linear phase FIR filter with an even symmetric weight vector. As such, the numerator roots must appear either on the unit circle, or if off and real, in reciprocal pairs, and if off and complex, in reciprocal conjugate quads. Thus, for appropriate choice of the filter weight(s), the zeros of the transfer function can be placed on the unit circle and can be distributed to obtain equal ripple stop-band response. In addition, due to the one pole difference between the two paths, the numerator must have a root at ± 1. When the two paths are added, the numerator roots are located in the left half plane, and, when subtracted, the numerator roots are mirror imaged to the right half plane forming low-pass and high-pass filters, respectively.

For the example shown in Equation (11.19), the numerator roots are on the unit circle for a limited range of α_0. If we compare the numerator with the sequence obtained from Pascal's triangle, we can determine the parameter α_0 to obtain a Butterworth filter, with repeated zeros at -1, the half sample rate. This value is $\alpha_0 = 1/3$. For values of α_0 greater than 1/3, two of the roots leave -1 and migrate as conjugate roots along the unit circle until they reach $\pm j1$ where they meet and annihilate the poles migrating away from the origin along the y-axis. As the zeros start their journey around the unit circle, the transition bandwidth is reduced while a stop-band side lobe rises between the zeros. A value of α_0 can be found to realize any side lobe level. For this example, the filter contains three poles and three unit-circle zeros, a total of five non-trivial roots for the cost of one multiply per input point.

The attraction of this class of filters is the unusual manner in which the transfer function zeros are formed. The zeros of the all-pass subfilters reside outside the unit circle (at the reciprocal of the stable pole positions) but migrate to the unit circle as a result of the sum or difference of the two paths. The zeros occur on the unit circle because of destructive cancelation of spectral components delivered to the summing junction via the two distinct paths, as opposed to being formed by numerator weights in the feed forward path of standard bi-quadratic filter. The stop-band zeros are a windfall. They start as the maximum phase all-pass zeros formed concurrently with the all-pass denominator roots by a single shared coefficient and migrate to the unit circle in response to addition of the path signals.

We now examine a filter containing a recursive all-pass stage in path-0 and a delay and recursive all-pass filter in path-1. This transfer function is shown in the following equation:

$$H(Z) = \frac{1 + \alpha_0 Z^2}{Z^2 + \alpha_0} \pm \frac{1}{Z} \frac{1 + \alpha_1 Z^2}{Z^2 + \alpha_1}$$

$$= \frac{\alpha_0 Z^5 \pm \alpha_1 Z^4 + (1 + \alpha_0 \alpha_1) Z^3 \mp (1 + \alpha_0 \alpha_1) Z^2 + \alpha_1 Z + \alpha_0]}{Z(Z^2 + \alpha_0)(Z^2 + \alpha_1)}.$$

(11.20)

This transfer function has five poles on the imaginary axis in the interval ± 1, and for appropriate selection of the two coefficients, five unit-circle zeros on the left half of the unit circle (for the sum of the paths) and the mirror reflection zeros on the right half of the unit circle (for the difference of the paths). As in the previous example, coefficients of the numerator can be compared to that of the appropriate Pascal triangle coefficient list to place all zeros at –1. This occurs when $\alpha_0 = 0.1055728$ and $\alpha_1 = 0.5278640$. For increased values, the roots migrate along the unit circle toward $\pm j1$, reducing the transition bandwidth and forming stop-band side lobes. By judicious choice of the two weights, the filter can be designed for equal-ripple side lobes of any level. For instance, we obtain a stop-band edge at 0.370 with –60 equal-ripple side lobes when $\alpha_0 = 0.1413486$ and $\alpha_1 = 0.5899948$. These weights correspond to a half-band elliptic filter with restricted pole positions.

Figure 11.20 presents the pole-zero diagram for the two-path filter formed with these weights. In this example, the path-0 is formed with one first-order section in Z^2 and path-1 is formed with one first-order section in Z^2 and a single delay Z^{-1}. The composite filter thus contains five poles and five zeros and requires one coefficient for path-0 and one coefficient for path-1. The design routine *Tony_des2* computed weights for the fifth-order polynomial to obtain –60 dB equal-ripple side lobes. The –3 dB pass-band edge is located at a normalized frequency of 0.25 and the stop-band edge that achieved the desired –60 dB stop-band ripple is located at a normalized

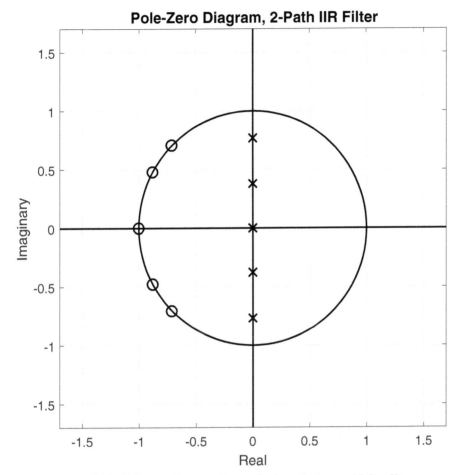

Figure 11.20   Pole-zero diagram of two path, five-pole, two-multiplier filter.

frequency of 0.370. The coefficients of the filter are listed here in decreasing order of Z:

Path-0	**Polynomial Coefficients:**		
Filter-0	[1	0	0.1413486]

Path-1	**Polynomial Coefficients:**		
Filter-1	[1	0	0.5899948]

The roots presented in Figure 11.20 describe the low-pass filter formed from the two-path filter. Figure 11.21 presents the phase slopes of the two

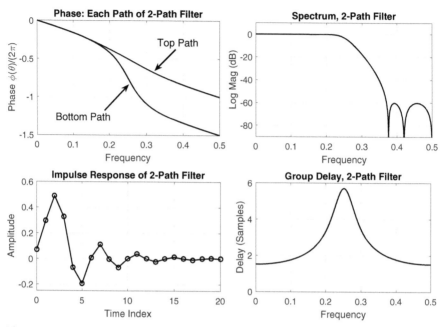

Figure 11.21   Two-path phase slopes, frequency, impulse, and phase responses of two-path, five-pole, two-multiplier filter.

paths of this filter as well as the frequency response, impulse response, and group delay response. We note that the zeros of the spectrum correspond to the zero locations on the unit circle in the pole-zero diagram. The low-frequency group delay matches the delay to the peak time response and we note the expected increased group delay at the filter band edge due to the dominant poles near the unit circle.

Figure 11.22 presents the pole-zero diagram for a higher-order two-path filter. In this example, the path-0 is formed with two first-order sections in Z^2 and path-1 is formed with two first-order sections in Z^2 and a single delay Z^{-1}. The composite filter thus contains nine poles and nine zeros and requires two coefficients for path-0 and two coefficients for path-1.

The *Tony_des2* design routine was used to compute weights for the ninth-order filter with -60 dB equal-ripple stop band. The -3 dB pass-band edge is located at a normalized frequency of 0.25 and the stop-band edge that achieved the desired 60 dB stop-band attenuation is located at a normalized frequency of 0.284. This too is an elliptic filter with constraints on the

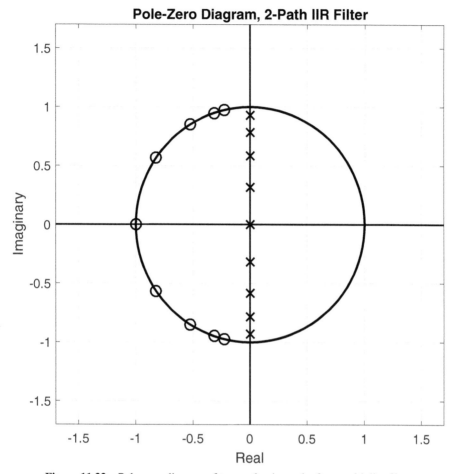

Figure 11.22 Pole-zero diagram of two-path, nine-pole, four-multiplier filter.

pole positions. The coefficients of the filter are listed here in decreasing order of *Z*:

Path-0	**Polynomial Coefficients**		
Filter-0	[1	0	0.101467517]
Filter-2	[1	0	0.612422841]

Path-1	**Polynomial Coefficients**		
Filter-1	[1	0	0.342095596]
Filter-3	[1	0	0.867647439]

Figure 11.22 presents the roots of the low-pass filter formed by the sum of the all-pass transfer functions of the two-path filter. The high-pass roots of the same filter would have the same poles, but the zeros would be reflected about the imaginary axis appearing in the low-frequency band on the unit circle. Figure 11.23 presents the phase slopes of the two paths of this filter as well as the frequency response, impulse response, and group delay response. We note that the zeros of the spectrum correspond to the zero locations on the unit circle in the pole-zero diagram. The low-frequency group delay matches the delay to the peak time response and we note the expected larger group delay at the filter band edge due to the dominant poles being nearer the unit circle.

Figure 11.24 presents a more detailed view of the phase response for each path of the nine-pole, two-path filter. The dashed lines represent the phase response of the two paths when the filter coefficients are set to zero. In this case, the two paths default to two delays in the top path and three delays in the bottom path. Since the two paths differ by one delay, the phase shift difference is precisely 180° at the half sample rate. When the coefficient of

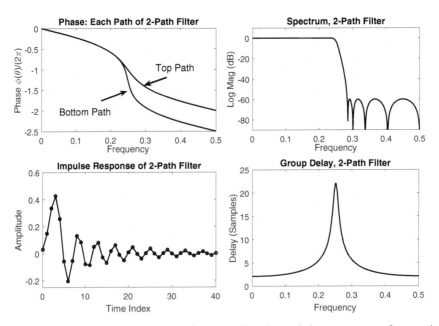

Figure 11.23   Two-path phase slopes, frequency, impulse, and phase responses of two-path, nine-pole, four-multiplier filter.

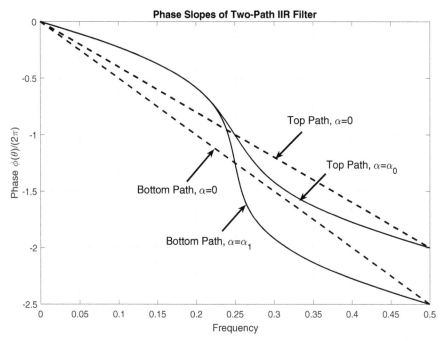

Figure 11.24 Phase response of both paths of two-path, nine-pole, four-multiplier filter.

the filters in each path is adjusted to their design values, the phase response of both paths assumes the bowed lazy-S curve described in Figure 11.7. Note that at low frequencies, the two phase curves exhibit the same phase profile and that, at high frequencies, the two-phase curves maintain the same 180° phase difference. Thus, the addition of the signals from the two paths will lead to a gain of 2 in the band of frequencies with the same phase and will result in destructive cancelation in the band of frequencies with 180° phase difference. These two bands, of course, are the pass band and stop band, respectively. We note that the two phase curves differ by exactly 180° at only four distinct frequencies as well as the half sample rate: these frequencies correspond to the spectral zeros of the filter. Between these zeros, the filter exhibits stopband side lobes that, by design, are equal ripple.

11.2.2 Two-Path Half-Band Filters: Linear Phase

We can modify the structure of the two-path filter to form filters with approximately linear phase response by restricting one of the paths to be pure delay. We accomplish this by setting all the coefficients of the filter in the upper leg to be zero. This forces the all-pass filters in this leg to revert to their default responses of pure delay with poles at the origin. As we pursue the solution to the phase matching problem in the equal-ripple approximation, we find that the all-pass poles must move off the imaginary axis. In order to keep real coefficients for the all-pass filters, we make one minor change in the all-pass filter structure. The change is to permit the cascade filters in the lower path to contain first- and second-order filters in Z^2. We lose degrees of freedom in the filter design process when we set the phase slope in one path to be a constant. Consequently, when we design an equal-ripple group delay approximation to a specified performance, we will need additional all-pass sections. To meet the same out-of-band attenuation and the same stop-band band edge as the non-linear phase design of the previous section, our design routine, *lineardesign*, determined that we require two first-order filters in Z^2 and three second-order filters in Z^2. This means that eight coefficients are required to meet the specifications that the non-linear phase design only required four coefficients. Path-0 (the delay only path) requires 16 units of delay, while the all-pass coefficient list is presented next in decreasing powers of Z along with its single delay element formed a 17th-order denominator.

Path-0	**Polynomial Coefficients**			
Delay	[zeros(1,16)	1]		

Path-1	**Polynomial Coefficients:**			
Filter-0	[1	0	0.832280776]		
Filter-1	[1	0	−0.421241137]		
Filter-2	[1	0	0.67623706	0	0.23192313]
Filter-3	[1	0	0.00359228	0	0.19159423]
Filter-4	[1	0	−0.59689082	0	0.18016931]

Figure 11.25 presents the pole-zero diagram of the linear phase all-pass filter structure that meets the same spectral characteristics as those outlined in the previous section. We first note that the filter is non-minimum phase due to the zeros outside the unit circle. We also note the near cancelation of the right half plane pole cluster with the reciprocal zeros of the non-minimum phase zeros. Figure 11.26 presents the phase slopes of the two filter paths, the filter frequency response, the filter impulse response, and the filter group delay. We first note that the phase slopes of the paths are linear; consequently, the group delay is also linear over most of the filter pass bands. Due to the non-minimum phase zeros of this filter, the group delay is greater than that for the previous filter. The constant group delay matches the 16-sample time

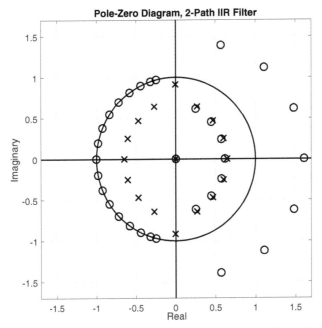

Figure 11.25 Pole-zero diagram of 2-path, 33-pole, 8-multiplier filter.

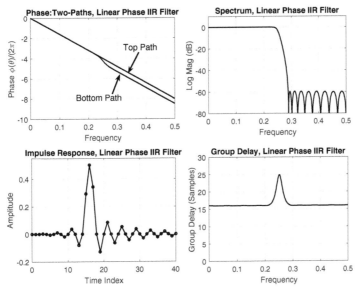

Figure 11.26 Two-path phase slopes, frequency, impulse, and phase responses of 2-path, 33-pole, 8-multiplier filter.

delay to the peak of the impulse response. The group delay increases at the filter band edge due to the dominant pole on the imaginary axis near the unit circle. Of course, the spectral zeros of the frequency response coincide with the transfer function zeros on the unit circle.

11.3 Comparison of Non-Uniform and Equal Ripple Phase Two-Path Filters

We have discussed two forms of the two-path recursive all-pass filter. In the case of non-uniform phase, the two paths contain one or more all-pass filters formed by first-order polynomials in Z^2, while in the case of linear phase, one path is limited to pure delay and the other contains one or more all-pass filters formed by first- and second-order polynomials in Z^2. In the following discussion, we distinguish roots at the origin from roots not at the origin. We say a root at the origin is inactive since it only affects phase and that a root off the origin is active since it affects both phase and magnitude. In the non-uniform phase filter, each coefficient contributes a second-order polynomial to the filter. The second path of this filter also contains an extra delay that contributes a pole at the origin. The total number of poles in the composite filter is $2n + 1$, where n is the number of coefficients. The number of zeros in the filter equals the number of poles with all the zeros active and, in fact, on the unit circle. In the uniform phase filter, each coefficient in the second path contributes a second-order polynomial in Z^2 that forms two active poles and a pair of inactive poles due to the matching delays in the first path. There is one more delay in the second path than in the first. Thus, the total number of poles in this composite filter is $4n + 1$, where n is the number of coefficients.

The delays in the first path are responsible for a large number of active zeros as the phases of the two paths interact by destructive cancelation.

A final comparison of the efficiency of the two forms of two-path recursive filters is to be found in Figures 11.27 and 11.28. Both figures present sets of curves showing the trade between out-of-band attenuation and transition bandwidth for two-path half-band filters for a range of filter complexity. The family of curves is indexed both by number of coefficients, m, and the degree, n, of the composite filter.

As a comparison, suppose we require a half-band filter with transition bandwidth of 3.0% and with 60 dB out-of-band attenuation. We find from Figure 11.27 that we can meet these requirements with a 9th degree, four-coefficient non-uniform phase filter and, from Figure 11.28, that we can meet

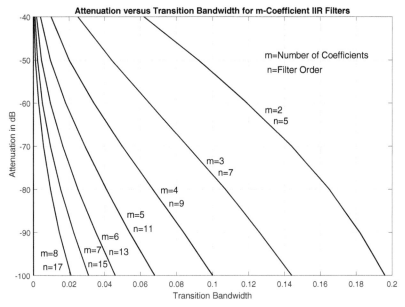

Figure 11.27 Variation of out-of-band attenuation versus transition bandwidth for non-uniform phase two-path filters containing 2−8 coefficients.

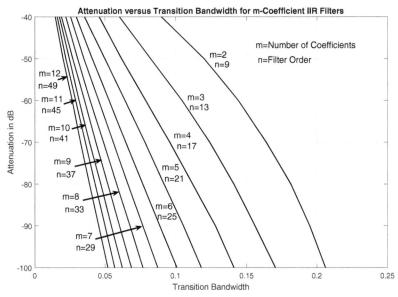

Figure 11.28 Variation of out-of-band attenuation versus transition bandwidth for uniform phase two-path filters containing 2−12 coefficients.

these requirements with a 33rd degree, eight-coefficient linear phase filter. To complete this section, we include for comparison, the pole-zero plot and the frequency response of a linear phase 65-tap FIR filter that meets the same specifications as the linear phase Infinite duration Impulse Response (IIR) filter.

The pole-zero diagram of the FIR filter is presented in Figure 11.29 and the impulse response and frequency response of the same filter is presented in Figure 11.30. Figure 11.31 presents for comparison the impulse response of the linear phase IIR and FIR filters. Note the shorter delay to the main

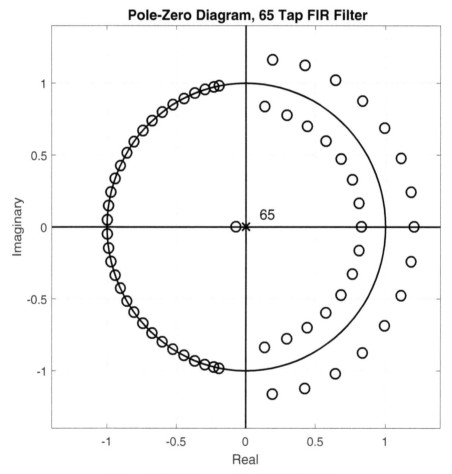

Figure 11.29   Pole-zero plot of 65-tap FIR filter.

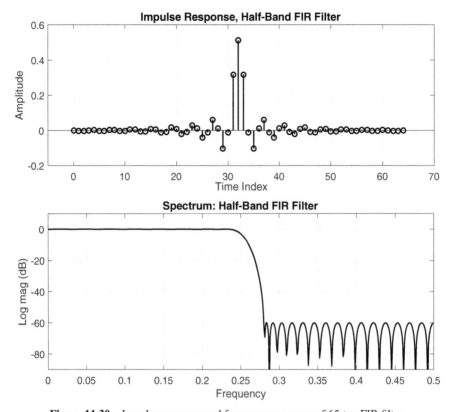

Figure 11.30 Impulse response and frequency response of 65-tap FIR filter.

lobe response of the IIR filter. Figure 11.32 presents the frequency response of the linear phase IIR and FIR filters. As can be seen, these filters have the same pass-band and stop-band edges as well as the same out-of-band attenuation level. Finally, Figure 11.33 presents a comparison of the in-band magnitude ripple of the two filters as well as the group delay of the equal ripple approximation to linear phase IIR filter. We expect that the group delay response of the two filters differs in that the FIR group delay is constant and the IIR is approximately equal ripple. We now observe that the amplitude response of the two filters differs in a similar fashion. The FIR amplitude is equal ripple while the IIR appears to be constant. In fact, the IIR ripple response is merely very small, for this example, on the order of 5-μdB. Note that the FIR filter requires 65 multiplies per output (32 if we take advantage of symmetry), while the IIR filter requires 8 multiplies. Now that we have

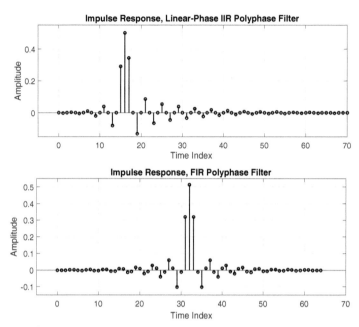

Figure 11.31   Comparison of impulse responses of linear phase IIR and FIR filters.

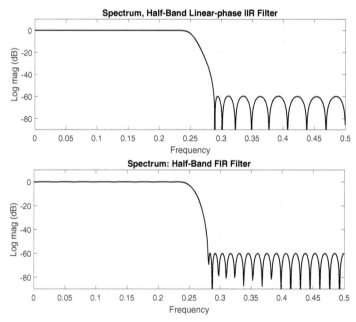

Figure 11.32   Comparison of frequency responses of linear phase IIR and FIR filters.

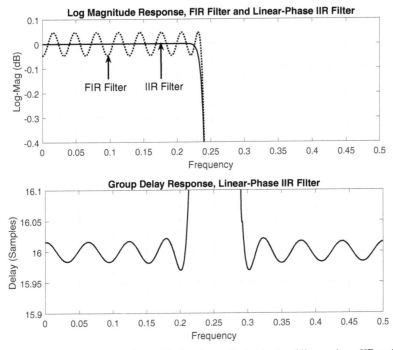

Figure 11.33 Detailed comparison of in-band magnitude ripple of linear phase IIR and FIR and of group delay of equal ripple linear phase IIR filter.

highlighted the relative performance of the linear phase IIR and FIR filters, we leave the FIR filter and return our attention to the primary topic, all-pass recursive filter.

11.4 Pass-Band and Stop-Band Response in Half-Band Filters

The all-pass networks that formed the half-band filter exhibit unity gain at all frequencies. These are lossless filters affecting only the phase response of signals they process. This leads to an interesting relationship between the pass-band and stop-band response of the half-band filter and, in fact, for any of the two-path filters discussed in this chapter. Examining Figure 11.17, we note that we have access to complementary filters: the low-pass and the high-pass versions of the half-band filter, the frequency responses of which are shown in Figure 11.34 where pass-band and stop-band ripples have been denoted by $\delta 1$ and $\delta 2$, respectively.

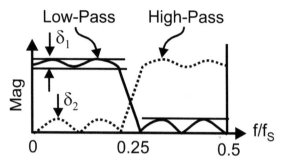

Figure 11.34   Magnitude response of low-pass and high-pass half-band filter.

The transfer functions of the low-pass and the high-pass filters are shown in Equation (11.21), where $P_0(Z)$ and $P_1(Z)$ are the transfer functions of the all-pass filters in each of the two paths. The power gain of the low-pass and high-pass filters is shown in Equation (11.22). When we form the sum of the power gains, the cross terms in the pair of products cancel and we obtain the results shown in Equation (11.23).

$$
\begin{aligned}
H_{\text{LOW}}(Z) &= 0.5 \cdot [P_0(Z) + Z^{-1}P_1(Z)] \\
H_{\text{HIGH}}(Z) &= 0.5 \cdot [P_0(Z) - Z^{-1}P_1(Z)]
\end{aligned}
\tag{11.21}
$$

$$
\begin{aligned}
|H_{\text{LOW}}(Z)|^2 &= H_{\text{LOW}}(Z) \cdot H_{\text{LOW}}(Z^{-1}) \\
&= 0.25 \cdot [P_0(Z) + Z^{-1}P_1(Z)] \cdot [P_0(Z^{-1}) + Z\,P_1(Z^{-1})] \\
|H_{\text{HIGH}}(Z)|^2 &= H_{\text{HIGH}}(Z) \cdot H_{\text{HIGH}}(Z^{-1}) \\
&= 0.25 \cdot [P_0(Z) - Z^{-1}P_1(Z)] \cdot [P_0(Z^{-1}) - Z\,P_1(Z^{-1})]
\end{aligned}
\tag{11.22}
$$

$$
\begin{aligned}
|H_{\text{LOW}}(Z)|^2 + |H_{\text{HIGH}}(Z)|^2 &= 0.25 \cdot [2|P_0(Z)|^2 + 2|P_1(Z)|^2] \\
&= 1.
\end{aligned}
\tag{11.23}
$$

Equation (11.23) tells us that at any frequency, the magnitude square of the low-pass gain and the magnitude square of the high-pass gain are equal to unity. This is a consequence of the filters being lossless. Energy that enters the filter is never dissipated, a fraction of it is available at the low-pass output and the rest of it is available at the high-pass output. This property is the reason the complementary low-pass and high-pass filters cross their 3-dB points. If we substitute the gains at peak ripple of the low-pass and high-pass filters into Equation (11.23), we obtain Equation (11.24) which we can rearrange and solve for the relationship between δ_1 and δ_2. The result is interesting. We learn here that the in-band ripple is approximately half the square of the out-of-band ripple. Thus, if the out-of-band ripple is –60 dB or 1-part in a 1000, then the in-band ripple is half of 1-part in 1,000,000, which is on the order of 5 μ-dB (4.34 μ-dB). The half-band recursive all-pass filters exhibit a very small in band ripple.

$$[1 - \delta_1]^2 + [\delta_2]^2 = 1$$
$$[1 - \delta_1] = \sqrt{1 - \delta_2^2} \cong 1 - 0.5\,\delta_2^2 \qquad (11.24)$$
$$\delta_1 \cong 0.5\,\delta_2^2.$$

11.5 Transforming Half-Band to Arbitrary-Bandwidth

In the previous section, we examined the design of two-path half-band filters formed from recursive all-pass first-order filters in the variable Z^2. We did this because we have easy access to the weights of this simple constrained filter, the constraint being stated in Equation (11.24). If we include a requirement that the stop-band be equal ripple, the half band-filters we examine are elliptic filters that can be designed from standard design routines. The filter so designed is our prototype that will be transformed to form other filters with specified (arbitrary) bandwidth and center frequency. In this section, elementary frequency transformations performed with our all-pass filters are introduced and their impact on the prototype architecture as well as on the system response is reviewed. In particular, frequency transformations that permit bandwidth tuning of the prototype are introduced first. Additional transformations that permit tuning of the center frequency of the prototype filter are also discussed.

11.5.1 Low-Pass to Low-Pass Transformation

We now address the first transformation to perform a transformation from the low-pass half-band filter to a low-pass arbitrary-bandwidth filter. Frequency transformations occur when an existing all-pass subnetwork in a filter is replaced by another all-pass subnetwork. In particular, we now examine the transformation shown in Equation (11.25).

$$\frac{1}{Z} \Rightarrow \frac{1 + bZ}{Z + b}; \quad b = \frac{1 - \tan(\theta_b/2)}{1 + \tan(\theta_b/2)}, \quad \theta_b = 2\pi\frac{f_b}{f_s} \qquad (11.25)$$

This is the generalized delay element we introduced in the initial discussion of first-order all-pass networks. In many applications, we can physically replace each delay in the prototype filter with the all-pass network and then tune the prototype by adjusting the parameter b. We have fielded many designs in which we perform this substitution. Some of these designs are cited in the chapter references. For the purpose of this paper, we perform the substitution algebraically in the all-pass filters comprising the two-path half-band filter and, in doing so, generate a second structure for which we will develop and present an appropriate architecture.

We substitute Equation (11.25) into the first order, in Z^2, all-pass filter introduced in Equation (11.18) and rewritten in the following equation:

$$G(Z) = H(Z)|_{Z \Rightarrow \frac{Z+b}{1+bZ}}$$

$$G(Z) = \frac{1 + \alpha Z^2}{Z^2 + \alpha}\bigg|_{Z \Rightarrow \frac{Z+b}{1+bZ}} . \tag{11.26}$$

After performing the indicated substitution and gathering terms, we find the form of the transformed transfer function as shown in the following equation:

$$G(Z) = \frac{1 + c_1 Z + c_2 Z^2}{Z^2 + c_1 Z + c_2}; \quad c_1 = \frac{2b(1 + \alpha)}{1 + \alpha b^2}, \quad c_2 = \frac{\alpha + b^2}{1 + \alpha b^2} \tag{11.27}$$

As expected, when $b \Rightarrow 0$, $c1 \Rightarrow 0$, and $c2 \Rightarrow \alpha$, the transformed all-pass filter reverts back to the original first-order filter in Z^2. The architecture of the transformed filter, which permits one multiplier to form the matching numerator and denominator coefficient simultaneously, is shown in Figure 11.35. Also shown is a processing block $G(Z)$ that uses two coefficients c_1 and c_2. This is seen to be an extension of the one-multiply structure presented in Figure 11.12. The primary difference in the two architectures is the presence of the coefficient and multiplier associated with the power of Z^{-1}. This term, formerly zero, is the sum of the polynomial roots and, hence, is minus twice the real part of the roots. With this coefficient being non-zero, the roots of the polynomial are no longer restricted to the imaginary axis.

The root locations of the transformed, or generalized, second-order all-pass filter are arbitrary except that they appear as conjugates inside the unit circle, and the poles and zeros appear in reciprocal sets as indicated in Figure 11.36.

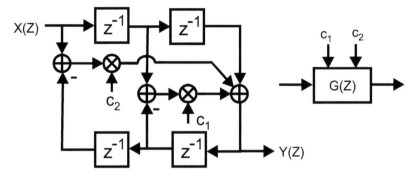

Figure 11.35 Block diagram of general second-order all-pass filter.

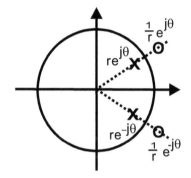

Figure 11.36 Pole-zero diagram of generalized second-order all-pass filter.

The two-path prototype filter contained one or more one-multiply first-order recursive filters in Z^2 and a single delay. We effect a frequency transformation on the prototype filter by applying the low-pass to low-pass transformation shown in Equation (11.25). Doing so converts the one-multiply first-order in Z^2 all-pass filter to the generalized two-multiply second-order all-pass filter and converts the delay, a zero-multiply all-pass filter to the generalized one-multiply first-order in Z all-pass filter. Figure 11.37 shows how applying the frequency transformation affects the structure of the prototype. Note that the five-pole, five-zero half-band filter, which is implemented with only two multipliers, now requires five multipliers to form the same five poles and five zeros for the arbitrary-bandwidth version of the two-path network. This is still significantly less than the standard cascade of first- and second-order canonic filters for which the same five-pole, five-zero filter would require 13 multipliers.

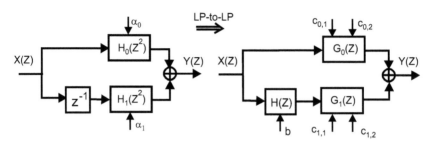

Figure 11.37   Effect on architecture of frequency transformation applied to two-path half-band all-pass filter.

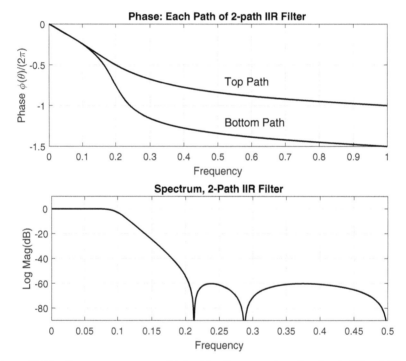

Figure 11.38   Frequency response obtained by frequency transforming half-band filter to normalized frequency 0.1.

Figure 11.38 presents the frequency response obtained by applying the low-pass to low-pass frequency transformation to the prototype two-path, two-multiply, half-band filter presented in Figure 11.21. The band-edge was moved from normalized frequency 0.25 to normalized frequency 0.1.

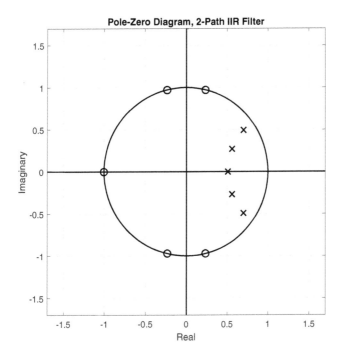

Figure 11.39 Pole-zero diagram obtained by frequency transforming half-band filter to frequency 0.1.

Figure 11.39 presents the pole-zero diagram of the frequency transformed prototype filter. The five poles have been pulled off the imaginary axis, and the five zeros migrated around the unit circle to form the reduced-bandwidth version of the prototype.

Figure 11.40 presents the response obtained by applying the low-pass to low-pass frequency transformation to move the band-edge from the normalized frequency 0.25 to normalized frequency 0.02. Figure 11.41 presents the pole-zero diagram of this frequency transformed prototype filter.

11.5.2 Low-Pass to Band-Pass Transformation

In the previous section, we examined the design of two-path, arbitrary-bandwidth low-pass filters formed from recursive all-pass first- and second-order filters as shown in Figure 11.36. We formed this filter by a transformation of a prototype half-band filter. We now address the second

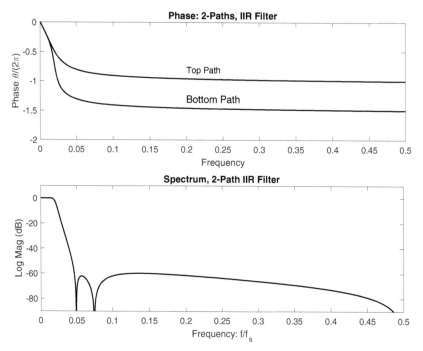

Figure 11.40 Frequency response obtained by frequency transforming half-band filter to frequency 0.02.

transformation, one that performs the low-pass to band-pass transformation. As in the previous section, we can invoke a frequency transformation wherein an existing all-pass subnetwork in a filter is replaced by another all-pass subnetwork. In particular, we now examine the transformation shown in the following equation:

$$\frac{1}{Z} \Rightarrow -\frac{1}{Z}\frac{1 - cZ}{Z - c}; \quad c = \cos(\theta_C), \quad \theta_C = 2\pi\frac{f_C}{f_S}. \qquad (11.28)$$

This, except for the sign, is a cascade of a delay element with the generalized delay element we introduced in the initial discussion of first-order all-pass networks. We can physically replace each delay in the prototype filter with this all-pass network and then tune the center frequency of the low-pass prototype by adjusting the parameter c. For the purpose of this paper, we perform the substitution algebraically in the all-pass filters comprising the two-path pre-distorted arbitrary-bandwidth filter and, in doing so generate

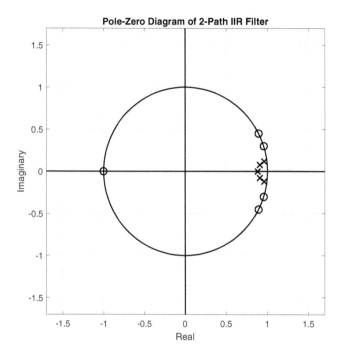

Figure 11.41 Pole-zero diagram obtained by frequency transforming half-band filter to frequency 0.02.

yet a third structure for which we will develop and present an appropriate architecture.

We substitute Equation (11.28) into the second-order all-pass filter derived in Equation (11.27) and rewritten in the following equation:

$$
\begin{aligned}
F(Z) &= G(Z)|_{Z \Rightarrow \frac{Z(Z-c)}{(cZ-1)}} \\
&= \frac{(b^2 + \alpha)Z^2 + 2b(1 + \alpha)Z + (1 + \alpha b^2)}{(1 + \alpha b^2)Z^2 + 2b(1 + \alpha)Z + (b^2 + \alpha)}\Bigg|_{Z \Rightarrow \frac{Z(Z-c)}{(cZ-1)}}.
\end{aligned} \tag{11.29}
$$

After performing the indicated substitution and gathering up terms, we find the form of the transformed transfer function as shown in the following equation:

$$
F(Z) = \frac{1 + d_1 Z + d_2 Z^2 + d_3 Z^3 + d_4 Z^4}{Z^4 + d_1 Z^3 + d_2 Z^2 + d_3 Z + d_4}
$$

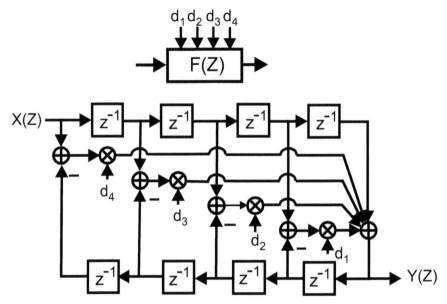

Figure 11.42   Block diagram of general fourth-order all-pass filter.

$$d_1 = \frac{-2c(1+b)(1+\alpha b)}{1+\alpha b^2} \qquad d_2 = \frac{(1+\alpha)(c^2(1+b)^2+2b)}{1+\alpha b^2}$$

$$d_3 = \frac{-2c(1+b)(\alpha+b)}{1+\alpha b^2} \qquad d_4 = \frac{\alpha+b^2}{1+\alpha b^2}.$$

(11.30)

As expected, when we let c $\Rightarrow$ 0, d1 and d2 $\Rightarrow$ 0, while d2 $\Rightarrow$ c1 and d4 $\Rightarrow$ c2, the weights default to those of the prototype, arbitrary-bandwidth, filter. The transformation from low-pass to band-pass generates two spectral copies of the original spectrum, one each at the positive and negative tuned center frequency. The architecture of the transformed filter, which permits one multiplier to simultaneously form the matching numerator and denominator coefficients, is shown in Figure 11.42. Also shown is a processing block $F(Z)$ which uses four coefficients d_1, d_2, d_3, and d_4. This is seen to be an extension of the two-multiply structure presented in Figure 11.35.

We have just described the low-pass to band-pass transformation that is applied to the second-order all-pass networks of the two-path filter. One additional transformation that requires attention is the low-pass to band-pass transformation that must be applied to the generalized delay or bandwidth-transformed delay from the prototype half-band filter. We substitute Equation (11.28) into the first-order all-pass filter derived in Equation (11.25) and

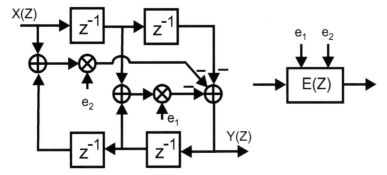

Figure 11.43 Block diagram of low-pass to band-pass transformation applied to low-pass to low-pass transformed delay element.

rewritten in the following equation:

$$E(Z) = \left.\frac{1+bZ}{Z+b}\right|_{Z \Rightarrow \frac{Z(Z-c)}{(cZ-1)}}$$

$$= \frac{(cZ-1)+bZ(Z-c)}{Z(Z-c)+b(cZ-1)} = \frac{-1+c(1-b)Z+bZ^2}{Z^2-c(1-b)Z-b}.$$

$$(11.31)$$

As expected, when c $\Rightarrow$ 1, the denominator goes to $(Z + b)(Z - 1)$, while the numerator goes to $(1 + bZ)(Z - 1)$ so that the transformed all-pass filter reverts to the original first-order filter. The distributed minus sign in the numerator modifies the architecture of the transformed second-order filter by shuffling signs in Figure 11.35 to form the filter shown in Figure 11.43. Also shown is a processing block $E(Z)$, which uses two coefficients e_1 and e_2.

In the process of transforming the low-pass filter to a band-pass filter, we convert the two-multiply second-order all-pass filter to a four-multiply fourth-order all-pass filter and convert the one-multiply low-pass to low-pass filter to a two-multiply all-pass filter. The doubling of the number of multiplies is the consequence of replicating the spectral response at two spectral centers of the real band-pass system. Note that the five-pole, five-zero arbitrary low-pass filter now requires 10 multipliers to form the 10 poles and 10 zeros for band-pass version of the two-path network. This is still significantly less than the standard cascade of first- and second-order canonic filters for which the same 10-pole, 10-zero filter would require 25 multipliers. Figure 11.44 shows how the structure of the prototype is affected by applying the low-pass to band-pass frequency transformation.

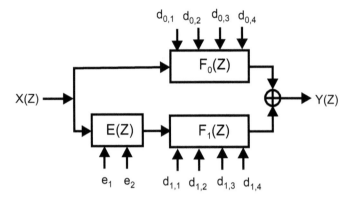

Figure 11.44   Effect on architecture of low-pass to band-pass frequency transformation applied to two-path arbitrary-bandwidth all-pass filter.

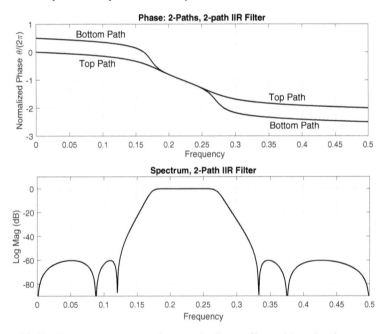

Figure 11.45   Frequency response of two-path all-pass filter subjected to low-pass to low-pass and then low-pass to band-pass transformations.

Figure 11.45 presents the frequency response obtained by applying the low-pass to band-pass frequency transformation to the prototype two-path, four-multiply, low-pass filter presented in Figure 11.38. The one-sided bandwidth was originally adjusted to a normalized frequency of 0.1 and is

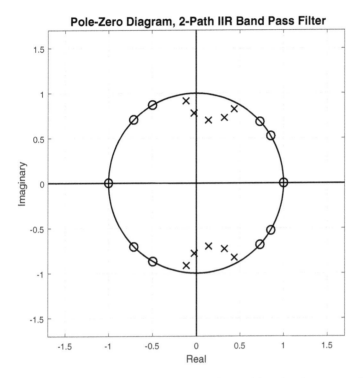

Figure 11.46   Pole-zero plot of two-path all-pass filter subjected to low-pass to low- pass and then low-pass to band-pass transformation.

now translated to a center frequency of 0.22. Figure 11.46 presents the pole-zero diagram of the frequency-transformed prototype filter. The five poles defining the low-pass filter have been pulled to the neighborhood of the band-pass center frequency. The five zeros have also replicated, appearing both below and above the pass-band frequency.

11.6 Multirate Considerations of Recursive Half-Band Filters

The half-band filters can be used in a number of multirate applications to perform up-sampling and down-sampling with imbedded resampling or in a variety of iterated filter applications. Since the half-band filter is formed with polynomials in Z^2, we can invoke the noble identity to slide a 2-to-1 down sampler from the filter output to the filter input or a 1-to-2 up sampler from the filter input to the filter output. This permits us to operate the half-band filter at the lower of the two rates when there is a rate change. The five multiplications required to obtain one data point from a five-coefficient down-sampling half-band filter are distributed over two input samples so that the workload for the half-band filter is 5/2 multiplies per input. Figure 11.47 illustrates the use of the noble identity applied to a 2-to-1 down sampler and Figure 11.48 illustrates the same relationship applied to a 1-to-2 up sampler. In each case, the resampler is slid through the all-pass filters, changing them from polynomials in Z^2 to polynomials in Z. The resampler cannot move through the delay. Recognizing the interaction of the delay and two resampling switches, we replace the three elements with a commutator that

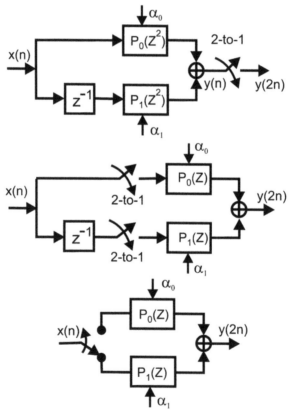

Figure 11.47   Half-band, all-pass polyphase filter with 2-to-1 down sampler moved from output to input and replaced with commutator.

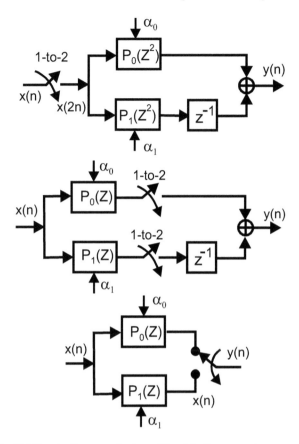

Figure 11.48 Half-band, all-pass polyphase filter 1-to-2 up sampler moved from input to output and replaced with commutator.

delivers successive inputs to the two-path filter for the down sampler or accepts successive outputs from the two paths of the up sampler.

The resampling half-band filter can be applied iteratively in a cascade of half-band filters to obtain sample rate changes that are powers of 2. Thus, three successive stages of half-band filtering can be cascaded to obtain efficient 8-to-1 down-sampling or 1-to-8 up-sampling. The efficiency of the half-band cascade suggests filter applications in which we use the cascade to down-sample, perform desired filtering at a reduced rate, and then up-sample with a second cascade. The following example demonstrates the significant savings due to applications of the half band, two-path recursive all-pass filter.

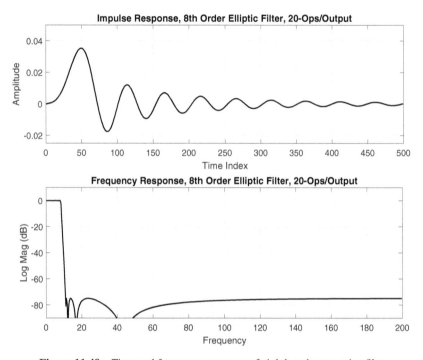

Figure 11.49   Time and frequency response of eighth-order recursive filter.

EXAMPLE 11.1. Applications of Half-Band Filters

We examine a low-pass digital filter that meets the following specifications:

Sample rate	=	400 kHz
Pass-band	=	8 kHz
Stop-band	=	12 kHz
In-band ripple	<	0.1 dB
Stop-band attenuation	>	75 dB

We first try the brute force approach and seek a conventional recursive filter solution. We find from a standard filter design package, such as MATLAB, that an eighth-order elliptic filter meets these specifications. Figure 11.49 presents the impulse response and the frequency response of the elliptic filter.

Figure 11.50 shows a standard implementation of the eighth-order filter as a cascade of four second-order canonic filters sometimes called biquads. Each filter requires five multiplies: two for the feedback path and three for

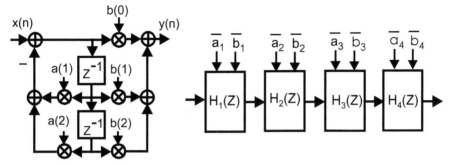

Figure 11.50   Cascade of four second-order canonic filters to form eighth-order elliptic filter.

the scaled feed forward path. The eighth-order filter requires four of these segments; hence, the filter requires 20 multiplies per input–output data pair. This filter is our reference design against which we compare other designs that offer reduced computation.

Our first alternate design uses the two-path all-pass recursive filter initially designed as a half-band with the desired stop-band attenuation and then transformed to the desired low-pass bandwidth by the low-pass to low-pass transformation discussed in Equation (11.27). The attraction of this option is that the system poles are formed by the filter coefficients and the system zeros are free by virtue of the destructive cancelation that occurs as the two paths are summed. Using the MATLAB design routine, *tony_des2*, we learn that a ninth-order two-path filter will meet the design specifications. Note that the two-path filter must always contain an odd number of roots. The structure of this filter is shown in Figure 11.51 where the segments $G(Z)$ and $H(Z)$ are the forms introduced in Figure 11.37. Note that this filter only requires nine multiplies to forms its nine poles and nine zeros. This represents a savings of more than a half relative to our cascade biquads. Figure 11.52 presents the time and frequency response of this two-path filter. Note that the extra zero resides at the half sample rate, a characteristic of the two-path filter.

The next alternate design takes advantage of the fact that the bandwidth of the desired filter is a small fraction of the sample rate. We implement this filter by first reducing the sample rate with a cascade of half-band filters, build the filter that meets the desired specification at the reduced rate, and then increase the sample rate with another cascade of half-band filters. The structure of this cascade is shown in Figure 11.53, where we see that the input filter chain contains three resampling half-band filters that reduce the sample rate 8-to-1 from 400 to 50 kHz. The first stage is a third-order one-multiply

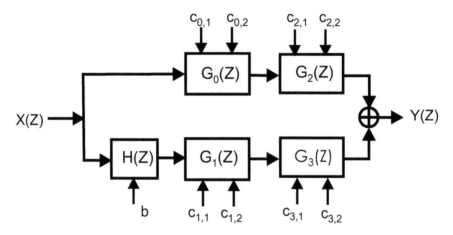

Figure 11.51 Structure of ninth-order two-path recursive all-pass filter.

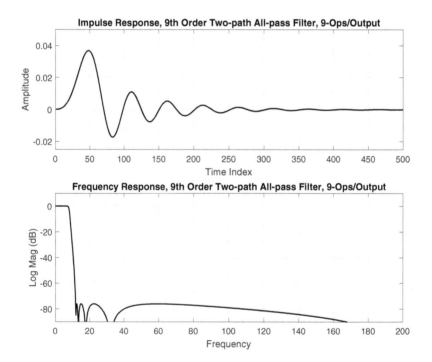

Figure 11.52 Time and frequency response of ninth-order two-path recursive filter.

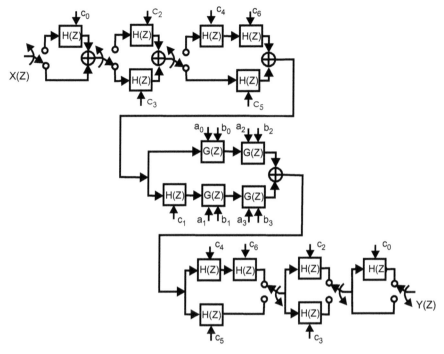

Figure 11.53 Structure of low bandwidth filter as a cascade of down-sampling half-band filters, the low-pass filter, and a cascade of up-sampling half-band filters.

filter, the second stage is a fifth-order two-multiply filter, and the third stage is a seventh-order three-multiply filter. The low-pass filter that performs the bandwidth control is a ninth-order two-path filter that now operates at 50 kHz rather than the original 400 kHz. The output filter chain contains the same filter set as the input chain but is applied as an up-sampling filter in the opposite order.

The easiest way to determine the workload for this resampling filter is to deliver eight input samples to the chain and determine the workload for the entire chain and then divide by the number of input samples. Starting at the input of the entire cascade, we note the first filter will perform four multiplies and deliver four samples to the next stage. The second stage will also perform four multiplies and deliver two samples to its next stage. The third stage will perform three multiplies and deliver one sample to the central filter. This filterer performs nine multiplies and delivers one sample to the start of the up-sample chain. The up-sample chain performs the same

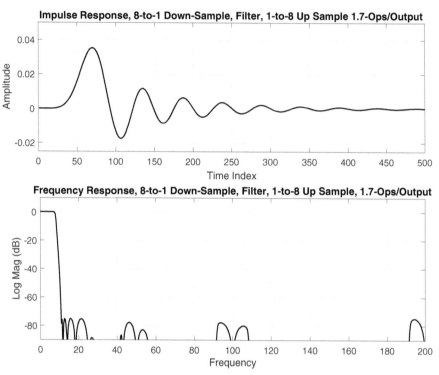

Figure 11.54   Time and frequency response of 8-to-1 down-sampled and 1-to-8 up-sampled ninth-order two-path recursive filter.

workload as the down-sampling chain for which we conclude that the total workload is (4 + 4 + 3 + 9 + 3 + 4 + 4) or 31 multiplies. These 31 multiplies are amortized over eight input–output sample pairs for a workload of 3.875 or approximately four multiplies per input sample. The budget can be visualized as approximately 1.5 multiplies per input point to reduce the sample rate, approximately 1 multiply per sample point to filter the input signal, and approximately 1.5 multiplies per output point to raise the sample rate to the original input rate. Comparing filter workloads, we recall that the original filter requires 20, the two-path requires 9, and the cascade-resampling filter requires 4 multiplies per sample point, which translates into successive savings of about a factor of 2.

Figure 11.54 presents the impulse response and frequency response of the composite resampling chain of filters. The group delay of the chain is seen to be greater than either of the earlier filter options. Figure 11.55

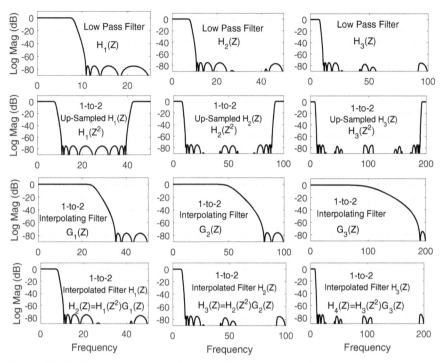

Figure 11.55 Frequency response for first of three segments of successive filtering: zero-packed bandwidth limiting filter and interpolating filter.

is a set of spectra showing the spectral transformations performed by the filtering chain. This explicitly shows the spectra obtained by zero insertion up-sampling followed by filtering to remove the spectral replicate due to the up-sampling operation. The figure presents the frequency response of the filter sequence starting at the low-pass central filter and then that of the zero-packed up-sampled version, the half-band filter following the up-sampling, and the result of passing through the half-band filter. The sequence starts in the upper left and proceeds by successive columns to the lower right with the figure at the end of one column becoming the start of the next column. Remember, the process we describe in Figure 11.53 does not separately zero-pack and filter, so these spectra are not actually available in the processing chain but are shown to offer insight to the sequence of operations.

11.7 Hilbert-Transform Filter Variant of Two-Path All-Pass Filter

A trivial variant of the two-path all-pass filter leads to a filter with wide application. This is the Hilbert-transform filter that is formed by a trivial modification to the basic two-path half-band filter. We can translate the half-band filter centered at baseband to another half band filter centered at the quarter sample rate. The simple transformation is related to the modulation theorem for Z-transforms. The theorem states that the Z-transform of a heterodyned time series is the Z-transform of the original time series with the variable Z replaced with a phase rotated Z. This relationship is illustrated in Equations (11.32)−(11.34). In Equations (11.32) and (11.33), we note the Z-transforms of sequences $h(n)$ and of the heterodyned sequence $h(n)\exp(j\theta_0 n)$. Here, we see that the phase rotation of the time series heterodyne has the same exponent as the position index exponent, and we are free to associate the phase rotator with the delay rather than with the time samples. This is shown in Equation (11.34).

$$H(Z) = \sum_{n=0}^{\infty} h(n)Z^{-n}$$

$$= h(0) + h(1)Z^{-1} + h(2)Z^{-2} + \cdots + h(k)Z^{-k} + \cdots \tag{11.32}$$

$$G(Z) = \sum_{n=0}^{\infty} h(n)e^{j\theta n}Z^{-n}$$

$$= h(0) + h(1)\,e^{j\theta}Z^{-1} + h(2)\,e^{j2\theta}Z^{-2} + \cdots + h(k)\,e^{jk\theta}Z^{-k} + \cdots \tag{11.33}$$

$$G(Z) = H(Z)|_{Z\Rightarrow e^{-j\theta}Z} \cdot$$

$$= H(e^{-j\theta}Z) \tag{11.34}$$

If we heterodyne the impulse response of a filter to the quarter sample rate, the phase rotator is $\exp(j\,\pi/2)$, which we can write more compactly as the operator j. Thus, we can replace each Z^{-1} in a Z-transform with jZ^{-1} to reflect the heterodyne of the time series. We now apply this transformation to the transfer function of the half band filter. Since the filter is formed by polynomials in Z^{-2}, the substitution is equivalent to replacing each Z^{-2} with $-Z^{-2}$. The effect of this transformation on the all-pass sections in the two-path filter is shown in Equation (11.35). Except for a sign change on the transfer function, the phase substitution on the delays Z^{-2} is equivalent to a sign change applied to the coefficients of the all-pass stages. The negative sign of the transfer function is canceled if there is an even number of stages in a filter path.

The one exception to the filter coefficient absorbing the sign change occurs at the single delay in the lower leg of the two-path filter. Here we must associate the *j* operator with the delay. This means that we declare the second leg of the two-path filter to be the imaginary part of the time series. This means, in turn, that the summation performed at the output of the two paths is no longer a true sum, but rather an association, pairing the real and imaginary outputs as a complex sample of the filter.

$$H(Z^2) = \frac{1 + \alpha Z^2}{Z^2 + \alpha} \Rightarrow G(Z) = H(-Z^2) = -\frac{1 - \alpha Z^2}{Z^2 - \alpha} \qquad (11.35)$$

Figure 11.56 shows how the transformation is applied to the signal flow representation of the half-band filter. Note the sign changes and the assignment of the second path to the imaginary output component. Here, we continue to invoke the option of down-sampling the output of the half-band filter by a factor of 2-to-1 in concert with the filter's bandwidth reduction. Since the filter still contains polynomials in Z^2, we can use the noble identity to interchange the order of filtering and down-sampling which results in the compact form shown in the bottom right of the figure.

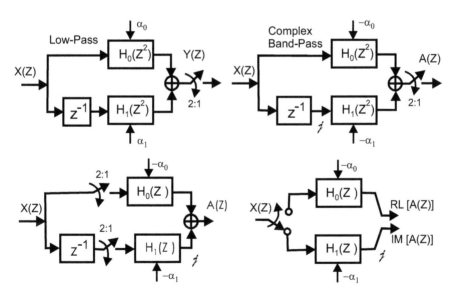

Figure 11.56 Low-pass half-band filter to positive frequency half-band filter.

The Hilbert-transform filter has always held the position of being our first exposure to a polyphase filter structure. In fact, it is a two-path filter, the precursor of the general M-path structure. It uses a combination of frequency-dependent phase shift and time domain phase rotation to obtain destructive cancelation of signals over selected frequency bands. The Hilbert transform cancels the negative frequency segment of the spectra arranging for a broadband frequency-dependent 90° phase shift between the two arms of the filter. The time domain phase shift applied through the j operator rotates the entire spectral response another 90° to effect a 180° phase difference along the two paths over the negative frequency band.

We have two basic options on how to achieve the frequency-dependent 90° phase shift. The minimum resource option places recursive first-order all-pass polynomials in Z^2 in both arms and uses the lazy-S phase contours to obtain the desired phase difference between the two arms. While this option realizes the desired phase difference between the two arms, the phase slope between input and output of the filter is not linear. There are many applications in which linear phase is not a requirement. Automatic gain control is one and digital receivers containing an equalizer are another. If linear phase is a requirement, we expend additional resources, restrict the upper path to contain only delay elements, and have the lower path supply the desired phase difference.

Figure 11.57 illustrates the phase structure of a non-uniform and a uniform phase filter designed for the same 5% transition bandwidth and 60 dB out-of-band attenuation. The two filters require four coefficients and seven coefficients respectively to meet these specifications. Figure 11.58 presents the pole-zero diagrams for the two options. We see here the effect of the center frequency transformation on the system roots. Replacing Z^{-2} with $-Z^{-2}$ in the path filters rotated the pole positions 90°, which is easily seen in the left figure where the poles have been moved from Y-axis to the X-axis. Similarly, applying the j operator to the lower path rotated the zero positions the same 90°. Compare the pole-zero diagram of Figure 11.58 with Figure 11.22.

Figure 11.59 presents the frequency and phase response of the two Hilbert-transform variants. The in-band phase profiles are inherited from the corresponding phase profiles of the two paths as seen in Figure 11.57. We easily see the stop-band zero locations. When used in the resampling mode, the four-coefficient, nine-pole/zero non-uniform phase filter requires two multiplies per input point to cancel the negative frequencies, while the seven-coefficient, 27-pole/zero uniform phase filter requires 3.5 multiplies per input point. For comparison, consider a 147-tap true half-band FIR filter

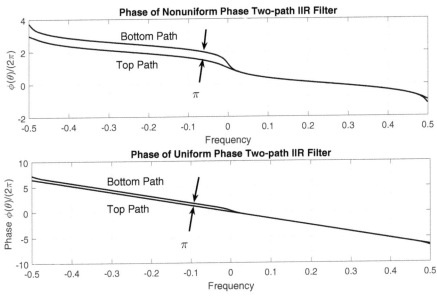

Figure 11.57 Path phases of two-path recursive all-pass, non-uniform phase, and uniform phase Hilbert-filters.

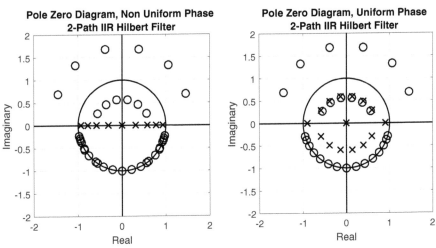

Figure 11.58 Pole-zero diagrams for non-uniform and uniform phase Hilbert-transform filters.

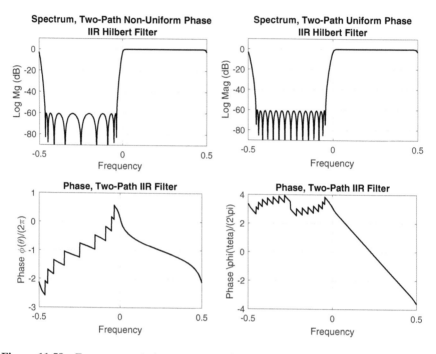

Figure 11.59   Frequency and phase response of non-uniform and uniform phase two-path recursive all-pass Hilbert-transform filters.

that would be required to meet the same specification. When partitioned into a 2-to-1 polyphase filter, the upper path would contain the even-indexed sample points which implements the delay to the center tap of the filter, while the lower path would contain the 74 symmetric odd-indexed samples. The filter would require 18 multiplies per input point if the architecture used the coefficient symmetry or 36 multiplies per input point if not. Thus, the best linear phase FIR, the linear phase IIR, and the non-linear phase filters would require 18, 3.5, and 2 multiplies, respectively, per input point to implement comparably performing Hilbert-transform filters.

11.8 M-Path Recursive All-Pass Filters

This chapter started with a discussion of M-path recursive all-pass filters. We then introduced recursive all-pass filter structures and demonstrated their use in a rich subset of M-path filters, two-path filters, and variants related to them.

We now return to the M-path filter that can be thought of as an extension of the two-path structure. We presented the standard structure of the M-path filter in Figure 11.2. In the case of non-linear phase, the paths contain a delay and one or more all-pass filters formed as first-order polynomials in Z^{-M}. In the case of linear phase, the first path is only delay, and the remaining paths contain a delay and one or more all-pass filters formed as first- and second-order polynomials in Z^{-M}. The polynomials in Z^{-M} enable the use of the noble identity, the interchange of the resampler, and the M-stage filter prior to the filtering operation rather after.

As a specific example to demonstrate the characteristics of the recursive all-pass M-path filters, we have designed and programmed a pair of eight-path filters with the same pass band and attenuation characteristics of the composite three-stage half-band filter discussed and presented in Figure 11.49. The specifications, recast in normalized frequency units, that the filter must satisfy are that pass-band frequency is 0.0625 (1/16), stop-band frequency is 0.09375 (1.5/16), pass-band ripple of less than 0.1 dB, and stop-band attenuation of at least 75 dB. The design routine for the non-uniform phase filter determined that these specifications required two stages in paths 0, 1, 2, 3, 4, and 5, and one stage in paths 6 and 7, each stage being a first-order polynomial in Z^{-8}. Thus, 14 coefficients define this eight-path filter. Remarkably, since these coefficients describe 14 first-order polynomials in Z^{-8}, the composite filter contains 112 pole-zero pairs. In like fashion, the design routine for linear phase filter determined that these specifications required 24 delays in path 0 and both a first-order and a second-order stage in paths 1, 2, 3, 4, 5, 6, and 7. These stages are polynomials in Z^{-8}. Here, 21 coefficients define the eight-path filter. Since these coefficients describe the equivalent of 24 first-order polynomials in Z^{-8}, the composite filter contains 192 pole-zero pairs.

The poles of the polynomials in Z^{-8} have eight-fold symmetry around the unit circle, and as we traverse the unit circle, we observe rapid changes in phase as we pass these poles. The spectral regions in the vicinity of the poles represent transition bands. These phase change regions are easily seen in Figure 11.60, which presents the phase profiles of the eight paths of the two eight-path filters. Between the transitions, the separate phase profiles differ by constant phase differences. In the five intervals seen in the positive frequency span, the phase differences are 0°, 45°, 90°, 135°, and 180°. The three unseen intervals in the negative frequency span have phase differences of –45°, –90°, and –135°. When the signals in these eight paths are summed, they constructively add in the 0° difference frequency interval (the pass-band) and

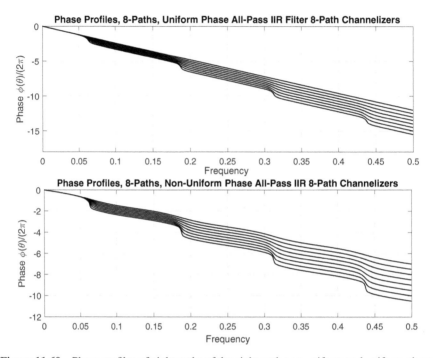

Figure 11.60   Phase profiles of eight paths of the eight-path non-uniform and uniform phase polyphase filter.

destructively add in the remaining intervals (the stop bands). The transition regions between the bands represent transition regions, either transitioning from pass band to stop band or from stop band to stop band.

Figure 11.61 shows the pole-zero diagrams for the pair of eight-path filters, and Figure 11.62 shows a detailed section of the unit circle near the first set of poles. We see the zeros on the unit circle that define the multiple stop bands as well as the pole clusters on the eight roots of $-\alpha$ formed by the first-order polynomials in Z^{-8}. We also see the cluster of zeros alongside the pole clusters shielding the stop-band intervals from the pole clusters. Figures 11.63 and 11.64 present the frequency response, a pass-band detail of the frequency response, the impulse response, and the pass-band phase response for the two versions of the eight-path recursive filter. The first characteristic that strikes our attention is the multiple transition bands between successive stop-band intervals. We explained this as the traversal of the pole clusters near the unit circle at the roots of the first-order polynomials

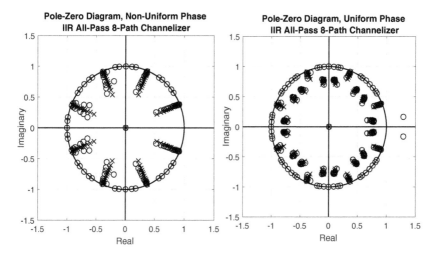

Figure 11.61 Pole-zero plots of composite eight-path recursive non-uniform and uniform phase filters.

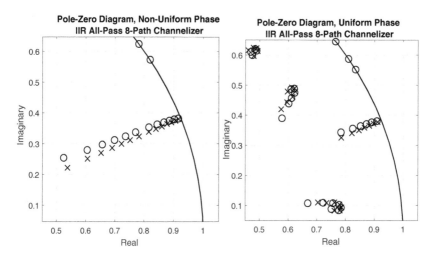

Figure 11.62 Details of pole-zero plots of composite eight-path filters.

in Z^{-8}. When the output of these filters is maximally decimated (8-to-1), the multiple transition bands alias to a common interval, the transition interval from the pass band to stop band.

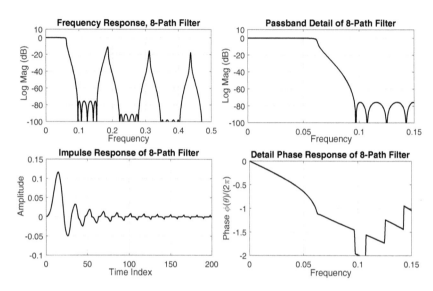

Figure 11.63   Frequency, impulse, and phase responses of non-uniform phase eight-path filter.

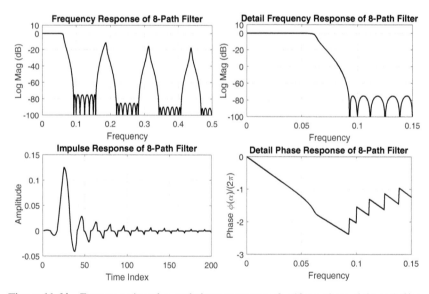

Figure 11.64   Frequency, impulse, and phase responses of uniform phase eight-path filter.

11.9 Iterated Half-Band Filters

The final architecture we consider for the all-pass structures is that of an iterated recursive filter. The iterated filter concept is useful when the filter bandwidth is a small fraction of the signal sample rate. Rather than designing the filter at the input sample rate, we design it at a fraction of the input rate, which often leads to shorter filters and always results in filters exhibiting reduced sensitivity to finite arithmetic. This reduced sensitivity is the result of using lower Q-poles in the recursion process. The implementation of the iterated filter involves designing the filter at a reduced sample rate and then zero-packing its impulse response and interpolating up to the desired higher sample rate with post-processing filters. This iterative filter up-samples the filter impulse response to the higher sample rate rather than down-sampling the data to the reduced sample rate as we did earlier.

To demonstrate the iterative filter process, let us again consider a filter meeting the specifications outlined in the previous section. These specifications, in normalized frequency units, are that pass-band frequency of 0.0625 (1/16), stop-band frequency of 0.09375 (1.5/16), pass-band ripple less than 0.1 dB, and stop-band attenuation at least 75 dB. We will perform the design for a sample rate one-fourth of the original sample rate but operate the filter at the system rate by 1-to-4 zero-packing and then remove the spectral replicates with two half-band filters. The structure of the iterative design is shown in Figure 11.65. We determined that a seventh-order input filter was required to satisfy the designed goals when cast as a half-band filter with pass band at 0.25 (1/4) and stop band at 0.375 (1.5/4). This filter requires three coefficients. Since the half-band filter is designed as polynomials in Z^{-2}, the 1-to-4 zero-packing converts it to polynomials in Z^{-8}. The second-stage filter is also a seventh-order, three-coefficient half-band filter that is

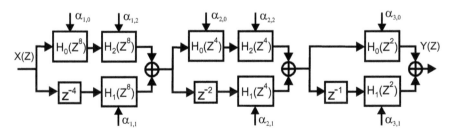

Figure 11.65   Iterated filter, up-sampled 1-to-4 by zero-packing, and then filtered to suppress spectral replicates.

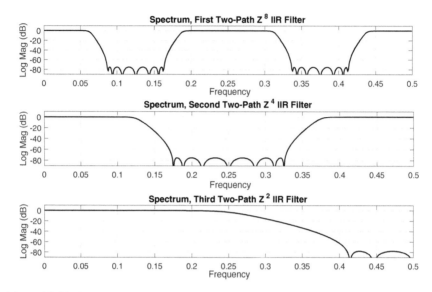

Figure 11.66   Spectra of 1-to-4 zero-packed first stage, 1-to-2 zero-packed second stage, and non-zero-packed third stage of iterated filter.

zero-packed 1-to-2. Finally, the third stage is a fifth order, two-coefficient half-band without zero-packing. The spectral responses of these filters are shown in Figure 11.66 and the composite spectra at successive filter outputs are shown in Figure 11.67. The spectra shown in each panel are the output spectra from the previous stage, which is the input to the current stage and the spectral response of the current stage.

11.9.1 Final Comparisons

We can now compare the workload to reduce the bandwidth and sample rate by a factor of 8-to-1 for various resampling options. Our reference design was a standard eighth elliptic filter implemented as a cascade of four biquadratic stages that required 20 multiplies per input data point. We then considered a two-path tuned half-band filter that required a ninth-order half-band prototype that required nine multiplies per data point, resulting in more than a 2-to-1 savings. We then examined the option of a cascade of three half-band filtering stages and found the total workload to be approximately 4.0 multiplies per input. Continuing with other option, we looked at the eight-path resampling filters. The non-uniform phase eight-path

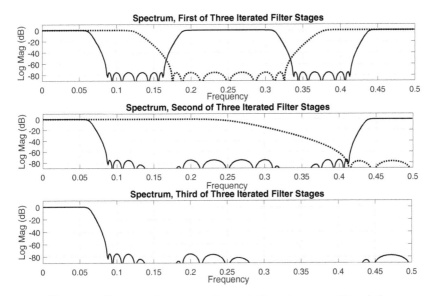

Figure 11.67 Spectral response of cascade filters in iterated filter chain.

filter requires 14 multiples per eight input data samples which is equivalent to 1.8 multiplies per input while the uniform phase eight-path filter requires 21 multiplies per eight input data samples which is equivalent to 2.7 multiplies per input.

For comparison, an eight-path polyphase partitioned FIR filter with the same spectral characteristics requires 13 multiplies per input sample point.

Finally, we looked at the iterated half-band filter option that required eight multiplies per input point. We conclude that the eight-path non-uniform phase and the eight-path uniform phase polyphase implementation are ranked first and second in the most efficient of the options for a single channel-filtering task. The eight-path implementations become even more efficient when the polyphase partition is used as a channelizer. Suppose, for instance, five channels centered baseband $(-2/8, -1/8, 0, 1/8, 2/8)$ are to be down-converted by a set of phase rotators following the path processing. The phase rotators are nearly trivial for this example but could be applied with an eight-point fast Fourier transform with three of the terms discarded. Now the 1.8 or 2.7 multiplies per input point is amortized over five-output points for a filter processing 0.36 or 0.54 multiplies per input per channel.

References

Ansari, R. and B. Liu "Efficient Sampling Rate Alteration Using Recursive (IIR) Digital Filters", IEEE Trans. On Acoustics, Speech, and Signal Processing, Vol. 31, Dec. 1983, pp. 1366-1373.

Drews, W. and L. Gaszi, "A New Design Method for Polyphase Filters Using All-Pass Sections", IEEE Trans. On Circuits and Systems, Vol. 33, Mar. 1986, pp 346-348.

harris, fred. "On the Design and Performance of Efficient and Novel Filter Structures Using Polyphase Recursive All-Pass Filters", Keynote Presentation ISSPA-92 (International Symposium on Signal Processing) Gold Coast, Australia 16–21 August 1992

harris, fred, and Eric Brooking, "A Versatile Parametric Filter Using an Imbedded All-Pass Sub-Filter to Independently Adjust Bandwidth, Center Frequency, and Boost or Cut", 95th Audio Engineer- ing Society (AES) Convention, New York, New York, 7-10 October 1993

harris, fred, Bob Caulfied, and Bill McKnight, "Use of All-pass Networks to Tune the Center Fre- quency of Sigma Delta Modulators", 27th Asilomar Conference on Signals, Systems, and Com- puters, Pacific Grove, CA., 31 October to 3 November 1993

harris, fred, Itzhak Gurantz, and Shimon Tzukerman "Digital (T/2) Low pass Nyquist Filter Using Recursive All-pass Polyphase Resampling Filter for Signaling Rates 10 kHz to 10 MHz", 26th

Annual Asilomar Conference on Signals, Systems, and Computers, Pacific Grove, CA., 26–28 October 1992

harris, fred, Maximilien d'Oreye de Lantramange and A.G. Constantinides, "Design and Implementa- tion of Efficient Resampling Filters Using Polyphase Recursive All-Pass Filters", 25th Annual Asilomar Conference on Signals, Systems, and Computer, Pacific Grove, CA., 4–6 November 1991

Jovanovic-Dolecek, Gordana. *Multirate Systems: Design and Applications*, London, Idea Group, 2002. Renfors, M. and T. Saramaki, "Recursive N-th Band Digital Filters, Parts I an II", IEEE Trans. CAS, Vol. 34, pp. 24-51, January 1987.

Vaidyanathan, P. P. *Multirate Systems and Filter Banks,* Englewood Cliff, NJ, Prentice-Hall Inc., 1993.

Valenzuela R. A. and Anton Constantinides, "Digital Signal Processing Schemes for Efficient Interpo- lation and Decimation", IEE Proceedings, Dec 1983

Problems

11.1 Verify that any transfer function of the form shown in the equation is an all-pass network. That is, $|H(\theta)| = 1$ for all θ. Sketch the location of the numerator and denominator roots for $M = 1, 2, 3,$ and 4.

$$H(Z) = \frac{(1 - \alpha Z^M)}{(Z^M - \alpha)}.$$

11.2 Group delay is the phase slope $d(\phi)/d(\theta)$ of a transfer function. Determine the group delay of a first-order all-pass network shown here. Plot the group delay for various values of the parameter

$$H(Z) = \frac{(1 - \alpha Z)}{(Z - \alpha)}.$$

11.3 Form a transfer function $H(Z)$ shown next as the sum of the two trivial all-pass networks $P_0(Z)$ and $P_1(Z)$. Comment on the root locations and determine the scale factor required to obtain a unity gain filter for arbitrary value of the parameter α.

$$P_0(Z) = \frac{1 - \alpha Z}{Z - \alpha}, \quad P_1(Z) = 1$$
$$H(Z) = P_0(Z) + P_1(Z).$$

11.4 Form a transfer function $H(Z)$ shown next as the sum of the two trivial all-pass networks $P_0(Z)$ and $P_1(Z)$. Comment on the root locations as functions of the parameters α and β and determine the scale factor required to obtain a unity gain filter for arbitrary value of the parameters α and β.

$$P_0(Z) = \frac{1 - \alpha Z}{Z - \alpha}, \quad P_1(Z) = \beta$$
$$H(Z) = P_0(Z) + P_1(Z).$$

11.5 Form a transfer function $H(Z)$ shown here as the sum of the two trivial all-pass networks $P_0(Z)$ and $P_1(Z)$. Determine and plot the root locations as a function of the parameter α.

$$P_0(Z) = \frac{1 - \alpha Z^2}{Z^2 - \alpha}, \quad P_1(Z) = \frac{1}{Z}$$
$$H(Z) = P_0(Z) + P_1(Z).$$

11.6 Form a transfer function $H(Z)$ shown here as the sum of the two all-pass networks $P_0(Z)$ and $P_1(Z)$. Determine the range of α_0 and α_1 for which the numerator roots are confined to the unit circle.

$$P_0(Z) = \frac{1 - \alpha_0 Z^2}{Z^2 - \alpha_0}, \quad P_1(Z) = \frac{1}{Z}\frac{1 - \alpha_1 Z^2}{Z^2 - \alpha_1}$$

$$H(Z) = P_0(Z) + P_1(Z).$$

11.7 Use the MATLAB design program *tony_des* to design a recursive half-band low-pass filter with stop-band normalized frequency edge equal to 0.30 and with at least 60 dB stop-band attenuation. Design an IIR Tchebyschev-II filter with same 3-dB edge, stop-band edge, and stop-band attenuation. Form and compare the frequency response of the two filters. Comment on their relative implementation complexity.

11.8 Use the MATLAB design program *tony_des* to design a recursive half-band low-pass filter with stop-band normalized frequency edge equal to 0.30 and with at least 60 dB stop-band attenuation. Design a Remez-based FIR filter with same 3-dB edge, stop-band edge, and stop-band attenuation. Form and compare the frequency response of the two filters. Comment on their relative implementation complexity.

11.9 Use the MATLAB design program *Lineardesign* to design a recursive linear phase half-band low-pass filter with stop-band normalized frequency edge equal to 0.30 and at least 60 dB out-of-band attenuation. Design a Remez-based FIR filter with same 3-dB edge, stop-band edge, and stop-band attenuation. Form and compare the frequency response of the two filters. Comment on their relative implementation complexity.

11.10 Use the MATLAB design program *tony_des* to design a recursive low-pass filter with pass-band and stop-band normalized frequency edges equal to 0.1 and 0.2, respectively, and with at least 60 dB stop-band attenuation. Design an IIR Tchebyschev-II filter with same 3-dB edge, stop-band edge, and stop-band attenuation. Form and compare the frequency response of the two filters. Comment on their relative implementation complexity.

11.11 Use the MATLAB design program *tony_des* to design a recursive low-pass filter with pass-band and stop-band normalized frequency edges equal to 0.1 and 0.2, respectively, and with at least 60 dB stop-band attenuation. Design a Remez-based FIR filter with same 3-dB edge, stop-band edge, and stop-band attenuation. Form and compare the frequency response of the two filters. Comment on their relative implementation complexity.

11.12 We want to design a low-pass filter with the following specifications:

Sample rate	100 kHz		
Pass-band	0-to-5 kHz	In-band ripple:	<0.1 dB
Stop-band	8-to 50 kHz	Stop-band atten.	60 dB

Use *Tony_des* to design a recursive non-uniform phase filter that satisfies this specification.

We now implement a cascade of two resampling half-band filters and a final spectral shaping filter to achieve the same filter specifications. Design the pair of half-band 2-to-1 resampling filters and the final output stage. What is the relative work between the two implementations?

12

Cascade Integrator Comb Filters

3-stage M-to-1 Down-sampling CIC Filter

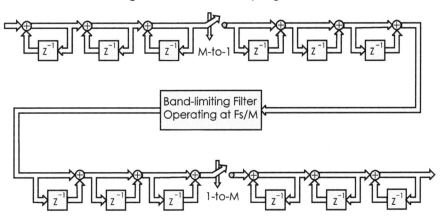

3-stage 1-to-M Up-sampling CIC Filter

Digital filters are formed by a standard set of resources, memory or delays, summing junctions, multipliers, and resamplers. Architectures that combine these resources build a variety of filtering systems. When faced with two or more contending filtering architectures to solve a given problem, we compare their relative cost as well as their relative performance. In the early days of digital signal processing (DSP), the number of multiplies and data transfers per data sample were standard cost measures. Another comparative cost is sensitivity to finite precision coefficients and required register and accumulator widths. A delight to Application Specific Integrated Circuit (ASIC) designers, and to a lesser extent to Field Programmable Gate Array (FPGA) designers, is a class of filters that does not require multipliers, requires few data transfers, and further exhibits an easy to derive relationship between filter specifications and accumulator widths. These filters are variants of a structure based on the sliding average filter. This chapter presents these filters and discusses their performance under finite arithmetic.

12.1 A Multiply-Free Filter

A filter with a rectangle-shaped impulse response is called a boxcar or a sliding average filter. It is a simple finite impulse response (FIR) filter with unit-valued coefficients that performs a filtering task without multiplies. The quality of the filtering is actually not very good since the spectral response of the boxcar filter, as shown in Equation (12.1) and illustrated in Figure 12.1, exhibits only 13 dB attenuation. The simplicity of one form of implementation is so attractive that we are drawn to this filter even though the filtering performance is not very good. We will simply have to find a way to improve its performance.

$$H(\theta) = \frac{\sin(\theta\frac{M}{2})}{\sin(\theta\frac{1}{2})}. \tag{12.1}$$

The FIR filter implementation of the boxcar filter is shown in Figure 12.2.

Conceptually, the filter computes an output as follows: an input arrives and the data in the register shifts one place to the right to accommodate the new arrival. The filter forms the sum of the contents of the registers and outputs this sum. This is shown in the following example:

$$y(n) = \sum_{k=0}^{M-1} x(n-k). \tag{12.2}$$

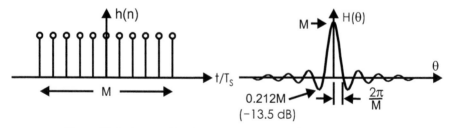

Figure 12.1   Boxcar filter impulse response and frequency response.

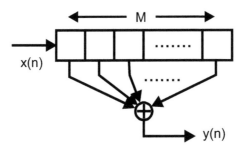

Figure 12.2 FIR filter implementation of boxcar filter.

The process is repeated upon the arrival of each new input sample. If we think about it, this is not a very efficient implementation of the filter. When the new input arrives, the shift to the right of the register contents discards the sample that had arrived M samples ago to make room for the latest input sample. The new output sum differs from the previous sum by the addition of the new data point and the removal of the M-sample-old data point. A recursive form of the boxcar filter can be implemented by altering the previous sum by the known difference. This form of the sum is shown in the following equation:

$$y(n) = \sum_{k=0}^{M-1} x(n-k) = x(n) - x(n-M) + \sum_{k=0}^{M-1} x(n-1-k) \quad (12.3)$$
$$= [x(n) - x(n-M)] + y(n-1).$$

This form of the filter is known as the cascade integrator comb (CIC). The block diagram of this filter is shown in Figure 12.3. The integrator is the "I" in the name CIC, and it is easily identified as the recursive accumulator. By default, the M units of delay and the subtraction preceding the integrator must be the comb filter, the second "C" in the name CIC. This section has a simple impulse response of 1 at time index 0 followed by a -1 at time index M. The Z-transform of this impulse response is shown in Equation (12.4). The reason this filter is called a comb filter can be traced to its frequency response, which as shown in Equation (12.5), is a sinusoid. The magnitude response of the filter, shown in Figure 12.4, has the appearance of a rectified sine wave with M periodic zeros spanning the frequency axis. The periodic zeros remind us of the teeth of a comb, and hence the name, comb filter.

$$C(Z) = 1 - Z^{-M} \quad (12.4)$$

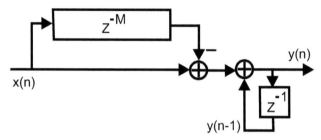

Figure 12.3   Cascade integrator comb filter.

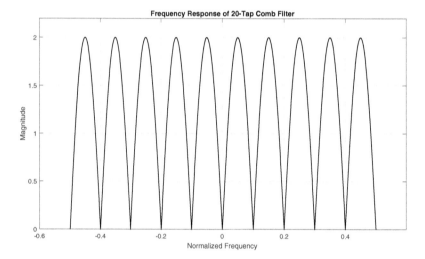

Figure 12.4   Frequency response of 10-tap comb filter.

$$C(\theta) = 1 - e^{-jM\theta}$$
$$= e^{-j\frac{M}{2}\theta}[e^{+j\frac{M}{2}\theta} - e^{-j\frac{M}{2}\theta}] \tag{12.5}$$
$$= 2j\, e^{-j\frac{M}{2}\theta} \sin(\frac{M}{2}\theta).$$

Note from Figure 12.3 that in this form of the filter, only two summations are required to form the filter output, while in the direct implementation, indicated in Figure 12.2, M additions are required to form the filter output. This is valid for any M. We are thus able to replace 100 adds with 2 adds if we implement the boxcar filter as a CIC filter. Both implementations require the same amount of memory, but modifications yet to come will reduce the memory required for the CIC.

An alternate derivation of the CIC structure is available from the Z-transform of the boxcar filter's impulse response. This is shown in Equation (12.6), which we recognize as the first M terms of the geometric series as shown in Equation (12.7).

$$H(Z) = \sum_{n=0}^{M-1} Z^{-n} \tag{12.6}$$

$$H(Z) = 1 + Z^{-1} + Z^{-2} + \cdots + Z^{-(M-2)} + Z^{-(M-1)}. \tag{12.7}$$

The closed form of this series is shown in the following equation:

$$H(Z) = \frac{1 - Z^{-M}}{1 - Z^{-1}}. \tag{12.8}$$

Equation (12.8) can be cast as a polynomial in Z as in the following equation:

$$H(Z) = \frac{1}{Z^{(M-1)}} \frac{Z^M - 1}{Z - 1}. \tag{12.9}$$

In this form of the Z-transform, we can easily see that the zeros, the numerator roots, are uniformly distributed around the unit circle at the M roots of unity and that the zero at $z = 1$ is canceled by the pole, a denominator root, at the same location. This pole zero cancelation is certainly valid in the ratio of polynomial representation of the filter but may not be in the following partition.

Returning to Equation (12.8), we perform a questionable step and then examine the consequences of that maneuver. We treat the numerator and denominator of Equation (12.8) as if they reside in different filters. Normally, we are free to do this. What we have to question here is: are we able to separate the numerator and denominator when they contain canceling roots? Technically, we cannot place canceling roots in separate filters. I once had a student who got angry with me for separating them. We will shortly see why we are not permitted do so and then find a work around. The separated form of the filter is shown in the following equation:

$$H(Z) = [1 - Z^{-M}] \cdot \left[\frac{1}{1 - Z^{-1}} \right]. \tag{12.10}$$

An obvious, but, as just mentioned, perhaps incorrect, conclusion to be drawn from this partitioned form of the boxcar filter is that the filter can be implemented as the cascade of two filters, an integrator and a comb. This is the structure shown in Figure 12.3, a structure arrived at from a different perspective.

Figure 12.5 presents two filter structures that we now compare to the original boxcar filter. The two structures are reordered versions of the comb filter and the integrator. This reordering is standard fare for linear systems since linear time invariant systems commute. Shown by each of the three filters is its impulse response. The impulse response of the tapped delay line version of the boxcar is simple and obvious.

The impulse response of the comb and integrator cascade is formed in two phases. We first see the comb response, which is a sequence of two impulses separated by M samples. The first impulse applied to the integrator circulates around the integrator, outputting the same unit-valued samples. When the second impulse arrives at the integrator, it cancels the circulating first impulse and the integrator response goes to zero. The composite response of the cascade filters matches the response of the boxcar. No surprise yet!

The impulse response of the integrator and comb cascade is also formed in two phases. We first see the integrator response, which is a recirculating copy of the input impulse that is a step or DC response. This step is delivered

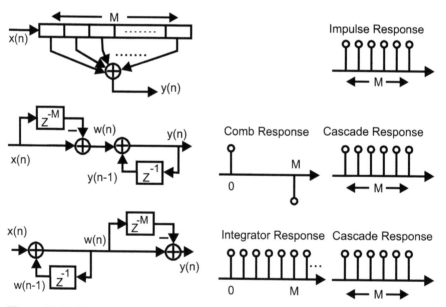

Figure 12.5 Structures and associated impulse responses of boxcar, cascade comb, and integrator, and cascade integrator and comb.

directly to the output and matches the boxcar response. M samples into the response, and an M unit delayed version of the output step arrives at the output summing junction where they subtract from the direct path output of the integrator and set each output sample to zero. This output sequence also matches the response of the boxcar filter. If an observer joined us now and examined only the input and output of this filter, he/she would see zero input and zero output and might conclude that the filter is at rest, a standard attribute of a stable system. In fact, it is not; it is circulating the original input impulse in the integrator and delivering pairs of samples via the comb filter to sum to zero and give the impression the system is at rest. What we are witnessing is the effect of an internal state of the system, the integrator that is neither observable nor controllable from external ports. Remember that this state did not exist in the original boxcar filter; it is the denominator root canceled by a zero of the numerator.

Figure 12.6 presents the same filter structures we just examined, but this time, we compare their responses to an input step. Shown by each of the three filters is its step response. The step response of the tapped delay line version of the boxcar contains two distinct intervals. The first interval of length M is a ramping transient response that lasts until the filter is filled. The second interval is the steady-state response to the step of constant level M.

The step response of the comb and integrator cascade is also formed in two phases. We first see the comb response, which is a square wave sequence of length-M samples. The square wave forms the transient ramp in the integrator. At the end of M samples, the comb response goes to zero and the integrator stops accumulating inputs. At this point, the integrator output has reached steady state and circulates and outputs its constant level of M. The composite response of the cascade filters matches the response of the boxcar. Still no surprise yet!

The step response of the integrator and comb cascade is also formed in two phases. We first see the integrator response, which is a ramp sequence formed by integrating the input step. This ramp is delivered directly to the output as the start of the boxcar response. M samples into the response, an M unit delayed version of the output ramp arrives at the output summing junction where they subtract from the direct path delivered by the integrator to form a constant difference of amplitude M for each output sample. This output sequence also matches the response of the boxcar filter. If an observer joined us now and examined only the input and output of this filter, he/she would see a finite input and a finite output and conclude that the filter is stable, an alternate definition of a stable system. In fact, it is not stable in this

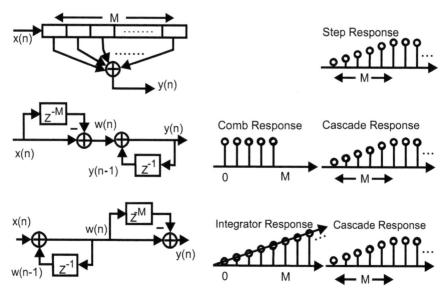

Figure 12.6 Structures and associated step responses of boxcar, cascade comb, and integrator, and cascade integrator and comb.

sense since the integrator is on the way to infinity. The difference between the ramp and the *M* unit delayed ramp formed in the comb filter is a constant even as the ramp climbs without bound. Here again, we are witnessing the effect of an internal state of the system that is neither observable nor controllable. Any finite state machine that tries to implement this structure is bound to overflow its accumulator.

12.2 Binary Integers and Overflow

We have identified a concern about register overflow in a CIC filter. To help understand the overflow, we now review a number of ways binary numbers are used to represent integers. Equation (12.11) presents the standard notation for binary numbers.

$$N = \sum_{i=0}^{b-1} a_i 2^i; \quad a_i = 0, 1. \tag{12.11}$$

Standard notation assigns the least significant bit (LSB) to the right so that Equation (12.11) can be written specifically as in the following equation:

$$N = a_{b-1} 2^{b-1} + a_{b-2} 2^{b-2} + \cdots + a_2 2^2 + a_1 2^1 + a_0 2^0. \tag{12.12}$$

Alternatively, we can represent the binary number by an ordered set, with position in the ordered set implying the appropriate power of 2. This is the standard notation we use in decimal notation. This notation form is shown in the following equation:

$$\text{Implied Weight}: \quad 2^{b-1} \ 2^{b-2} \ 2^{b-3} \ \ldots \ 2^2 \ 2^1 \ 2^0$$
$$N = a_{b-1} \ a_{b-2} \ a_{b-3} \ldots a_2 \ a_1 \ a_0.$$
(12.13)

An example of the binary representation of a set of decimal integers with implied powers of 2 is shown in Table 12.1.

The integers identified in Table 12.1 are all positive. If we require both positive and negative representations of our integers, we have to add a sign modifier. The form of a binary number with a sign modifier is shown in the following equation:

$$\text{Implied Weight}: \quad s \quad 2^{b-1} \ 2^{b-2} \ 2^{b-3} \ \ldots \ 2^2 \ 2^1 \ 2^0$$
$$N = a_b \ a_{b-1} \ a_{b-2} \ a_{b-3} \ \ldots \ a_2 \ a_1 \ a_0.$$
(12.14)

In *sign-magnitude* notation, the sign bit is a multiplier +1 or –1. In *offset-binary* notation, the sign bit is an added term $a_b 2^b$ and the number is decoded by subtracting 2^b from its binary representation. In *2's-complement* notation, the sign bit is an added $-a_b 2^b$. Examples of 4-bit binary numbers in the three binary representations are shown in Table 12. 2.

Figure 12.7 presents the overflow behavior of a 2's-complement binary counter. The overflow is, as expected, periodic. The unique behavior of the overflow is that the difference between points in the counter (or on circle) is correct even if the counter has experienced an overflow. It is well known that intermediate overflows of a 2's-complement accumulator leads to the correct answer as long as the accumulator is wide enough to hold the correct answer.

Table 12.1 Decimal integers and their 3-bit binary representation.

Decimal	Binary
0	0 0 0
1	0 0 1
2	0 1 0
3	0 1 1
4	1 0 0
5	1 0 1
6	1 1 0
7	1 1 1

Table 12.2   Decimal integers and their signed 4-bit binary representation.

Decimal	Sign-magnitude	Offset-binary	2's-Complement
+7	0 1 1 1	1 1 1 1	0 1 1 1
+6	0 1 1 0	1 1 1 0	0 1 1 0
+5	0 1 0 1	1 1 0 1	0 1 0 1
+4	0 1 0 0	1 1 0 0	0 1 0 0
+3	0 0 1 1	1 0 1 1	0 0 1 1
+2	0 0 1 0	1 0 1 0	0 0 1 0
+1	0 0 0 1	1 0 0 1	0 0 0 1
+0	0 0 0 0	1 0 0 0	0 0 0 0
−0	1 0 0 0	N/A	N/A
−1	1 0 0 1	0 1 1 1	1 1 1 1
−2	1 0 1 0	0 1 1 0	1 1 1 0
−3	1 0 1 1	0 1 0 1	1 1 0 1
−4	1 1 0 0	0 1 0 0	1 1 0 0
−5	1 1 0 1	0 0 1 1	1 0 1 1
−6	1 1 1 0	0 0 1 0	1 0 1 0
−7	1 1 1 1	0 0 0 1	1 0 0 1
−8	N/A	0 0 0 0	1 0 0 0

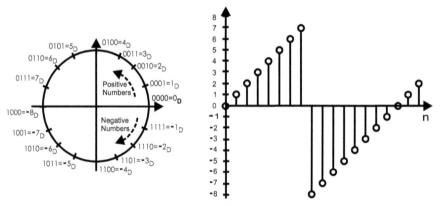

Figure 12.7   Overflow behavior of 2's-complement binary counter.

As an example, let us examine the step response of a CIC with 4 units of delay in the comb filter. Here the number 1 is added to the accumulator on each clock cycle and the comb filter following the accumulator forms the difference between the current input and the input formed 4 units ago. The sequence of outputs values is presented in Table 12.3 where we see that the output of the comb filter is correct in spite of the overflowing accumulator.

If the accumulator is sufficiently wide and if the CIC filter is performed with 2's-complement arithmetic, the CIC output will be correct in spite of

Table 12.3 Comb filter operating on output of overflowing accumulator.

$w(n)$	math	$w(n-4)$	$=$	$y(n)$
0	–	0	=	0
1	–	0	=	1
2	–	0	=	2
3	–	0	=	3
4	–	0	=	4
5	–	1	=	4
6	–	2	=	4
7	–	3	=	4
−8	–	4	=	−12 = 4
−7	–	5	=	−12 = 4
−6	–	6	=	−12 = 4
−5	–	7	=	−12 = 4
−4	–	−8	=	4
−3	–	−7	=	4
−2	–	−6	=	4
−1	–	−5	=	4

the internal overflow of the integrator. By sufficiently wide, we mean that the accumulator width must be the sum of the number of input bits and the number of bits required to accommodate the growth M of the M-tap prototype filter upon which the CIC is based. For instance, with 10-bit input data and a 100 tap boxcar filter with a gain of 100, we require 7 bits of growth for a bit field width of 17 bits. The accumulator must have 17 or more bits to implement the 100-tap boxcar filter as a CIC filter. Bit width is addressed in more detail in a later section.

12.3 Multistage CIC

We now address the problem that the boxcar filter or its surrogate, the CIC filter, is really not a good filter. We improve performance of the boxcar by forming a filter as a cascade of multiple boxcar filters. It is common to use 3−5 cascade stages with many applications using 3 or 4 stages. We can examine the impulse response of a multiple stage boxcar to gain insight of how a set of identical relatively poor performing filters work together to become a pretty good filter. The transfer function for a K-stage boxcar filter is shown in Equation (12.15) and the corresponding frequency response is shown in Equation (12.16). The two parameters that define this filter are M, the length of each subfilter, and K, the number of filters in the cascade.

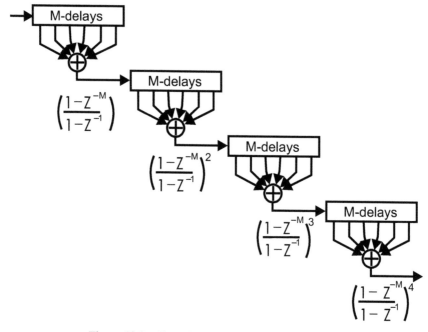

Figure 12.8   Cascade of four length-*M* boxcar filters.

$$H_K(Z) = \left[\frac{1 - Z^{-M}}{1 - Z^{-1}} \right]^K \tag{12.15}$$

$$H_K(\theta) = \left[\frac{sin(\theta \frac{M}{2})}{sin(\theta \frac{1}{2})} \right]^K . \tag{12.16}$$

For ease of understanding, we examine a specific cascade filter. The form of a four-stage cascade is shown in Figure 12.8. Here we see that the transfer functions from the input to the output of successive filter stages are increasing integer powers of the first stage transfer function $(1 - Z^{-M})/(1 - Z^{-1})$.

Figure 12.9 presents the impulse response observed at the outputs of the four successive length-10 boxcar filters. For ease of comparison, each response has been normalized to unity peak amplitude. As expected, the successive outputs are a rectangle, a triangle formed by convolving two rectangles, a piecewise quadratic formed by convolving three rectangles, and a piecewise cubic formed by convolving four rectangles. These are successively smoother sequences and are expected to exhibit successively lower side-lobe levels. Figure 12.10 presents the frequency responses of the

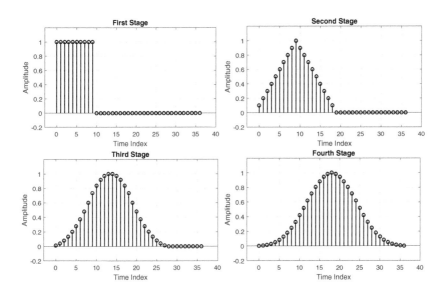

Figure 12.9 Scaled impulse responses of four-stage, 10-tap boxcar filter.

impulse responses shown in Figure 12.8. Here too, the spectra have been normalized for unity gain. The most obvious feature of these spectra is the reduction in maximum side-lobe level as well as rate of side-lobe attenuation in successive filters. The maximum side-lobe level of the first filter is –13 dB and since the spectra of successive filters are multiples of the first spectrum, the level of the maximum side lobe in the successive filters are multiples of –13 dB. The second, and more important feature, of these spectra can be seen in Figure 12.11 which presents the main-lobe response and a detailed look at the spectra at the first zero crossing. Here we see the effect of the repeated zeros on the ability of the filter to suppress spectra centered at normalized frequency 1/10 or 1/M for the general case. We noted the additional rejection capabilities of the higher order spectral zeros in our discussion on arbitrary interpolators.

Figure 12.12 shows how the CIC filter with two, three, and four stages are able to suppress spectral copies of a baseband signal, originally oversampled 4-to-1, that has been up sampled again by 1-to-10 zero-packing. We can easily see that the higher order filter with higher order zeros at the spectral copies are better able to suppress the undesired spectral copies.

An effect we observe in Figure 12.10 and hint in Figure 12.11 is the frequency-dependent gain reduction of the filter's main-lobe response. This

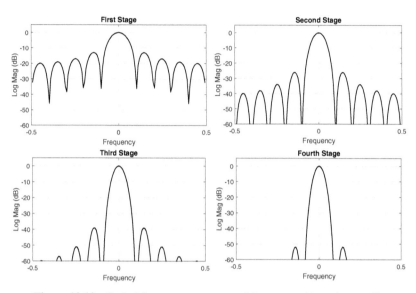

Figure 12.10   Scaled frequency responses of four-stage, 10-tap boxcar filter.

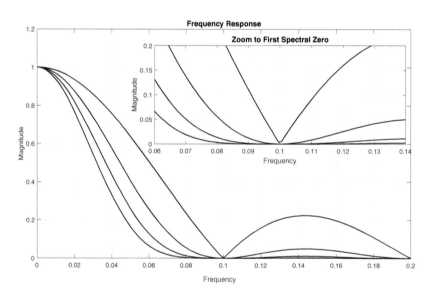

Figure 12.11   Zoom to zero of frequency responses of four-stage, length-10 boxcar.

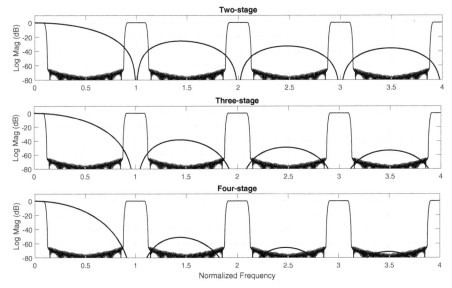

Figure 12.12 Suppression spectral replicas by two-, three-, and four-stage CIC filter.

nonuniform pass-band gain will distort a baseband spectrum being processed by this filter. The main-lobe gain reduction usually limits the input signal bandwidth to be less than 25% of the main-lobe width. For this example, input bandwidth would be less than 0.025 or 1/4 of 1/10, the 1/10 being the main-lobe width of the 10-tap boxcar filter. The spectral distortion due to the main lobe is corrected by embedding the inverse response in an FIR filter preceding or following the CIC. Additional comment on the correction process is postponed until we examine a system using the CIC with an embedded resampling operation.

12.4 Hogenauer Filter

The CIC filter is used in both up sampling and down sampling applications. The CIC contains two subfilters, the comb and the integrator, which can be applied in either order. If the filter is applied to an up sampling task, the comb filter is placed at the input to the CIC filter. On the other hand, if the CIC is applied to a down sampling task, the comb filter is placed at the output of the CIC filter. This ordering is established to permit the application of the noble identity and thus enable the reordering of the resampling switch and the comb filter. This reordering is illustrated in Figure 12.13 for the

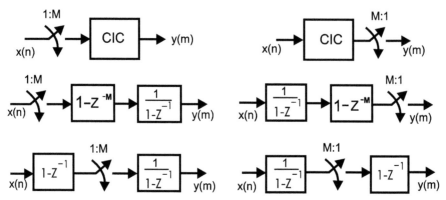

Figure 12.13   Order and reordering of resampling switch and subfilters of CIC filter for up sampling and for down sampling.

two cases of up sampling and down sampling. Note that the comb filter becomes a differentiator at the low data rate after it is reordered with the resampling switch. When the CIC absorbs the resampling switch, the filter structure is known as a Hogenauer filter. Remember that when ordering the three components in a resampling mode, the integrator always operates at the higher of the two rates, the differentiator operates at the lower rate, and the resampler resides between the pair.

A CIC filter with any number of stages can be similarly converted to a Hogenauer filter by first ordering all the integrators on one side of the filter and the comb filters on the other side, then applying the noble identity to interchange the resampling switch and the multiple comb filters. This reordering for a three-stage down sampling CIC filter is shown in Figure 12.14. The up sampling version of this same process is similar except that the derivatives are at the input and the integrators are at the output.

12.4.1 Accumulator Bit Width

We observed earlier that in spite of overflowing accumulators, the composite CIC filter would compute correct filter outputs, provided the additions were performed with 2's-complement arithmetic and provided the bit field width of the accumulator exceeded the word width required by the output sequence. The required bit width is the number of bits in the input data words plus the number of bits required to accommodate the maximum filter gain. This is shown in Equation (12.17). The gain of a K-stage filter with length-M comb filters is the product of the DC gains of each prototype boxcar filter as shown

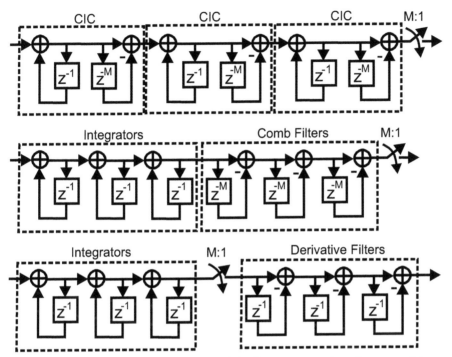

Figure 12.14 Down sampled three-stage CIC filter, rearranged, and converted to a Hogenauer filter.

in Equation (12.18). Substituting Equation (12.18) in Equation (12.17), we determine the bit-width required of the CIC accumulators as indicated in Equation (12.19).

$$b_{\text{ACCUM}} = b_{\text{DATA}} + \text{CEIL}[\text{Log}_2(\text{GAIN})] \qquad (12.17)$$

$$\text{GAIN} = M^K \qquad (12.18)$$

$$b_{\text{ACCUM}} = b_{\text{DATA}} + \text{CEIL}[K \cdot \log_2(M)]. \qquad (12.19)$$

As an example, suppose we have a two-stage CIC with length-20 comb filters processing 7-bit input. The gain of the filter is 20*20 or 400 for which we require 9 bits to provide for the signal growth through the filter. The required accumulator width is 9 bits of growth added to the 7 bits of input data or 16 bits. Figure 12.15 presents the time series at the outputs of a CIC's two 16-bit integrators and two comb filters. The input to the CIC is a cosine of amplitude 63 with period equal to 1000 samples. Note the overflow of the second integrator and the recovery from the overflow by the two comb filters.

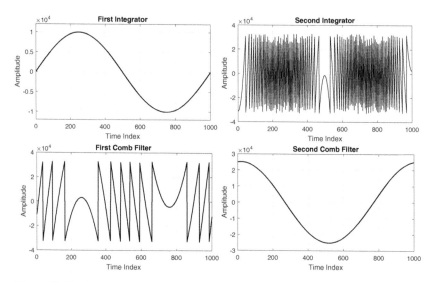

Figure 12.15   Two-stage CIC filter, 20-tap comb, 16-Bit accumulator, and 7-bit input.

For comparison with a properly operating CIC, we examine the performance of an improperly operating filter. Figure 12.16 presents the time series from a similar two-stage CIC filter operating on the same input signal except that the accumulator bit width of this CIC filter is 15 bits. As in the previous example, the second accumulator overflows and the two comb filters attempt to reverse the overflow. We observe that the output of the second comb filter has not successfully recovered from the overflow and that there is a residual overflow of a single bit at its output. When the input signal level is reduced by a factor of 2, the filter successfully recovers from the internal overflow since the reduced input level results in a reduced output level supportable by the 15-bit accumulator width.

12.4.2 Pruning Accumulator Width

In the previous section, we identified the bit growth required in a CIC to successfully recover from internal accumulator overflow. The bit growth reflects the filter gain between input and output of the filter. In an up sampling filter, this growth occurs in successive integrator stages, and the integrators at the beginning of the chain do not require the same bit width as does the final integrator. We can prune the most significant bits (MSBs) of accumulators to the level corresponding to the maximum gain of each integrator. In a down

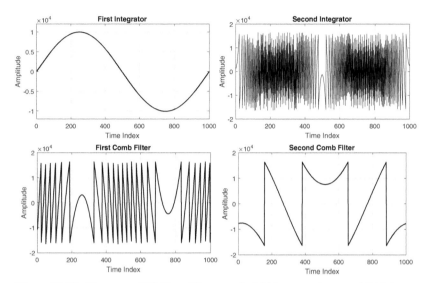

Figure 12.16 Two-stage CIC filter, 20-tap comb, 15-bit accumulator, and 7-bit input.

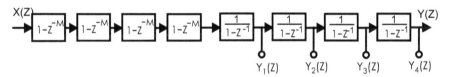

Figure 12.17 Four-stage up sampling CIC filter with integrator outputs identified.

sampling filter, the growth appears immediately in the first integrator stage and all subsequent integrators and comb filters must honor the MSB of the first integrator stage. Scaling applied at the output of the CIC to remove the filter processing gain discards the lower order bits of the CIC process. We can prune lower order bits early in the filtering chain to the bit position in any stage that cannot grow beyond the LSB of the output word. We now examine the two cases of MSB pruning for up sampling CIC filters and of LSB pruning for down sampling CIC filters.

12.4.2.1 Up Sampling CIC

Figure 12.17 is a block diagram of a four-stage CIC up sampling filter. The output of each integrator in the chain is identified and is available at the indicated tap points.

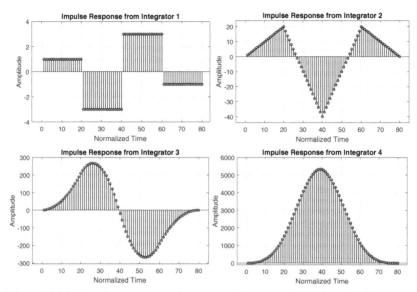

Figure 12.18   Impulse responses at integrators of four-stage CIC filter with 20-tap comb.

The transfer functions from the input to each integrator output are identified in Equation (12.20). Here, at each position, the included integrators are coupled with a matching comb filter with the unmatched or excess comb filters listed separately as a preprocessor to the equivalent boxcar filters.

$$\frac{Y_k(Z)}{X(Z)} = H_k(Z) = (1 - Z^{-M})^{(4-k)} \left[\frac{1 - Z^{-M}}{1 - Z^{-1}} \right]^k. \tag{12.20}$$

Figure 12.18 presents the impulse response from the input to the four integrator positions when there are 20 delays in each comb filter. Note the successively larger output levels at the outputs of the successive integrators.

To determine the maximum signal gain from the input to $y_k(n)$, the output of integrator k, we apply the input sequences $x_k(n)$ satisfying the following equation:

$$x_k(n) = \text{sign}[y_k(n)]. \tag{12.21}$$

The sequences that maximize the output levels at each of the integrator stages are shown in Figure 12.19.

The sequence that maximizes the output at a selected integrator lead to nonmaximum output levels at the remaining integrators. This can be seen in Table 12.4 which lists the maximum level obtained at each integrator for the

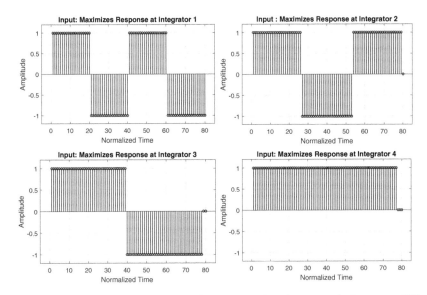

Figure 12.19 Input sequences to maximize output from integrators in four-stage CIC.

Table 12.4 Maximum integrator response levels for sequences that maximize selected integrator response.

	Sequence Int-1	Sequence Int-2	Sequence Int-3	Sequence Int-4
Integrator 1	**−160**	109	80	40
Integrator 2	858	**−1068**	640	267
Integrator 3	−6821	9848	**−10,680**	5340
Integrator 4	73,725	113,496	141,626	**160,000**

four input sequences identified in Figure 12.19. The bold entries in the table correspond to the maximum level that integrator can exhibit for sequences that result in stable responses at the final output stage. Figure 12.20 presents the integrator responses for the input sequences shown in Figure 12.19 designed to maximize the amplitude response for the selected integrator.

Each comb filters in the processing sequence, with one addition per stage, has a gain of 2 with a cumulative gain at the kth stage of 2^k. The impulse weights of a comb filter cascade are the terms of Pascal's triangle with alternating signs. These terms are shown in Figure 12.21. To maximize the sum at the output of each comb filter, as we did for the integrators, we set the input sequence to the sign of the comb filter impulse response. To also probe the response of the integrator chain while probing the comb filters, we

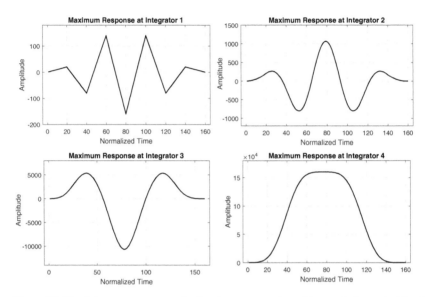

Figure 12.20   Integrator responses for input sequence selected to maximize output level.

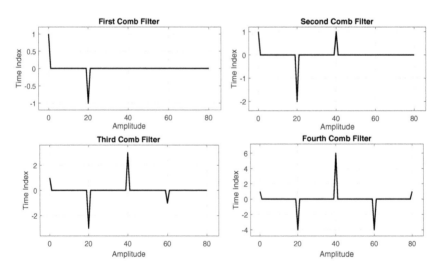

Figure 12.21   Impulse responses at comb filters of four-stage CIC filter with 20-tap comb.

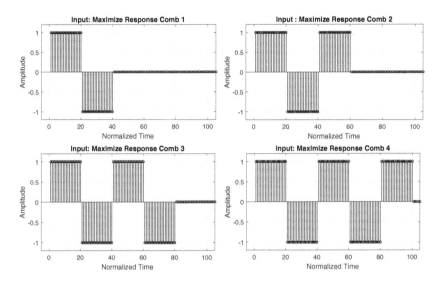

Figure 12.22 Input sequences to maximize output from comb filters in four-stage CIC.

extend the samples for the m-sample interval between samples of the comb filter response. The probe signals to test both the comb and the integrator segments of the CIC are shown in Figure 12.22. We note that the successive sequences use the previous sequence as a prefix and simply add one more segment. Consequently, we will find that the sequence that maximizes the output for comb filter k also maximizes the output for comb filter $k - 1$.

Table 12.5 lists the maximum level obtained at each comb filter and at each integrator output for the four input sequences identified in Figure 12.21. The bold entries in the table correspond to the predicted maximum levels that comb filter can exhibit; we also see the peak values of the integrators obtained while maximizing the output of the comb filters. For ease of comparison, we also list from Table 12.4, the maximum values of the integrator outputs. The bit width results determined from the two sets of probes applied to the up sampling CIC filter of Figure 12.17 are tabulated in Table 12.6.

The bit growth per processing stage can be visualized by the graph of bit field width shown in Figure 12.23. Each stripe in the graph indicates the number of bits entering and the number of bits leaving the indicated process. For the example just concluded, we can see how the total of 18-bit growth from input to output is distributed between the comb filters and the integrators.

Table 12.5　Maximum comb response levels for sequences that maximize selected comb response.

	Sequence Comb-1	Sequence Comb-2	Sequence Comb-3	Sequence Maximum Comb-4	Level
Comb-1	2	2	2	2	2
Comb-2	3	4	4	4	4
Comb-3	6	7	8	8	8
Comb-4	10	14	15	**16**	**16**
Integrator-1	120	140	−160	160	**160**
Integrator-2	720	900	855	855	**1068**
Integrator-3	−8020	6821	−6821	6821	**10,680**
Integrator-4	73,710	73,725	73,725	73,725	**160,000**

Table 12.6　Maximum comb and integrator response and required bit field width.

	Maximum gain	Growth bits
Comb-1	2	1
Comb-2	4	2
Comb-3	8	3
Comb-4	16	4
Integrator-1	160	9
Integrator-2	1068	10
Integrator-3	10,680	14
Integrator-4	160,000	18

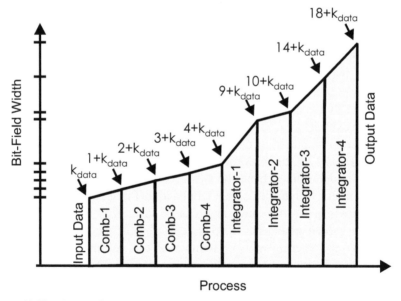

Figure 12.23　Graphs showing required bit field at input and output of each process in CIC.

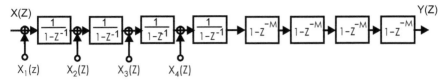

Figure 12.24   Four-stage down sampling CIC filter with integrator inputs identified.

12.4.3 Down Sampling CIC

Figure 12.24 is a block diagram of a four-stage CIC down sampling filter. Pruning of this process will be performed by discarding lower order bits in each accumulator. The discarded bits will be treated as additive noise to each integrator. Our interest here is the noise gain from each integrator and from each comb filter to the output port. The input of each integrator in the chain is identified and is available at the indicated input points from which we will determine the noise gains.

The transfer functions from the input of each integrator to the CIC output are identical to the transfer functions identified in Equation (12.20) and repeated here as Equation (12.22) to reflect the change of input and output variables. Here at each position, the integrators included in the signal path are coupled with a matching comb filter with the unmatched or excess comb filters listed separately as post-processor to the equivalent boxcar filters.

$$\frac{Y_k(Z)}{X(Z)} = G_k(Z) = \left[\frac{1 - Z^{-M}}{1 - Z^{-1}}\right]^k (1 - Z^{-M})^{(4-k)}. \tag{12.22}$$

Figure 12.25 presents the impulse response from the four-integrator input positions to the output when there are 20 delays in each comb filter. Note the successively reduced output levels at the output when there are a reduced number of integrators between the selected input and the CIC output.

The noise power measured at the output due to noise inserted at the input to integrator k is shown in the following equation:

$$\sigma^2{}_{\text{OUT}(k)} = \sigma_k^2 \sum_{n=0}^{N-1} g_k^2(n) = \sigma_k^2 N G_k. \tag{12.23}$$

The standard deviation of the output noise due to the gain from each integrator is listed in Table 12.7. In a similar manner, noise inserted at the input to the comb filters can be formed using the comb filter impulse response. The comb filter impulse responses are the same as those shown in Figure 12.21.

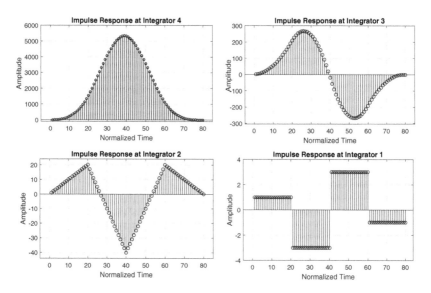

Figure 12.25   Impulse responses from integrators of four-stage CIC filter with 20-tap comb.

Table 12.7   Noise gain from each CIC stage to output.

	SQRT(Noise gain)	Noise bit growth
Integrator-1	24,785	14.6
Integrator-2	1462.4	10.5
Integrator-3	146.3	7.2
Integrator-4	20	4.3
Comb-1	8.4	3.1
Comb-2	4.5	2.2
Comb-3	2.5	1.3
Comb-4	1.4	0.5

Table 12.7 shows the noise gain for each of the integrator and comb filter stages in the CIC.

The total noise contributed to the output by the separate noise sources injected at each stage due to pruning LSB bits in each of the $2K$ filter segment is shown in the following equation:

$$\sigma^2_{\text{TOTAL}} = \sum_{k=0}^{2K-1} \sigma^2_{\text{OUT}(k)} = \sum_{k=0}^{2K-1} \sigma^2_k NG_k. \qquad (12.24)$$

It is reasonable to assign equal levels of noise contribution from each noise source.

Table 12.8 Bit locations below output LSB for register pruning.

	Bits below output LSB	Integer number of bits
Integrator-1	14.6 + 3 = 17.6	18
Integrator-2	10.5 + 3 = 13.5	14
Integrator-3	7.2 + 3 = 10.2	11
Integrator-4	4.3 + 3 = 7.3	8
Comb-1	3.1 + 3 = 6.1	7
Comb-2	2.2 + 3 = 5.2	6
Comb-3	1.3 + 3 = 4.3	5
Comb-4	0.5 + 3 = 3.5	4

For this condition, the noise contributed by each source must satisfy the following equation:

$$\sigma^2_{\mathrm{OUT}(k)} = \frac{1}{2K}\sigma^2_{\mathrm{Total}}. \tag{12.25}$$

The acceptable noise level that can be inserted at each noise source must account for the $1/2K$ of Equation (12.25). Since bits levels define the noise level at each insertion point, we convert the $1/2K$ to bits as is done in the following equation:

$$\text{Additional Number Bits} = \log_2(2K) \tag{12.26}$$

Our four-stage example, with eight noise terms, requires the noise terms to drop an additional 3 bits for the overall noise to be below the specified output LSB. When we incorporate the additional attenuation of the scaled noise in our example, we obtain the parameters listed in Table 12.8.

The LSB bit pruning per processing stage can be visualized by the graph of bit field width shown in Figure 12.26. Each stripe in the graph indicates the number of bits entering and the number of bits leaving the indicated process. For the example just concluded, we can see how the total of 18-bit growth occurring at the input stage permits the LSB pruning as we progress from input to output through the integrators and the comb filters.

12.5 CIC Interpolator Example

We now examine an example of the use of the Hogenauer (resampled CIC) filter. The example is part of a modulator that shapes and up samples in a polyphase filter and then interpolates to a higher output sample rate with a CIC filter. Figure 12.27 presents an example in which the shaping is performed by a 1-to-5 polyphase filter and the additional interpolation is

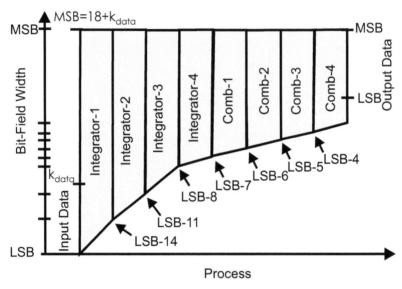

Figure 12.26   Graph showing required bit field at input and output of each process in CIC.

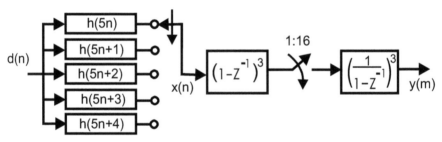

Figure 12.27   Cascade of 1-to-5 shaping filter and 1-to-16 CIC interpolating filter.

performed in a three-stage 1-to-16 CIC filter. The CIC is implemented as a Hogenauer filter with three resampled comb filters at the input, a 1-to-16 resampler, and three integrators at the output.

Figure 12.28 presents the time and frequency response of the output series obtained by applying an impulse to the input of the shaping filter. The time axis is normalized to input symbol rate of unity and the frequency axis is normalized to unit 3 dB bandwidth or unity symbol bandwidth.

We gain insight into the operation of the CIC filter by tracking the impulse response of the shaping filter through the CIC filter. By this process, we obtain the impulse response of the composite shaping and interpolating filter. Figure 12.29 follows the response from the shaping filter through the

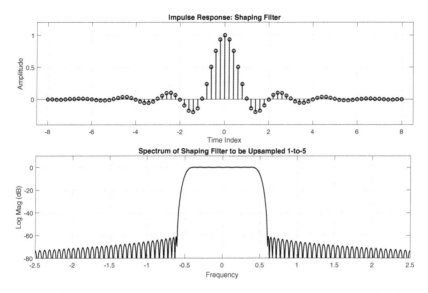

Figure 12.28 Time and frequency response of time series from 1-to-5 shaping filter.

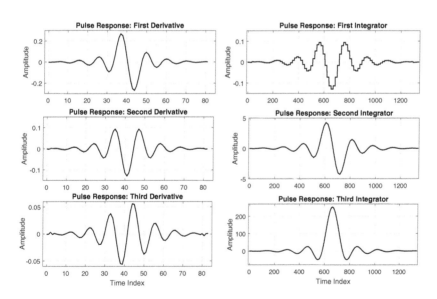

Figure 12.29 Shaping pulse response of three-stage 1-to-16 Hogenauer filter.

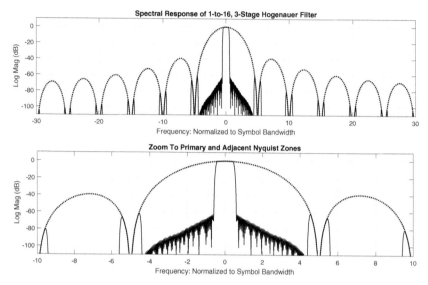

Figure 12.30   Spectrum of composite impulse response with CIC spectral overlay.

three input comb filters at the input rate and then follows the 1-to-16 up sampled sequence through the three output integrators. We can clearly see the equivalent ZOH effect of the 1-to-16 up sampling operation at the output of the first integrator.

Figure 12.30 presents the frequency response obtained from the output of the composite polyphase shaping filter and three-stage CIC interpolating filter. We can clearly see the effect of the CIC spectral zeros on the up sampled input time series. The maximum residual spectral level brackets the first spectral zeros of the CIC filter. This level is seen to be –60 dB, which is the target level of 60 dB attenuation. These spectral levels fall below the noise floor of a 10-bit digital-to-analog converter (DAC) and will be concealed by the system noise. An interesting note is that if the shaping filter were to operate as a 1-to-4 up sampling filter, the three-stage CIC would not obtain the desired 60-dB attenuation. Thus, the choice is to replace the three-stage CIC with a four-stage CIC or to raise the output rate of the shaping filter from 1-to-4 to 1-to-5. This second option is more cost effective since the addition of a fourth CIC stage would not only add another integrator but would also increase the bit field width of the other three integrators.

12.6 Coherent and Incoherent Gain in CIC Integrators

Figure 12.31 presents curves showing the maximum coherent gain between input and integrators of up sampling CIC filters of order from 2 through 5 for up sampling rates from 2 through 100. These curves were generated in a manner similar to the process described in the up sampling CIC example that examined the specific case of up sampling by 20 with a fourth-order CIC. The input sequences that probed the CIC are of the form shown in Figure 12.18. These sequences maximize the peak amplitude at each integrator. The curves contain the information required to determine the required bit width for each integrator in an up sampling CIC operating at a specific resampling rate.

Figure 12.32 presents curves showing the maximum incoherent gain between integrators and output of down sampling CIC filters of order from 2 through 5 for down sampling rates from 2 through 100. These curves were generated in a manner similar to the process described in the down sampling CIC example that examined the specific case of down sampling by 20 with a fourth-order CIC. A single impulse at each integrator probed the CIC to form the impulse response from which the sum of squares was computed to determine the incoherent gain. The curves contain the information required

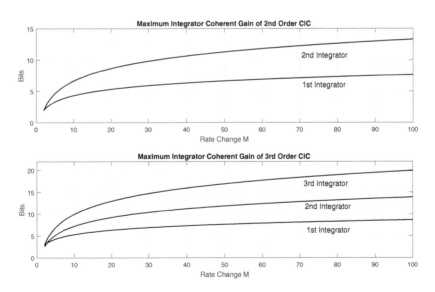

Figure 12.31(a)   Coherent bit growth of integrators of second- and third-order CIC filters as function of rate change *M*.

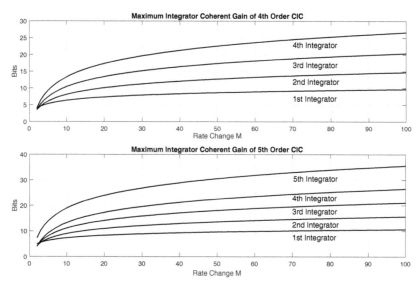

Figure 12.31(b)   Coherent bit growth of integrators of fourth- and fifth-order CIC filters as function of rate change *M*.

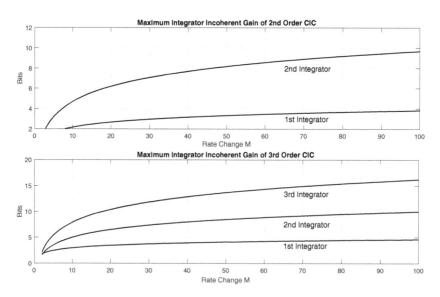

Figure 12.32(a)   Incoherent bit growth of integrators of second- and third-order CIC filters as function of rate change *M*.

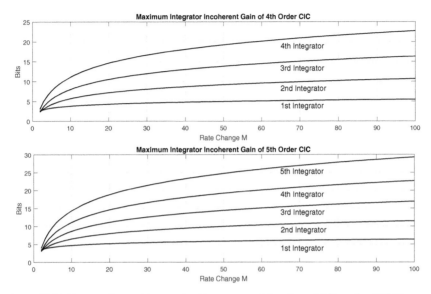

Figure 12.32(b)   Incoherent bit growth of integrators of fourth- and fifth-order CIC filters as function of rate change *M*.

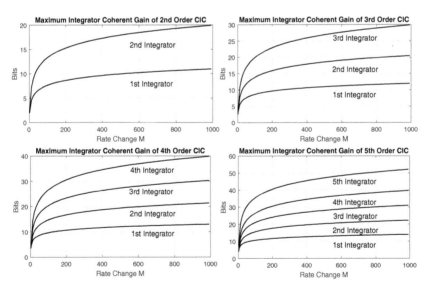

Figure 12.33(a)   Coherent bit growth of integrators of second-, third-, fourth-, and fifth-order CIC filters as function of rate change *M*.

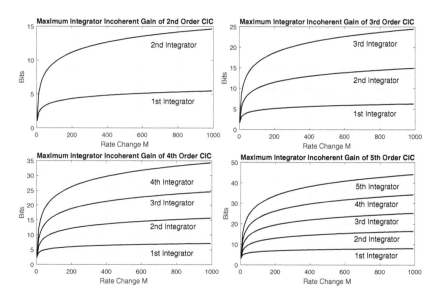

Figure 12.33(b)   Incoherent bit growth of integrators of second-, third-, fourth-, and fifth-order CIC filters as function of rate change M.

to determine the position in a bit field that can be pruned to reduce the bit width for successive integrators in a down sampling CIC operating at a specific resampling rate. Figure 12.33 presents the same curves shown in Figures 12.28 and 12.29 for a wider range of up and down sampling rates from 2 to 1000.

12.7 Equal Ripple Stopband Bifurcate Zeros

The original attraction of the CIC filter was that it recast an M-tap boxcar integrator as a cascade of an integrator filter and a comb filter that required only two additions. The second attraction was the Hogenauer transformation that moved the resampling operation to be between the integrator and the down sampled comb filter, now a derivative filter. Recasting the comb to a derivative reduced the M delays of the comb to 1 delay of the derivative. As we had indicated early in this chapter, the CIC is not a particularly good filter. Prior to the Hogenauer exchange, a single-stage CIC filter has a frequency response which is a sinc function with a DC gain of M and with $M - 1$ spectral zeros at multiples of f_S/M. The amplitude of its first spectral side lobe is $0.22M$ which is only 13 dB attenuation. We improved the performance of the boxcar by applying the filter K times. The K-fold convolution of its multiple stages forms a smooth continuous impulse response with a spectrum that is the Kth power of the sinc spectrum. The Kth power spectrum has K repeated zeros at the multiples of f_s/M frequencies. The multiple zeros set the first M derivatives of the local spectrum to zero which widens the width of the stopbands at these frequencies. What we will do in this section is separate the repeated zeros to obtain a wider bandwidth equiripple stopband.

We start this process with a two-stage CIC filter. The impulse response of the M-tap boxcar convolved with another M-tap boxcar is a triangle. The spectrum of the triangle has a DC gain of M^2, double zeros at multiples of f_S/M, and amplitude of its first spectral side lobe is $(0.22M)^2$, which is an improved 26 dB attenuation. The double zeros of the second-order CIC filter have a wider stopband bandwidth because the amplitude and first derivative are zero at the zero locations. We now separate the two zeros to obtain equal ripple and wider bandwidth stopband. We start by examining the repeated spectral zeros of the two-stage 10-tap filter's triangle spectrum shown in Figure 12.34. We can separate the two zeros at their points of tangency by lowering the frequency response curve by a small amount. We can lower the entire curve by subtracting a low level impulse from the peak amplitude of the time domain triangle. The low level impulse has a constant amplitude frequency response phase aligned with spectrum of the triangle.

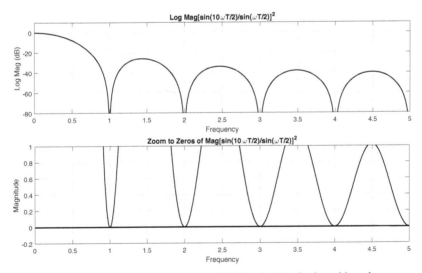

Figure 12.34 Spectrum of two stage 10-tap CIC filter's triangle shaped impulse response.

The spectrum obtained by subtracting 0.1 from the triangle center weight is shown in Figure 12.35. Here we see the bifurcation of the two zeros and the 0.1 magnitude stopband ripple between the zeros. The peak spectrum is now 99.9 due to the reduced amplitude of the triangle's center weight.

Figure 12.36 shows the pole-zero diagrams of the normal CIC filter and of the bifurcated CIC. The separated zeros bracket their original locations at

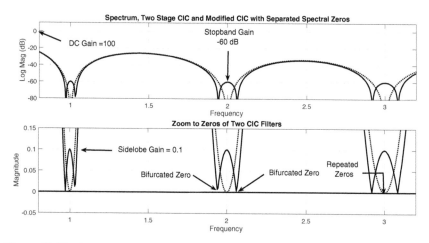

Figure 12.35   Spectrum of two-stage CIC filter and two-stage CIC with bifurcated spectral zeros.

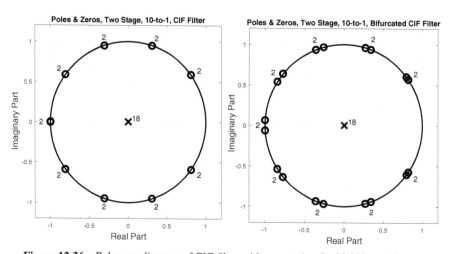

Figure 12.36   Pole zero diagram of CIC filter with repeated and with bifurcated zeros.

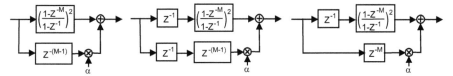

Figure 12.37 Block diagram of modified CIC filter. Reduce center weight by delaying input $M - 1$ samples. Scale and subtract delayed sample. Insert one delay in each branch of filter. Merge delays in lower path.

multiples of $2\pi/10$. Note that the spacing between the zeros becomes wider as we move further from DC at θ equal 0. Figure 12.37 shows the evolution of the bifurcated CIC to the precursor of the Hogenauer transformation. The first step in the process, shown in left most subplot, is to add the parallel path composed of the $M - 1$ delays with coefficient α and output summation. The delay aligns the input with the location of the center sample of the triangle impulse response of the cascade M-tap boxcar filter response. It is sensible that the coefficient α be a binary shift such as 2^{-2} or 2^{-3} to avoid a multiplication in the bifurcated filter. In the center subplot, we see the second step in the process, the inserted single delay in each path. This delay raises the number of delays to M so that we can invoke the noble identity and interchange the comb filter and companion delay with the M-to-1 resampling operation. The third step in the process, seen in the right most subplot, is to merge the new delay with the original $M - 1$ delays. The evolution of the Hogenauer variation continues in Figure 12.38. Here, in the upper subplot, the two-stage boxcar transfer function is partitioned into the cascade of two integrator filters and two length-M comb filters. In the lower subplot, we perform the noble identity interchange of the M-to-1 down sample with the Z^{-M} comb filters and companion delay to form the two derivative filters and its companion 1-tap delay. The α coefficient and output adder slide through the interchange. In the upper path of resampled and bifurcated CIC, the 1 unit input delay is folded into the input integrator. This fold means the output of the integrator is obtained from the internal delay register instead of its input.

Figure 12.39 presents the sequence of steps to convert a three-stage bifurcated boxcar filter into a Hogenauer resampled filter. We start, in the top subplot, by adding a third stage to the two-stage bifurcated boxcar filter. In the center subplot, we partition and redistribute the boxcar filters into cascade integrators and comb filters. We see that the third boxcar filter contributes an input integrator filter and an output derivative filter that brackets the two-stage bifurcated CIC. The bottom subplot shows the result of the noble identity

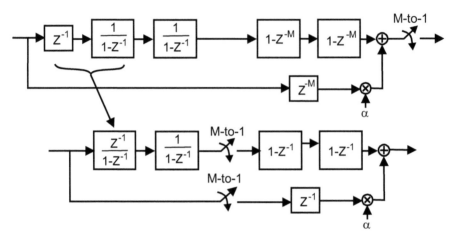

Figure 12.38 Upper subplot: partition boxcar transfer function ratio of polynomials to cascade integrators and comb filters. Lower subplot: interchange order of *M*-to-1 down sample and polynomials in *M* with the noble identity.

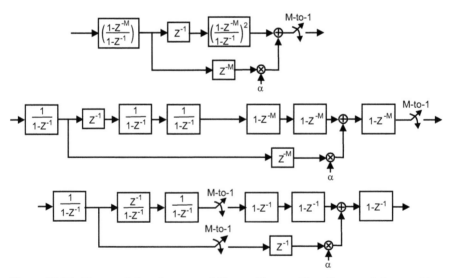

Figure 12.39 Upper subplot: three stage bifurcated boxcar filter. Center subplot: partition of three-stage boxcar transfer function ratio of polynomials to cascade integrators and comb filters. Bottom subplot: interchange order of *M*-to-1 down sample and polynomials in *M* with the noble identity.

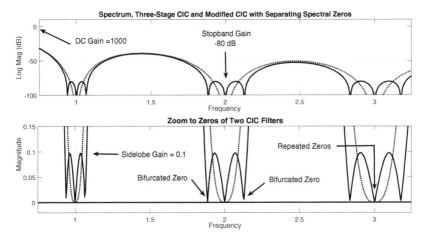

Figure 12.40 Spectrum of three-stage CIC filter and three-stage CIC with bifurcated spectral zeros.

filter and resample interchange. Figure 12.40 shows the spectrum of the three-stage bifurcated boxcar filter prior to the noble identity interchange. We see the addition gain introduced by the third stage, increasing DC gain from 100 to 1000. The equal ripple stopband gain was set to 0.1 for a stopband attenuation level to DC gain level ratio of -80 dB. We also see the widening of the stopband with the equal ripple stopband option. We suggest the reader read the cited paper to see additional bifurcated CIC options.

12.8 Compensation of CIC Main-Lobe Droop

The main-lobe spectral response of a multiple stage CIC filter exhibits a frequency-dependent droop which represents spectral distortion of any signal passing through it. It is common practice to correct for that droop with a compensating filter which applies a frequency-dependent gain to oppose the droop. The Taylor series of the CIC main-lobe response is quadratic independent of the number of stages. The number of stages only affects the coefficient of the quadratic term. The compensating filter can be designed with its Taylor series quadratic term equal size but opposite sign to the CIC's Taylor series. When we perform the product of the two spectra, the CIC filter, and the compensating filter, we cancel the quadratic terms and significantly improve the flatness of the cascade filter and its compensator. The simplest way to deliver a quadratic compensating spectrum is with a

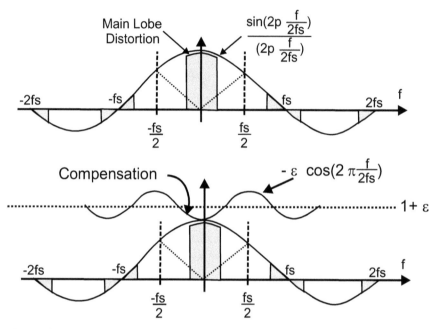

Figure 12.41   Upper subplot: main lobe of sinc spectrum with desired pass band distorted by main-lobe droop. Lower subplot: same spectra with compensating filter spectrum, an offset, and scaled opposing polarity cosine.

properly offset and scaled cosine. The upper subplot of Figure 12.41 shows the spectral sinc overlaying the fractional pass-band bandwidth with its quadratic droop distortion. The lower subplot shows the same spectrum with a compensating spectra formed by an offset cosine with opposing polarity quadratic correction. The offset positions the cosine to be tangential to the sinc's main lobe. Equation (12.27) shows the quadratic term in the Taylor series expansion centered on the $(\text{sinc})^K$ main lobe. Equation (12.28) shows the Taylor series expansion of the offset scaled cosine compensating filter. Finally, Equation (12.29) determines the value of required to compensate for the kth power of the CIC main-lobe quadratic droop. The form of compensating filter is shown in Equation (12.30). The cosine is embedded in the filter vector at its end points with half amplitudes required by Euler's representation. The DC

$$\left[\frac{\text{Sine}(\frac{\theta}{2})}{(\frac{\theta}{2})} \right]^K = \left[\frac{(\frac{\theta}{2}) - 1\frac{1}{3!}(\frac{\theta}{2})^3 + \cdots}{(\frac{\theta}{2})} \right]^K$$

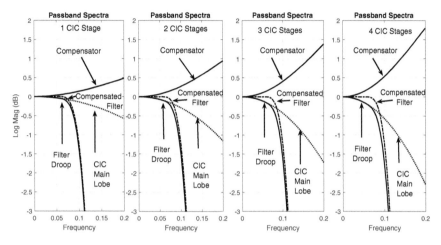

Figure 12.42 Spectra of one-, two-, three-, and four-stage CIC filters, their droop responses, their compensation filters, and their compensated responses.

$$= \left[1 - \frac{1}{6}(\frac{\theta}{2})^2 + \cdots\right]^K = \left[1 - \frac{K}{6}(\frac{\theta}{2})^2 + \cdots\right] \quad (12.27)$$

$$(1 + \varepsilon) - \varepsilon \cos(\theta) = (1 + \varepsilon) - (1 - \frac{1}{2!}\theta^2 + \cdots)$$

$$= 1 - \frac{\varepsilon}{2!}\theta^2 + \cdots \quad (12.28)$$

$$\frac{\varepsilon}{2}\theta^2 = \frac{K}{6}(\frac{\theta}{2})^2 = \frac{K}{24}\theta^2, \quad \therefore \varepsilon = \frac{K}{12} \quad (12.29)$$

$$c(Z, k) = \left[\frac{-\varepsilon(K)}{2} \quad \text{zeros}(1 : M - 1) \quad (1 + \varepsilon(K))\right.$$

$$\left. \text{zeros}(1 : M - 1) \quad \frac{-\varepsilon(K)}{2}\right].$$

offset term resides at the center of the filter and the $M - 1$ inserted zeros packed between the coefficients is needed to extend the offset cosine over all the Nyquist zones spanned by the CIC spectrum even though it is only compensating for the spectral copy in the primary Nyquist zone. Figures 12.42 and 12.43 show the spectra of one-, two-, three-, and four-stage CIC filters along with their spectral droop, their corresponding compensation filters designed and their compensated filter responses. The compensate

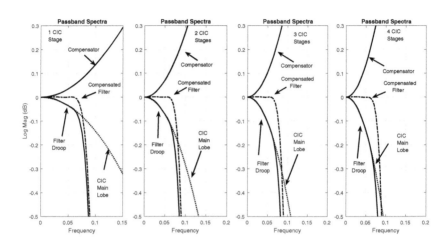

Figure 12.43 Finer detail spectra of one-, two-, three-, and four-stage CIC filters, their droop responses, their compensation filters and their compensated responses.

response are seen to be quite flat and upon closer scrutiny, we would be able to see the in-band ripple levels around the 0 dB reference levels of the compensated FIR filters that were processed by the four CIC filters. There are other techniques, such as filter sharpening, that work well with wider bandwidth segments of the CIC filter spectral response. We have chosen to not present them here, but we have cited one in our reference list.

References

Chu, S. and C.S. Burrus, "Multirate Filter Design Using Comb Filters", IEEE Trans. On Circuits and Systems, Vol. 31, Nov. 1984, pp. 913-924.

Crochiere, Ronald and Lawrence Rabiner, "Multirate Signal Processing", Englewood Cliff, NJ, Prentice-Hall Inc. 1983.

Fliege, Norbert, *Multirate Digital Signal Processing: Multirate Systems, Filter Banks, Wavelets*, West Sussex, John Wiley & Sons, Ltd, 1994.

Hentschel, Tim. *Sample Rate Conversion in Software Configurable Radios*, Norwood, MA, Artech House, Inc., 2002

Hogenauer, E.B. "An Economical Class of Digital Filters for Decimation and Interpolation", IEEE Trans. Acoustics. Speech Signal Proc., Vol. ASSP-29, April 1981, pp. 155-162

Jovanovic-Dolecek, Gordana. *Multirate Systems: Design and Applications*, London, Idea Group, 2002.

harris, fred and Jovanovic-Dolecek, Gordana, *Bifurcate Repeated Stop band Zeros of CIC Filter*, DSP 2016, Beijing, China, 16-18 October 2016

harris, fred, *Reduce Energy Requirements by Coupling a Poly-Phase Pre-Filter and CIC Filter in High-Performance Sigma-Delta A/D Converters*, ISCAS-2014, Melbourne, Australia, 1-5-June 2014.

Jovanovic-Dolecek, Gordana and harris, fred, *Design of CIC Compensator Filter in a Digital Receiver,* Elsevier DSP Journal, 2009.

Mitra, Sanjit. *Digital Signal Processing: A Computer-Based Approach*, 2nd ed., New York, McGraw-Hill, 2001.

Mitra Sanjit and James Kaiser, *Handbook for Digital Signal Processing*, New York, John Wiley & Sons, 1993.

P. P. Vaidyanathan, "Multirate Systems and Filter Banks", Englewood Cliff, NJ, Prentice-Hall Inc., 1993.

Problems

12.1 Program a 20-tap version of the three forms of the boxcar integrator shown in Figure 12.5 and determine the impulse response of all three versions. Note the state of the integrator in the two forms of the CIC.

12.2 Program a 20-tap version of the three forms of the boxcar integrator shown in Figure 12.5 and determine the step response of all three versions. Note the state of the integrator in the two forms of the CIC.

12.3 Program a 20-tap version of the three forms of the boxcar integrator shown in Figure 12.5 using integer arithmetic with a 5-bit 2's-compliment accumulator and with a 4-bit 2's-compliment accumulator and then determine the step response of all three versions. Note the state of the integrator in the two forms of the CIC for the two different width accumulators.

12.4 The spectrum of a P-stage CIC filter exhibits multiple zeros at multiples of $1/M$th of the sample rate. Determine the attenuation available at an offset of $1/(4M)$ from the zero at $1/M$; i.e. at $(3/4M)$. This is the frequency that aliases into the bandwidth of $1/(4M)$, the bandwidth of the final cascade of the CIC, and the 4-to-1 FIR filter following the CIC filter. In particular, how many stages of CIC are required to obtain 60, 80, 100, and 120 dB attenuation at the edge of the first Nyquist zone to alias to baseband?

12.5 A P-stage CIC filter of length M has a steady-state DC gain of M^P. For an $M = 100$ and $P = 5$, determine the width required of the accumulators when the input data is 16 bits. Repeat for $M = 1000$.

12.6 For the block diagram in Figure 12.17, imagine an input signal consisting of samples of a unity amplitude sine wave of normalized frequency 0.1. Determine the amplitude of the sinusoid observed at the four integrator output ports as a function of M. In particular, what is the set of amplitudes for $M = 100$? Repeat for normalized frequency of 0.01.

12.7 For the block diagram in Figure 12.24, imagine injecting an input signal consisting of samples of a unity amplitude sine wave of normalized frequency 0.1 in any of the four input ports of the integrator train. Determine the amplitude of the sinusoid observed at the output port as a function of M. In particular, what is the set of amplitudes for $M = 100$? Repeat for normalized frequency of 0.01.

12.8 Program the filter chain consisting of a 1-to-4 polyphase Nyquist filter followed by a four-stage CIC filter that will up sample the filter output by a factor of 20. The Nyquist filter has a 50% excess bandwidth and a side-lobe level 60 dB below pass-band response. Apply an impulse to the input and record and plot the time response of the filter chain at each stage as well as the frequency response at the final output. Repeat this programming task but replace the Nyquist filter with a 1-to-5 polyphase filter and a four-stage CIC filter that will up sample by a factor of 16. What is the peak output amplitude of the final output for the two filter implementations? Can either implementation be realized with a three-stage CIC? If so, which one, and what is the peak output amplitude for that case?

12.9 A CIC filter of length M does not have to be down sampled by M-to-1. An interesting option is to resample by $M/2$. When we operate the filter in this manner, the folding frequency for the filter output is f_s/M rather than $f_s/(2M)$. Program a third-order CIC filter of length $2M$ operating at an M-to-1 down sample rate. In this test, form a 1-to-4 polyphase Nyquist filter with 50% excess bandwidth and 60 dB side lobes. Zero-pack this signal 1-to-20 and process it with a third-order CIC filter of length 40. Examine the frequency response of the filter. Compare the response to a 1-to-20 zero-packed signal processed with a third-order CIC of length 20. Also examine the response of 1-to-20 zero-packed signal processed with a second-order CIC filter of length 40. Comment on the performance differences and the workload difference of the three options.

13

Cascade and Multiple Stage Filter Structures

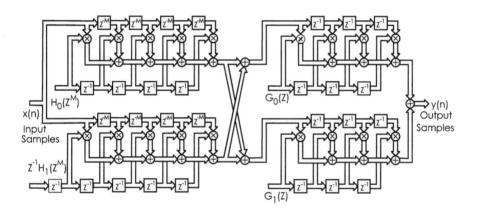

Often a cascade of two or more stages of filtering and resampling best serves a multirate signal-processing task. Various structures that emphasize this perspective have been developed. This chapter presents a number of systems in which cascade filtering is applied to the signal to obtain multiresolution partition of a signal or to obtain significant reduction in signal-processing workload.

13.1 Interpolated FIR Filters

The major emphasis of this text has been the application of resampling techniques to time series. We now switch perspectives and apply the resampling procedure to the filter processing time series. We know that the length of a finite impulse response (FIR) filter is proportional to the ratio of the filter sample rate to the filter's transition bandwidth. In many applications, transition bandwidth is small because the filter bandwidth is small so that

437

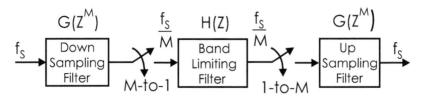

Figure 13.1   Resampling filtering option when ratio of sample rate to transition bandwidth is large.

narrow transition bandwidth is often synonymous with narrow pass-band bandwidth. Thus, we find that for many filters operating with a large ratio of sample rate to pass-band bandwidth, the filter length is prohibitively large. In other chapters, we addressed this problem by preprocessing the input time series with a resampling filter in which the signal bandwidth and the sample rate of the output time series is simultaneously reduced. The bandwidth-limiting filter, now designed to operate at the reduced sample rate, is shorter in proportion to the sample rate reduction. If it is necessary to preserve the original input sample rate, the output of the bandwidth-limiting filter is then up-sampled back to the original input sample rate. This cascade structure is shown in Figure 13.1 where $G(Z^M)$ performs the two resampling operations, while $H(Z)$ performs the desired bandwidth limiting function.

An alternate to down-sampling the input time series to the sample rate of the bandwidth-limiting filter is to raise the sample rate of the bandwidth-limiting filter. The filter is up-sampled by zero-packing its impulse response 1-to-M. The up-sampled filter now has the longer impulse response expected for operation at the higher sample rate but does not require any additional arithmetic operations. It does, however, require additional memory to hold the larger set of input data samples. The Z-transform of the up-sampled filter is $H(Z^M)$. The frequency response of the up-sampled filter has M spectral replicates at multiples of the baseband design sample rate of f_s/M. A second filter, following the zero-packed filter, preserves the baseband spectra while filtering out the multiple spectral replicates. This filter structure is shown in Figure 13.2, while the frequency responses observable at various positions in the cascade is shown in Figure 13.3.

The second filter interpolates the impulse response of the first. In the composite impulse response, the zeros in the impulse response of the first filter are replaced by interpolated samples. This occurs while the filtering

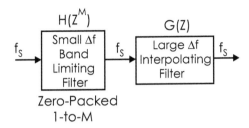

Figure 13.2 Interpolated FIR filter. Up-sampled prototype and interpolating clean-up filter.

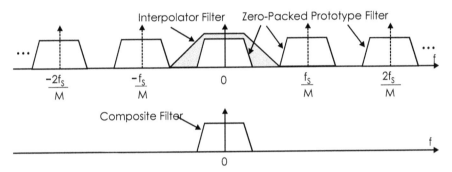

Figure 13.3 Spectral description of zero-packed, interpolating, and composite filters.

action of the second filter suppresses the spectral replicates of the first. By maintaining two separate filters, the composite workload of the two filters is the sum of the lengths of the prototype nonzero-packed filter and that of the following interpolating filter. In the filtering operation, the zero-packed filter with few active coefficients accomplishes the desired spectral bandwidth limiting process with fewer arithmetic operations but exhibits multiple pass bands. The second filter following the zero-packed filter passes the desired spectra while rejecting the replica spectral bands. We do pay a small penalty in that the cascade filter requires additional data register space to operate the pair of filters. The composite filter also exhibits additional group delay relative to the single filter realization as well as the in-band ripple of both filters.

13.1.1 Interpolated FIR Example

As an example of this process, consider the implementation of a digital filter with the following specifications:

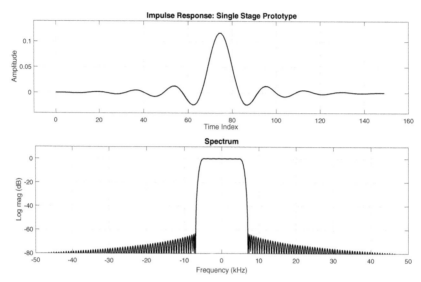

Figure 13.4   Impulse response and frequency response of prototype 150-tap filter.

Filter Type	Low Pass
Sample Rate	100 kHz
Pass Band	0–5 kHz
Stop Band	7–50 kHz
In-Band Ripple	0.1 dB pk-to-pk
Stop-Band Attenuation:	60 dB
Filter Length	150 Taps

A single-stage filter that meets these specifications requires 150 taps. Figure 13.4 presents the impulse response and frequency response of the single-stage filter.

The specifications for an interpolated two-stage filter designed to operate the input stage at one-fifth of the input rate, or 20 kHz followed by the interpolating filter operating at the input rate of 100 kHz, are listed next. As expected, the first stage filter length is one-fifth the length of the full sample rate prototype filter. The interpolating filter has a length of 40 taps, so the arithmetic workload for the cascade is 30 + 40 or 70 ops per input as opposed to the 150 ops per input for the single-stage prototype. Figure 13.5 presents the impulse response and frequency response of the two filter stages as well as the cascade of the two stages.

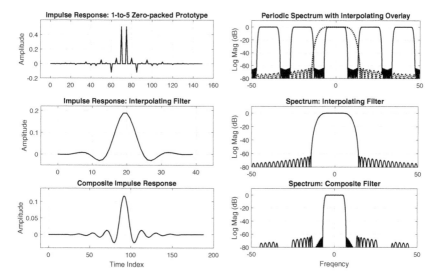

Figure 13.5 Impulse response of two cascade filters and composite interpolated response along with frequency responses of each impulse response.

	First Stage	**Second Stage**
Filter Type	Low Pass	Low Pass
Sample Rate	20 kHz	100 kHz
Pass Band	0–5 kHz	0-to-5 kHz
Stop Band	7–10 kHz	15–50 kHz
In-Band Ripple	0.08 dB pk-to-pk	0.02 dB pk-to-pk
Stop-Band Attenuation	60 dB	60 dB
Filter Length	30 Taps	40 Taps

As a final comparison, we can form a cascade of three filters that perform 5-to-1 down-sampling, band limiting, and 1-to-5 up-sampling of the form shown in Figure 13.1. The input and output stages of this cascade are polyphase partitions of the interpolator in the example just described, while the band limiting filter is the nonzero-packed version of the first-stage filter from the same example. Figure 13.6 presents one of the five impulse responses that can be observed at the filter output plus the frequency response of the three stages and the composite response of the three stages. The frequency responses have all been generated at the final output sample rate.

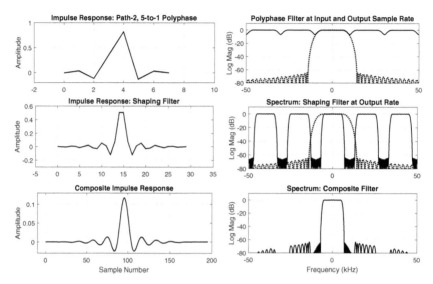

Figure 13.6   Impulse response of three cascade resampling filters and composite impulse response along with frequency responses of each impulse response.

When this third-option filter operates, for every five input points, the workload performed by the three stages are 40, 30, and 40 ops, respectively, for a total of 110 ops per five input samples. The workload per sample for this option is 22 ops per input sample. Interestingly, this option exhibits the smallest arithmetic workload even though it exhibits the longest composite impulse response. The length of the impulse response can be determined in the following manner: an applied impulse outputs an eight-point sequence from the first filter corresponding to one of the five legs of the input polyphase filter. This sequence then passes through the 30-tap filter to output a sequence of length 37 points. This 37-tap sequence in turn moves through the five legs of the output polyphase filter to output a sequence of length 224 points (37*5 + 40 − 1). Table 13.1 compares the arithmetic workload and the equivalent filter length of the three filter options examined in this example.

In this comparison, we note an observable truism about signal processing: to obtain the lowest processing burden, a signal-processing task should always be conducted at the lowest possible sample rate commensurate with the signal bandwidth!

Table 13.1 Comparison of three filter structures: coefficient lengths, arithmetic workload, and length of composite filter impulse response.

Option	Filter coefficients	Arithmetic workload	Impulse response length
Single stage	150 taps	150 ops/input	150 taps
Interpolated filter	30 and 40 taps	70 ops/Input	189 taps
5-to-1 and 1-to-5 Resampling filter	40, 30, and 40 taps	22 ops/input	224 taps

13.2 Spectral Masking Filters Based on Half-Band Filters

In the previous section, we discussed the task of controlling filter length where filter length is proportional to the ratio of sample rate to transition bandwidth. We also noted that narrow transition bandwidth is often associated with narrow pass-band bandwidth. The large ratio of sample rate to filter bandwidth permitted us to reduce the sample rate for the filter design and thus obtain a system design requiring a reduced number of coefficients. What design option is available when the ratio of sample rate to transition bandwidth is large but the pass-band bandwidth is too large to permit changes in sample rate? The primary design technique is known as *frequency masking*. This technique can be used to obtain narrow transition bandwidths for arbitrary pass-band or stop-band bandwidths.

The spectral masking design technique is an extension of the interpolated FIR (IFIR) filter design technique. For clarity of understanding, we first examine a restricted subset of this procedure that starts with a half-band filter and its complementary high-pass filter. Later we will extend the process to an arbitrary low-pass filter and its complementary high-pass filter. In this procedure, a half-band filter $H(Z)$ is designed with a desired transition bandwidth at $1/M$th of the desired sample rate. This filter is zero-packed 1-to-M to form the filter $H(Z^M)$. The zero-packed filter maintains the narrow transition bandwidth but exhibits a periodic pass band due to the zero-packing. The complementary filter of the zero-packed two-path polyphase filter also exhibits a periodic spectral pass band.

Masking filters following both filters eliminate selected subsets of their periodic pass bands. The signal paths with the two masked spectra are then added to form the composite spectrum. Equation (13.1) describes the Z-transform of the composite spectral masking filter, while Figure 13.7 presents the block diagram of the same filter. Figure 13.8 presents the spectral response of the separate filters and the response of them in cascade as well as the composite system.

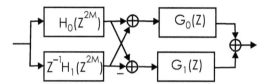

Figure 13.7   Block diagram of half-band spectral masking filter.

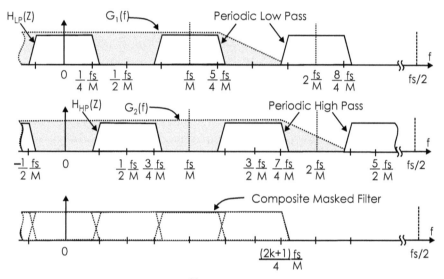

Figure 13.8   Spectra of top path, $H(Z^M)$ and masking filter $G_1(Z)$, spectra of bottom path, $H_C(Z^M)$, and masking filter $G_2(Z)$, and spectra of sum of two paths.

$$T(Z) = H(Z^M)G_0(Z) + H_C(Z^M)G_1(Z). \qquad (13.1)$$

We see that the frequency response of the composite filter has a pass-band edge that matches the upper edge of one of the spectral replicates and that the transition bandwidth matches that of the prototype low-pass filter. We also note from Figure 13.8 that the cutoff frequency of the composite masked filter is $f_s/4M$ above one of the spectral replicate center frequencies. The cutoff frequency is the term presented in Equation (13.2). Equation (13.3) reflects the constraint that the cutoff frequency must be less than the half sample rate. Combining these equations, we obtain Equation (13.4), which expresses all possible cutoff frequencies obtainable with masking filters for a range of M in a 1-to-M zero-packed half-band filter. Table 13.2 lists a set of frequencies that satisfies Equation (13.4) for a small range of up-sampling factor M. Here

Table 13.2 Possible ratios of 6 dB pass-band bandwidth-to-sample rate (f_c/f_s) for range of small integer (M) for 1-to-M zero-packing in masking filter.

	$k = 1$	$k = 2$	$k = 3$	$k = 3$	$k = 4$
$M = 2$	3/8	–	–	–	–
$M = 3$	3/12	5/12	–	–	–
$M = 4$	3/16	5/16	7/16	–	–
$M = 5$	3/20	5/20	7/20	9/20	–
$M = 6$	3/24	5/24	7/24	9/24	11/24

we see, for instance, that in a 1-to-5 up-sampled masking filter, we can obtain low-pass filters with band edge frequencies of 3/20, 5/20, 7/20, and 9/20 of the sample rate. As an example, if we have a sample rate of 100 kHz, we can design a masking filter based on a half-band prototype with a 6 dB cutoff frequency of 35 kHz.

$$f_C = \frac{2k + 1}{4M} f_S \tag{13.2}$$

$$2k + 1 < 2M \tag{13.3}$$

$$\frac{f_C}{f_S} = \frac{2k + 1}{4M}, \quad 2k + 1 < 2M. \tag{13.4}$$

We now illustrate the masking filter design process with an example that meets the following specification based on a 6 dB frequency equal to 7/20 of sample rate: from the listed specifications, we estimate the required filter length to be 151 taps.

Filter Type	Low Pass
Sample Rate	100 kHz
Pass Band	0–34 kHz
Stop Band	36–50 kHz
In-Band Ripple	0.1 dB pk-to-pk
Stop-Band Attenuation	60 dB
Filter Length	151 Taps

Figure 13.9 presents the impulse response and frequency response of the single-stage 151-tap filter designed to meet these specifications.

The specifications for a frequency masked two-path filter designed to operate the input stage at one-fifth of the input rate, or 20 kHz followed by the two masking filters operating at the input rate of 100 kHz, are listed next. We note the need for the stages to exhibit additional out-of-band attenuation

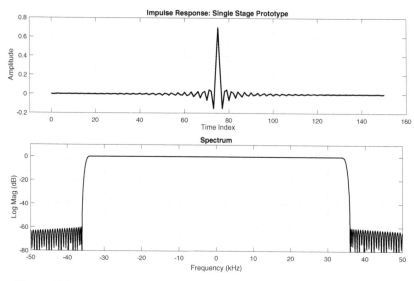

Figure 13.9   Impulse response and frequency response of prototype 151-tap filter.

to accommodate the increase in side-lobe levels that occur when summing the two paths. The half-band filter, selected to have a 2 kHz transition bandwidth centered at 5 kHz requires 37 taps. The masking filters with equal transition bandwidths are of the same length and require 35 taps to meet the listed specifications. The arithmetic workload for the cascade is 37 + 35 + 35 or 107 ops per input as opposed to the 151 ops per input for the single-stage prototype. The computational savings is approximately 30%. Figure 13.10 presents the impulse response and frequency response of the three filter stages as well as the composite response of the two paths.

	First Stage	**Mask-1**	**Mask-2**
Filter Type	Half Band	Low Pass	Low Pass
Sample Rate	20 kHz	100 kHz	100 kHz
Pass Band	0−4 kHz	0−26 kHz	0−36 kHz
Stop Band	6−10 kHz	35−50 kHz	45−50 kHz
In-Band Ripple	0.025 dB	0.025 dB	0.025 dB
Stop-Band Attenuation	66 dB	65 dB	65 dB
Filter Length	37 Taps	35 Taps	35 Taps

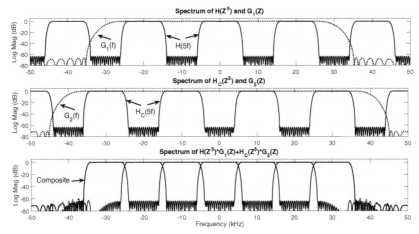

Figure 13.10   Spectra of up-sampled 1-to-5 half-band low-pass filter and first masking filter, of up-sampled 1-to-5 half-band high-pass filter and second masking filter, and of composite sum filter.

13.3 Spectral Masking Filters: Complementary Filters

A half-band filter was selected for the prototype in the previous section because it permitted us to determine a simple relationship for the possible band edge frequencies available in the masking structure. We can modify the design process slightly so that arbitrary band edge frequencies can be obtained from the same structure. The more general procedure starts with a prototype filter $H(Z)$ with bandwidth $f_c = \alpha f_s$. The parameter α is nominally between 0 and 0.5 but is usually near 0.25. This filter is designed with the desired transition bandwidth at $1/M$th of the sample rate and, as earlier, is zero-packed to form the filter $H(Z^M)$. The up-sampled filter maintains the narrow transition bandwidth but exhibits the periodic pass band due to the zero-packing. The filter must have an odd number of coefficients to be able to form its complement as the difference between a delay path and the zero-packed filter. The $(N - 1)/2$ units of delay is the midpoint of the filter $H(Z^M)$. The Z-transform of the complementary filter $H_C(Z^M)$ is shown in the following equation:

$$H_C(Z^M) = (Z^{-(N-1)/2} - 1)H(Z^M). \qquad (13.5)$$

We want the complementary filter to have the same stop-band attenuation as that of the prototype filter. To achieve this goal, the prototype filter must have equal-valued pass-band and stop-band ripple. This is accomplished by setting equal-valued penalty function weights in the Remez algorithm. The procedure now follows the steps derived in the previous example except that we have to select an appropriate resampling rate. We select this rate by comparing the ratios of Table 13.2 to the desired ratio of bandwidth to sample rate and selecting an up-sample rate corresponding to a close ratio. We then sketch the replicated spectra at this up-sample ratio and adjust the bandwidth of the baseband prototype to move the replica band edge to the desired cutoff frequency.

We now illustrate the general masking filter design process with an example that meets the specifications outlined in the previous example except that the band edge has been changed from 34 to 30 kHz.

Filter Type	Low Pass
Sample Rate	100 kHz
Pass Band	0–30 kHz
Stop Band	32–50 kHz
In-Band Ripple	0.1 dB pk-to-pk
Stop-Band Attenuation	60 dB
Filter Length	151 Taps

Again we find that a single-stage filter that meets these specifications requires 151 taps, but we seek a more efficient solution. We require a masking filter with a normalized band edge of 0.3. Examining the entries in Table 13.2, we find that the ratio 7/24 or 0.292 in the 1-to-6 up-sample is the closest entry with the ratio 5/16 or 0.312 in the 1-to-4 up-sample row a close second. Selecting the 1-to-4 up-sample, we form the sketch shown in Figure 13.11 and reduce the upper band edge at the first replicate from 31.25 to 30 by reducing the baseband bandwidth from 6.25 to 5.

The specifications for a shifted frequency mask two-path filter reflecting the shifted edges indicated in Figure 13.11 are listed next. We note the additional out-of-band attenuation required to accommodate the increase in side-lobe levels that occur when summing the two paths. Also note that the in-band ripple and out-of-band attenuation of the first stage correspond to equal values of ripple (0.56×10^{-3}). The masking filters now have different transition bandwidths and, hence, have different lengths of 25 and 33 taps. The time series output from the two paths must be time aligned by adding 4 units of delay to the output of the top path, which has the shorter filter. The arithmetic workload for this frequency-masked filter is 45 + 25 + 33 or 103 ops per input as opposed to the 151 ops per input for the single-stage

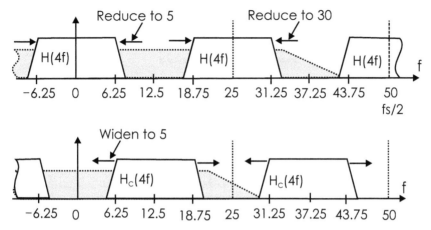

Figure 13.11 Initial spectral representation of 1-to-4 up-sampled baseband filter.

prototype. This represents a computational savings of approximately 32%. Figure 13.12 presents the impulse response and frequency response of the separate paths and of the composite filter.

	First Stage	**Mask-1**	**Mask-2**
Filter Type	Low Pass	Low Pass	Low Pass
Sample Rate	25 kHz	100 kHz	100 kHz
Pass Band	0−5 kHz	0−31 kHz	0−20 kHz
Stop Band	7−12.5 kHz	44−50 kHz	30−50 kHz
In-Band Ripple	0.005 dB	0.045 dB	0.045 dB
Stop-Band Attenuation	60 dB	60 dB	60 dB
Filter Length	45 Taps	25 Taps	33 Taps

13.4 Proportional Bandwidth Filter Banks

Proportional bandwidth filter banks are used to obtain a spectral partition in which the center frequencies and the bandwidths of the filters are equally spaced on a logarithmic scale or logarithmically spaced on a linear scale. Filter banks with this property are often called constant Q filter banks where Q in this application is the ratio of center frequency to bandwidth. Vibration analysis, audio analysis, and signal detection and classification of periodic signals with harmonic components are best performed in constant Q filter banks. The graphic equalizer of a home entertainment system has a constant

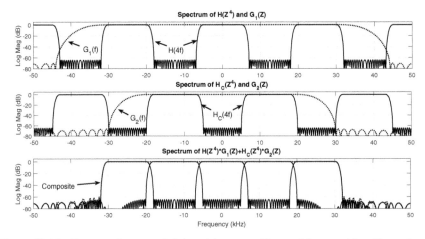

Figure 13.12   Spectra of up-sampled 1-to-4 low-pass filter and first masking filter, of up-sampled 1-to-4 complementary filter and second masking filter, and of composite sum filter.

Q filter bank. Audio instruments monitoring sound pressure level for noise abatement have constant Q filter banks with each filter spanning one-third or one-sixth of an octave.

13.4.1 Octave Partition

In this section, we examine the standard approach to proportional bandwidth filtering. A processing scheme is designed to decompose a single octave of bandwidth located near the quarter sample rate. With appropriate half-band filtering and sample rate reduction, the frequency band occupying the next lowest octave slides up to occupy the same spectral interval at the reduced sample rate. This octave shifting due to 2-to-1 resampling is illustrated in Figure 13.13.

Thus, a single octave processing scheme is applied to successively lower octaves as prefiltering and resampling iteratively delivers the sequence of octaves to the same fractional bandwidth at successively reduced sample rates. The signal flow of the multioctave process is shown in Figure 13.14. The suboctave processor extracts spectra from the top octave partitioning that octave with a bank of fractional octave filters. The filters used for the octave processing and for the half-band processing can be FIR or infinite impulse response (IIR). IIR filters are often used to perform the spectral partition when the application is spectral analysis, audio equalization, or audio compression.

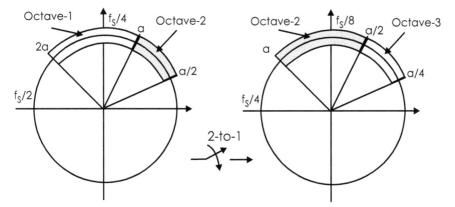

Figure 13.13 Spectral band, octave $K + 1$ (octave-2) sampled at f_s shifts to location of octave K (octave-1) when 2-to-1 down-sampled to $f_s/2$.

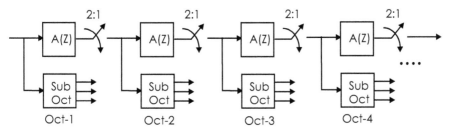

Figure 13.14 Multioctave processing iteratively filter and down-sample to access successive octaves.

FIR filters are used for image processing or other applications requiring linear phase.

An interesting feature of the multioctave processor is the aggregate workload to implement the system. Let us assume that the processing workload of the first stage, the input stage that forms the top octave and the first half-band filter, is N operations per input sample. The next stage requires the same workload, but operating at half the clock rate requires $N/2$ operations per input sample. The workload for the sequence of stages is the sum shown in Equation (13.6). Here we see that the workload for an arbitrarily large number of octaves is less than twice the workload of the first octave.

$$\frac{\text{OPS}}{\text{Input}} < \{N + N/2 + N/4 + N/8 + \cdots\} < 2N \qquad (13.6)$$

13.4.2 Proportional Bandwidth Filters

Suppose we have M filters spanning a frequency interval in equal increments on a log scale. We first address the center frequencies of the filter bank. Let f_0 be center frequency of the lowest frequency filter in the filter bank. The kth proportional bandwidth filter is located at the center frequency shown in the following equation:

$$f_k = f_0(1+r)^k. \tag{13.7}$$

We can solve for the parameter r if we evaluate Equation (13.7) with f_k set to the Mth center frequency and the exponent k set to $M - 1$ as shown in the following equation:

$$f_M = f_0(1+r)^M. \tag{13.8}$$

To design the filters, we have to identify the band edges of each filter as well as assure that the lower band edge of the first filter and the upper band edge of the Mth filter correspond to the bandwidth spanned by the filter bank. We modify Equations (13.7) and (13.8) to partition the inclusive interval into $2M$ increments and then use odd indices in the exponent to identify the band centers and even indices to identify band edges. The modified form is shown in the following equation:

$$f_{\text{Low-Edge}} = f_{\text{High-Edge}}(1+s)^{2M}. \tag{13.9}$$

After solving for the parameter s, we can use the expressions of Equation (13.10) to solve for the critical frequencies of the proportional bandwidth filter bank.

$$\left.\begin{array}{rcl} k\text{th Lower Band Edge} &=& f_{\text{Low-Edge}}(1-s)^{2k-2} \\ k\text{th Center Frequency} &=& f_{\text{Low-Edge}}(1-s)^{2k-1} \\ k\text{th Upper Band Edge} &=& f_{\text{Low-Edge}}(1-s)^{2k} \end{array}\right\} \quad k = 1, 2, \ldots, M. \tag{13.10}$$

13.4.2.1 Example Proportional Bandwidth Design

Let us examine a standard 15-band, International Organization for Standardization (ISO) 1/3 octave filter bank used as a graphic equalizer in a high-end audio system. For this design, the input sample rate is 48 kHz, with 15 frequency bands spanning the frequency interval 20–20 kHz with center frequencies listed here.

ISO Center Frequencies

25, 40, 63, 100, 160, 250, 400, 630, 1000, 1600, 2500, 4000, 6300, 10,000, 16,000.

If we apply Equation (13.10) to these requirements, we can solve for the term $(1 + s)$ and then verify the center frequencies of the filter bank. This is done in the following equations:

$$20,000 = 20 (1 + s)^{30}$$
$$30 \log_{10}(1 + s) = 1000 \quad\quad (13.11)$$
$$(1 + s) = 10^{\frac{3}{30}} = 1.2589$$

$$f_S = 20 (1.2589...)^{(2k-1)}, k = 1, 2, \ldots, M. \quad\quad (13.12)$$

Table 13.3 lists the center frequencies obtained by solving Equation (13.12) and the corresponding center frequencies identified in the ISO standard. We see that, applying standard rules for identifying parameters for instruments, the ISO frequencies have been adjusted so that in every five intervals, the values repeat in increments of 10 as in, 25, 250, and 2500 Hz. The table also lists the computed band edges of the proportional bandwidth filters without the shifted correction to match the ISO frequency list.

Starting at the higher frequency bands, we observe that bands 15 and 14 bracket 12 kHz, the quarter sample rate. These bands must be implemented at the full sample rate of 48 kHz and the remaining bands can be processed at the reduced rate of 24 kHz after a half-band filter and 2-to-1 down-sampling. Continuing up the list, we note that bands 13 and 12 bracket (or are close) to 6 kHz, the new quarter sample rate, and must be implemented at the 24 kHz sample rate. The "close to" consideration is used to avoid the condition that the filter transition bandwidth crosses the quarter sample rate. The remaining filters can be processed at a reduced rate after a second half-band filter and a second 2-to-1 down-sampling. In this manner, we can allocate where in the down-sample chain each channel filter has to be implemented. This partition was performed for the list of frequency edges listed in Table 13.3 and filters were designed to obtain the band edges at the selected sample rate.

To meet the requirements of all 15 filters at their corresponding sample rate, the filters were designed with a normalized unity sample rate. Examining the filters, we found that only four filters operating at their appropriate sample rate were required to implement all 15 filters. Table 13.4 presents the filters in the bank, the sample rate at which they operate, their (uncorrected) normalized frequency edges, and the implementation assignments. A fifth, half-band, filter was required to finish the design suite. Figure 13.15 presents the filter partition for the 15-band proportional bandwidth filter bank. As shown, the filter bank performs as a spectrum analyzer. To operate as a graphic equalizer, gain control would be applied to the filter outputs and

Table 13.3 Computed and ISO standard center frequencies of 15-band proportional bandwidth filter bank, and also, lower and upper edge of same bands.

Filter number	Computed frequency	ISO standard frequency	Lower edge	Upper edge
1	25.18	25	20	32
2	39.91	40	32	50
3	63.25	63	50	80
4	100.24	100	80	125
5	158.87	160	126	200
6	251.79	250	200	317
7	399.05	400	317	502
8	632.46	630	502	796
9	1002.38	1000	796	1262
10	1588.66	1600	1262	2000
11	2517.85	2500	2000	3170
12	3990.52	4000	3170	5024
13	6324.56	6300	5024	7962
14	10,023.74	10,000	7962	12,619
15	15,886.56	16,000	12,619	20,000

Table 13.4 Down-sampled sample rate and normalized band edge frequencies for 15-band proportional bandwidth filter bank.

Filter number	Sample rate	Lower edge	Upper edge	Filter type
1	93.75	0.2133	0.3413	C
2	187.5	0.1707	0.2667	B
3	375	0.1333	0.2133	D
4	375	0.2133	0.3333	C
5	750	0.1680	0.2667	B
6	1500	0.1333	0.2113	D
7	1500	0.2113	0.3347	C
8	3000	0.1673	0.2653	B
9	3000	0.2653	0.4207	A
10	6000	0.2103	0.3333	C
11	12,000	0.1667	0.2642	B
12	24,000	0.1321	0.2093	D
13	24,000	0.2093	0.3318	C
14	48,000	0.1659	0.2629	B
15	48,000	0.2629	0.4167	A

then the outputs would be up-sampled and merged by a filter bank of dual structures.

Figure 13.16 presents the frequency response of the first 10 proportional bandwidth filters in the filter bank, plus the four prototype filters identified

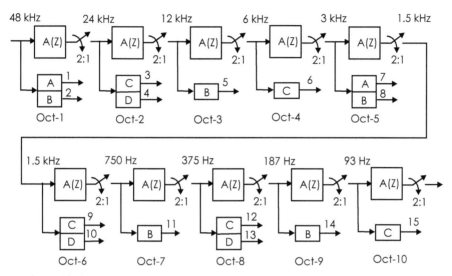

Figure 13.15   Filter partition for 15-band proportional bandwidth analysis filter bank.

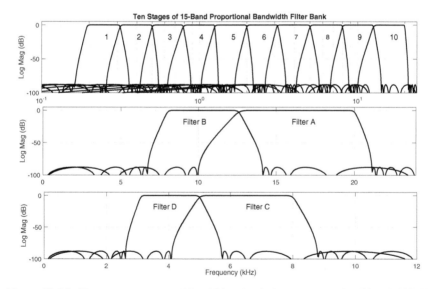

Figure 13.16   Frequency response of first 10 bands of 15-stage proportional bandwidth filter bank and four filter types distributed through half-band filter chain.

in Table 13.4 as filter types A, B, C, and D. Each of these absurdly good filters was designed as a two-path polyphase IIR filter of the type presented in Chapter 11. Each is a nine-pole band-pass filter requiring 19 multiplies per output sample. More traditional filter banks use three-pole Butterworth filters that in the polyphase IIR form would require six multiplies per output sample. The half-band filter is also a two-path recursive polyphase structure requiring four multiplies per output sample to form the nine poles of the low pass. The half-band filter operates at half rate with the resampling being performed prior to the filtering operation. An upper bound to the processing workload for the entire filter chain can be estimated by examining the workload for the first stage. This workload is seen to be

First Stage:	Half-Band	2 ops/input
	Filter A	19 ops/input
	Filter B	19 ops/input
	Total First Stage	40 ops/input

The workload for the entire chain must be less than twice this workload, which is less than 80 arithmetic operations per input sample to obtain the output of 15 filters. The actual count is 72 operations per input sample. The workload per filter is seen to less than five operations per input sample. Implementing the proportional bandwidth filters as third-order Butterworth prototypes would reduce the workload for the entire 15-band filter bank to approximately 26 operations per input sample. Amortized over the 15 filters, this is less than two operations per input sample per filter.

13.4.2.2 Fractional Bandwidth Design Example

This example examines a 1/12-octave proportional bandwidth spectrum analyzer. The design decomposes an octave band of frequencies spanning the normalized frequency interval of 3/16−6/16 into 12 proportional bandwidth filters. The same 12-band filter bank processes successive octaves after they are filtered and down-sampled 2-to-1 by a half-band filter. The chain is repeated 10 times to span the three decades from 20 Hz to 20 kHz. The 12 filters spanning the octave were implemented as two-path recursive polyphase structures requiring six multiplies to form each three-pole band-pass elliptic filter. The half-band filter was also designed as a two-path recursive polyphase structure implemented as a resampling filter requiring three multiplies

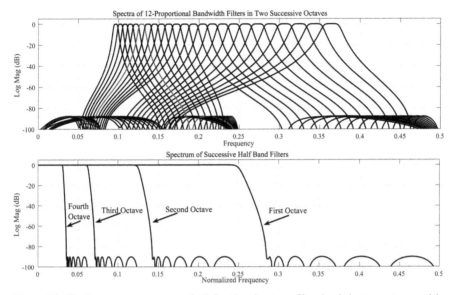

Figure 13.17   Frequency response of 12-fractional octave filter bank in top octave and in next lower octave and spectra of four successive half-band filter response.

per input sample. The workload required to process the first octave is seen to be

First Stage:	Half Band	3 ops/input
	12-Proportional Bandwidth Filters	72 ops/input
	Total First Stage	75 ops/input

The workload for the entire chain must be less than twice this workload, which is less than 150 arithmetic operations per input sample to obtain the output of the 12 filters per octave for each of the 10 octaves spanned by the process. A total of 120 filter outputs are formed by this process, and amortizing the workload over the 120 filters, we find that the workload per filter is 1.25 operations per filter per input sample. Figure 13.17 presents the frequency response of the 12-fractional octave filters spanning the top octave from 3/16 to 6/16 along with the response of the same filters in the next lower octave. Also shown is the frequency response of the first four half-band filters operating at their successively lower normalized sample rates of 1, 1/2, 1/4, and 1/8.

13.5 10-Channel Audiometric Filter Bank Example

An interesting example of a cascade multirate filter bank is the set of filters found in a modern hearing aid. The hearing aid function is to improve speech understanding and user comfort. At the simplest level, the hearing aid contains a microphone, and an analog-to-digital converter (ADC) that collects and digitizes acoustic signals. Its processor forms multiple reduced bandwidth time series from different frequency bands and applies various signal enhancement algorithms to them. The modified signals are recombined and delivered to a digital-to-analog converter (DAC), analog amplifier, and speaker which present the modified acoustic signal to the ear canal. Signal enhancements include channel-dependent amplification, frequency warping or displacement, per channel automatic gain control, channel noise suppression, acoustic feedback suppression, dynamic range compression, and more. Small physical sizes as well as small processing delays are important. Batteries are quite small and battery power is very limited.

Our interest is the audiometric channelizer. We run into the problem that there is no single standard defining these channelizers. The structures of the channelizers are considered company private and are not, in general, available to the community. One open source (Jim Kates) lists six center frequencies at [0.25, 0.50, 1.0, 2.0, 4.0, and 6.0] kHz and many audiology instruments and hearing aids use filters at these center frequencies. We have found a number of tabulated lists with additional center frequencies which we show graphically in Figure 13.18. The sample rate of the signal collection system is 32 kHz. The lists describe band-pass filters more or less centered at the listed frequencies. The two filters at low and high end of the lists are exceptions. The filter corresponding to the lowest and highest frequencies are low-pass and high-pass filters. We will use the frequency centers of the 10-channel channelizer, labeled Signia, for the remainder of this example and make no claims that our example reflects the actual performance of their filter bank implementation.

In-band ripple and stop-band attenuation levels of the filter are chosen from tabulated values found in an ANSI Standard for octave-band and fractional-octave-band analog and digital filters. For the example we present here, we use 0.05 dB in-band ripple and 70 dB stop-band attenuation, as used in the Sokolova filter bank paper.

An additional problem we face is that the community has not identified the bandwidths or crossover frequencies of adjacent frequency bands or the transition bandwidths of the filters. We could choose the crossover

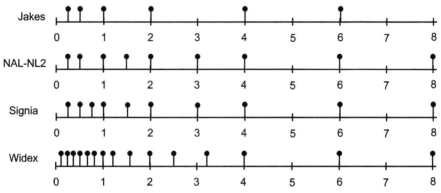

Figure 13.18 Graphical representation of center frequencies in audiometric filter banks with 6, 9, 10, and 15 channel filters.

frequencies to be the midpoints between adjacent centers on a linear frequency scale (algebraic mean) or on a logarithmic frequency scale (geometric mean). This would make sense if there was an algorithmic assignment of the center frequencies as there was for the graphic equalizer of the earlier section. Since the center frequencies were selected for convenience, we can do the same and select the crossover frequencies to balance the asymmetries around the center frequencies. This is what was done in the stylized spectral responses shown in Figure 13.19.

As we move toward lower frequencies, we notice that the spacing between channel centers is reduced by a factor of 2 after equally spaced filter pairs. Specifically, the top two filters are separated by 2.0 kHz, the next two by 1.0 kHz, and the next two by 0.5 kHz. This is a proportional bandwidth spacing of frequency centers for the list of frequencies {0.25, 0.5, 1, 2, 4, 8}. The thought occurs to us: "what would be the center frequencies of a proportional BW bank with centers separated by multiples of sqrt(2)?" For instance, the frequency between 4 and 8 would be sqrt(4*8) or 5.66, which is reasonably close to 6, the listed frequency between 4 and 8. Similarly, the frequency between 2 and 4 would be sqrt(2*4) or 2.83 which is reasonably close to 3, the listed frequency between 2 and 4. We decided to design the filter set based on the proportional BW for the top seven filters and resort to equal spacing for the bottom three filters. This option also supports the spacing and bandwidths suggested in Figure 13.19 but has the benefit of an algorithmic relationship which is easy to adjust by a simple common scaling factor.

Since the spacing between center frequencies at the low end of the spectrum is smaller than the spacing at the high end, their bandwidths must

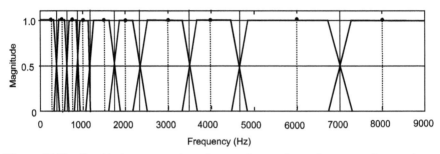

Figure 13.19   Graphical representation of center frequencies and crossover frequencies at the band edges in 10-channel audiometric filter bank.

also be smaller. Smaller bandwidths suggest narrower transition bandwidths, which in turn means longer filters. Longer filters mean additional signal-processing workload which would be a matter of concern for the small batteries in the hearing aid. We note that after every filter pair, we can reduce the sample rate by a factor of 2 because the bandwidth has been reduced by a factor of 2. We can then form the next filter pair with reduced bandwidth at reduced sample rate at half the workload of the previous filter pair. This is where the multirate filtering comes to play. Without the sample rate reduction, the reduced bandwidth filter, with commensurate reduce transition bandwidth would require successive double length filters to meet the filter specifications. Reducing the sample rate preserves the filter length but reduces the workload by operating at half the clock speed. The sample rate reduction is performed by a sequence of 2-to-1 down-sampling half-band filters and the original sample rate for each filter in the bank is retrieved by a sequence of 1-to-2 up-sampling half-band filters.

13.5.1 Signal Reconstruction in Synthesis Filter Bank

A sufficient condition to obtain perfect reconstruction of the composite output signal from the channelized components is that the transition bands of the adjacent filters are odd symmetric about their midpoints and have complementary amplitudes. In earlier reconstruction filter banks, we interleaved adjacent pairs of frequency bands with different bandwidths but with the same length, and hence equal transition bandwidths which automatically supported perfect reconstruction. On the other hand, the graphic equalizer which had a succession of narrower bandwidth filters with successfully narrower transition bandwidths was not required to form a perfect reconstruction composite signal, so we had no need to match transition bandwidths of adjacent channels in that filter bank. What is unique about the audiometric filter bank is that we are required to do so here with its output channels.

Many of the band-pass filters in the channelizer have a wider bandwidth filter to their right and a narrower bandwidth filter to their left. Since adjacent bandwidths change by the sqrt(2), each of the proportional bandwidth filters has the transition bandwidth on its left side narrower by sqrt(2) than the transition bandwidth on its right side. The lowest three frequency filters have the same frequency spacing and hence have the same bandwidth and same transition bandwidth on the left and right sides. These are simple to implement. Having different transition bandwidths on the two sides of band-pass filters requires a bit more work. We start the design process with two low-pass FIR filters designed with pass bands at the top two frequencies with bandwidth ratios sqrt(2), 8 kHz, and 5.657 kHz. Since the lower frequency has a bandwidth and transition bandwidth smaller than its right most neighbor by sqrt(2), then its length must be larger by the sqrt(2). For instance, if the top frequency filter has a length of 101 taps (always an odd number), then the next filter length must be 101*sqrt(2) or 143 taps. The two low-pass filters each have a complementary high-pass filter obtained from the difference of the center tap unity weight and the low-pass output. We can identify h1a and h1b as the low pass and high pass of the first filter and h2a and h2b as the low pass and high pass of the second filter. The cascade of these filters along with 2-to-1 down-samplers can form the impulse responses of all the channel filters of the proportional bandwidth filter bank. The signal flow diagram showing the operation of this cascade for the first six filters is shown in Figure 13.20. If we apply an impulse at the far left input point, we will see the impulse response g1 formed by the high-pass filter h1b which is the high-pass channel filter formed at the 32 kHz sample rate. The input low-pass filter response h1a is band-limited by the high-pass filter h2b to form the time response x2b. The response x2b is almost the band-pass filter centered near 6 kHz formed at the 32 kHz sample rate. A bit of house cleaning editing of this response forms the desired impulse response g2. We will discuss the editing shortly. The complementary output from h2a forms the time series x2a which is down-sampled 2-to-1 and then filtered with the high-pass filter h1b to form the time series x3b. The response x3b is almost the band-pass filter centered near 4 kHz. This sequence requires some editing to form the desired output g3.

The cascade of filter h1a low-pass with filter h2b high-pass forms, within an editing operation, the impulse response of the band-pass filters with asymmetric transition bandwidths. By design, the left transition bandwidth

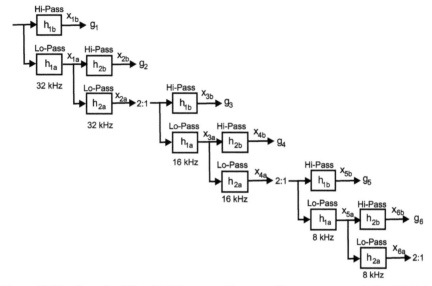

Figure 13.20   Cascade of filter h1 high-pass and low-pass filter responses with filter h2 high-pass and low-pass filter responses with alternate stage 2-to-1 down-sampling to form impulse responses of channelizer filters.

is narrower by sqrt(2) than the right transition bandwidth.. This is shown in Figure 13.21. Here we see the low-pass filter spectral response of the first filter in the cascade is altered by the high-pass filter spectral response of the second filter to form the desired band-pass filter spectral response. We see how the low-pass filter imparts its transition bandwidth to the right side of the band-pass spectral response, while the longer high-pass filter imparts its narrower transition bandwidth to the left side of the band-pass spectral response. The spectral response of the 10-channel audiometric filter bank is shown in Figure 13.22. Here we can clearly see the reduced transition bandwidths on the two sides of the seven successive reduced bandwidth channel filters.

We have two final comments on the channelizer filter design. The first comment directed toward the filter design algorithm for the two low-pass filters we have been calling h1a and h2a. We have already described why they are of different lengths and the lengths must be odd. We expect that their complementary transition bandwidths are odd symmetric about their half amplitude gain level to obtain perfect reconstruction. Interestingly, this happens automatically if the filters are designed as windowed sinc sequences but does not happen when the filters are designed with the FIRPM (or Remez)

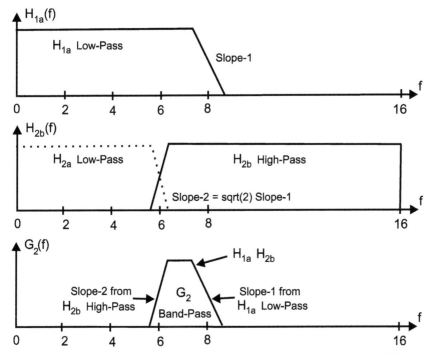

Figure 13.21 Spectral description of interaction between low-pass filter H1a and high-pass filter H2b to form band-pass filter with steeper transition bandwidth on left side than on right side.

algorithm. Thus, the design of the two low-pass filters h1a and h2a should use the windowed sinc. This is true for the seven upper frequency filters as well as for the three lower frequency filters. The final comment is the housecleaning editing operation referred to earlier. We noted that the channel filter impulse responses g2 shown in Figure 13.20 were obtained from the impulse response series x2b formed by the cascade of two filters h1a and h2b. The convolution of the two impulse response forms a sequence of length 1 less than the sum of their lengths. There is a ring up transient response as the two filters engage their weights and a ring down transient response as the two filters disengage their weights. Both sets of weights are quite small at the start and end of the transients. Small value weights times small value weights results in very small value output weights. These can be edited by selectively discarding them without changing the frequency response of the filter. If we remove too many of the weights, we will see increased levels of stop-band side-lobe levels, so we should monitor the spectrum of the edited weight set and stop removing

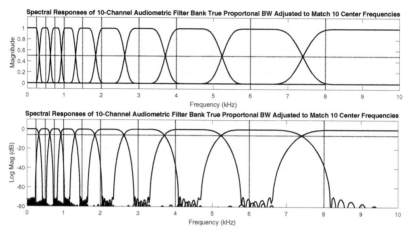

Figure 13.22   Spectral description of interaction between low-pass filter H1a and high-pass filter H2b to form band-pass filter with steeper transition bandwidth on left side than on right side.

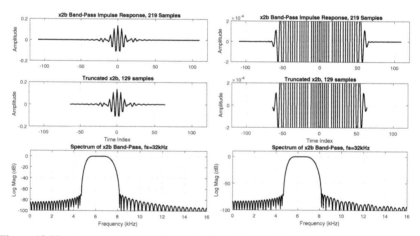

Figure 13.23   Impulse response of cascade filters h1a and h2b. Original 219-point sequence with zoom to lower level amplitudes. Edited to 129-point sequence with zoom to lower level amplitudes. Spectra of original sequence and edited sequence.

weights when the stop-band side lobes start to rise. What we do is go to the center weight and extract weights symmetrically about the center as shown in Figure 13.13. We can expect that the edited length will match the filter length that formed the spectrum with desired transition bandwidth, which was 129 points in this example. Figure 13.23 shows the 219-point time series x2b

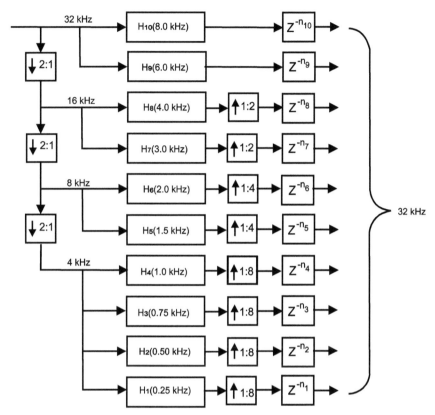

Figure 13.24 Block diagram of audiometric 10-channel channelizer. Impulse responses g1 through g10 of filters formed in Figure 13.20 and edited to reduced lengths as done in Figure 13.23 are inserted in this block diagram. The half-band filters in the three 2-to-1 down-sampling arms at the input port and the various up-sampling filters and inserted time delay at the output port have spectral requirements that are functions of the transition bandwidths of the filter bank spectra.

obtained for the first channel filter in Figure 13.20 and then edited down to the 129-point time series g2. As seen, the spectra of the two time series are essentially the same in their pass-band regions. Since the first two filters in the design chain have different lengths, the successive filters in the chain will have the same alternating lengths. If the center frequency is a binary fraction of the long filters center frequency, it will also have the long length, and if its center frequency is a binary fraction of the short filters center frequency, it will also have the short length. These lengths were 129 and 91 for the design illustrated in this example.

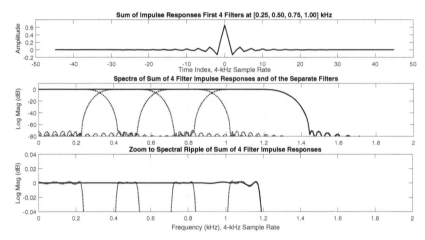

Figure 13.25 Sum of four lowest channel time responses and reconstructed spectrum response. Note extremely small level of ripple in reconstructed spectrum.

$$g2 = x2b(110 + (-64 : 64)). \tag{13.13}$$

Figure 13.24 presents the block diagram of the 10-channel audiometric channelizer. The top two channel filters are the high-pass filter and the first band-pass filter both operating at the 32 kHz input sample rate. All the filters are even symmetric and the filtering operation can fold the data samples about their midpoint to save half of the multiplies. The same input signal is filtered and down-sampled by the first half-band filter and presented to the next pair of channel filters operating at half sample rate of 16 kHz. This operation is repeated two more times till we reach the 4 kHz sample rate. The first filter in the last set of 4 is the last one with an asymmetric transition bandwidth. The next two filters are equally spaced and have the same bandwidth and symmetric transition bandwidth. These are designed as low-pass filters with the transition bandwidth of filter 4 and are translated to be cosine heterodynes to the spectral position that minimizes spectral reconstruction error. The last filter in this set is a low-pass filter of the same length, hence same transition bandwidth with a pass band adjusted to minimize spectral reconstruction error with its adjacent filter. A little bit of adjusting and nudging corners results in a remarkably small reconstruction error. The time and frequency response of the sum of the four low frequency filters is shown in Figure 13.25.

References

Crochiere, Ronald and Lawrence Rabiner. *Multirate Signal Processing*, Englewood Cliff, NJ, Prentice-Hall Inc., 1983.

Fliege, Norbert. *Multirate Digital Signal Processing: Multirate Systems, Filter Banks, Wavelets*, West Sussex, John Wiley & Sons, Ltd, 1994.

Jovanovic-Dolecek, Gordana. *Multirate Systems: Design and Applications*, London, Idea Group, 2002. Mitra, Sanjit. *Digital Signal Processing: A Computer-Based Approach*, 2nd ed., New York, McGraw-Hill, 2001.

Mitra, Sanjit and James Kaiser, *Handbook for Digital Signal Processing*, New York, John Wiley & Sons, 1993.

Vaidyanathan, P. P. *Multirate Systems and Filter Banks"*, Englewood Cliff, NJ, Prentice-Hall Inc., 1993.

Alice Sokolova, Dhiman Sengupta, Kuan-Lin Chen, Rajesh Gupta, Baris, Aksanli, fredric harris, Harinath Garudadri, *"Multirate Audiometric Filter bank for Hearing Aid Devices"* Asilomar Conference on Signals, Systems and Computers, 2020, Pacific Grove, California 1-4 November, 2020.

James M. Jakes, *Digital Hearing Aids*, Plural Publishing, San Diego, CA, 2008

Problems

13.1 Design a FIR filter with the following specifications:

Sample Rate	1.0 MHz		
Pass Band	$0-10$ kHz	Pass-Band Ripple	0.1 dB
Stop Band	$20-f_s/2$ kHz	Stop-Band Attenuation	60.0 dB

Now repeat the same design for a sample rate of 50 kHz. Note that the filter length is proportional to the sample rate.

Now zero-pack the shorter filter to increase its length to the high data rate filter. The zero-packing raises its sample rate and accesses the 20 spectral copies at the higher sample rate. Now design an interpolating filter to suppress the spectral copies to –60 dB. Determine and plot the frequency response of the cascade filter pair. Compute and compare the workload required to implement the single large filter and the cascade interpolated filter.

13.2 Design a FIR filter with the following specifications:

Sample Rate	1.0 MHz		
Pass Band	0–10 kHz	Pass-Band Ripple	0.1 dB
Stop Band	20–f_s/2 kHz	Stop-Band Attenuation	60.0 dB

Now repeat the same design for a sample rate of 500 kHz. Note that the filter length is proportional to the sample rate.

Design a half-band filter to lower the bandwidth and sample rate 2-to-1. Form a composite filter as a cascade of three filters, a half-band 2-to-1 filter, the low pass filter, and a half-band 1-to-2 filter. Determine the workload for the cascade and compare it to the workload of the single-stage filter.

13.3 Design a FIR filter with the following specifications:

Sample Rate;	1.0 MHz		
Pass Band:	0–10 kHz	Pass-Band Ripple	0.1 dB
Stop Band	20–f_s/2 kHz	Stop-Band Attenuation	60.0 dB

Now repeat the same design for a sample rate of 250 kHz. Note that the filter length is proportional to the sample rate.

Design a cascade of two half-band filters to lower the bandwidth and sample rate 4-to-1. Form a composite filter as a cascade of three filter sets, the two half-band 2-to-1 filters, the low-pass filter, and the two half-band 1-to-2 filters. Determine the workload for the cascade and compare it to the workload of the single-stage filter.

13.4 We are going to design a FIR filter with the following specifications:

Sample Rate	80 kHz		
Pass Band	0–30 kHz	Pass-Band Ripple	0.1 dB
Stop Band	35–40 kHz	Stop-Band Attenuation	60.0 dB

This is our reference design. We now design a FIR filter to up-sample and mask. The prototype filter has the following specifications:

Sample Rate	40 kHz		
Pass Band	0–10 kHz	Pass-Band Ripple:	0.1 dB
Stop Band	15–20 kHz	Stop-Band Attenuation	60.0 dB

This second filter is zero-packed 1-to-2 and partitioned into a two-path polyphase filter as shown in Figure 13.7. We form the low-pass and high-pass legs of this filter. Determine and plot the frequency response of the replicated low-pass and high-pass filters. Now design the masking filter to reject the high frequency replicate in the low-pass path. Be sure this filter has an odd number of taps so that we can align the lower leg with the filtered upper leg. Mask the upper leg with its second filter and form the sum with its output and the delayed lower leg. Verify that the masked filter satisfies the filter specification. Compare the two filters, the masked and the direct form. Compare the workload required to implement the two filters.

13.5 Design a set of three proportional bandwidth IIR filters that span an octave of frequency between $f_s/8$ and $f_s/4$. The filters will be third-order inverse Tchebyschev filters, sometimes referred to as Tchebyschev-II filters. Determine the normalized band edges of the filter set. These band edges will be the 3 dB crossover points of adjacent filters, and the filters will exhibit 60 dB out-of-band attenuation. Plot the frequency response of the three filters.

13.6 Design a set of three proportional bandwidth FIR filters that span an octave of frequency between $f_s/8$ and $f_s/4$. Determine the normalized band edges of the filter set. These band edges will be the 1 dB crossover points of adjacent filters, and the filters will have a transition bandwidth that is 60 dB down in the center of the next lowest frequency band and exhibit $1/f$ spectral side lobes thereafter. Plot the frequency response of the three filters.

13.7 Design a set of three proportional bandwidth FIR filters that span an octave of frequency between $3/16\ f_s$ and $3/8\ f_s$. Determine the normalized band edges of the filter set. These band edges will be the 1 dB crossover points of adjacent filters, and the filters will have a transition bandwidth that is 60 dB down in the center of the next lowest frequency band and exhibit $1/f$ spectral side lobes thereafter. Plot the frequency response of the three filters. Compare the filter length required to span this octave to filter length required to span the octave defined in problem 13.6. Explain the difference if any.

13.8 Design a pair of adjacent linear phase FIR low-pass filters for a proportional bandwidth filter bank as described in Section 13.5.1. The two filters h1a and h2a have pass-band edges at 6.8 and 4.8 kHz with transition bandwidths of 1 and 0.707 kHz. These ratios are sqrt(2). The filters have an odd number of coefficients to give us access to their complementary high-pass filters h1b and h2b. The input sample rate is 24 kHz and the stop-band attenuation is 80 dB. These filters coupled with a 2-to-1 down-sample, when operated as cascade low-pass and high-pass pairs, can form the four-filter frequency responses shown in the Figure P13.8. Implement the first three pairs of filters and show their time response and frequency response. Figure P13.8 shows the spectral response of the second and third pair. The first pair contains a high-pass filter and a band-pass filter.

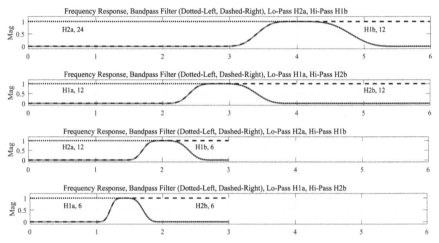

Figure P13.8   Spectra of four proportional bandwidth filters at successive sample rates of 24, 12, and 6 kHz.

14

Communication Systems Applications

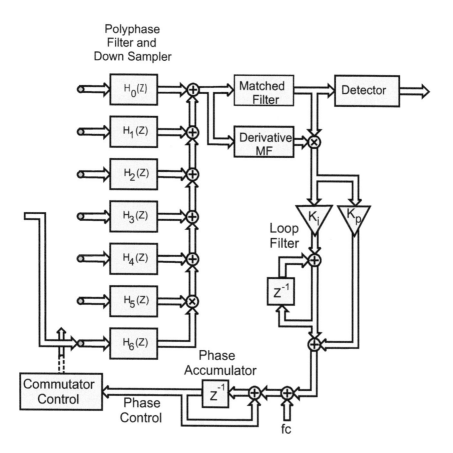

Communication systems make liberal use of multirate filters in several ways. Multirate processing finds application in shaping filters, channelizers, interpolators, efficient bandwidth and sample rate reduction schemes, anti-alias filtering, and in many other applications. This chapter deals with applications that fall under the other category. This chapter is a potpourri of applications we have found over many years of applying signal processing techniques to communication systems. Some of these applications are standard and are well known among practitioners but are included to bring the neophyte into the inner circle. Other applications are unique and will quicken the heart of even the most seasoned practitioner.

14.1 Conventional Digital Down Converters

A radio receiver down-converts and demodulates a narrowband radio frequency (RF) signal embedded in a block of frequencies assigned to a particular radio service. For instance, the commercial frequency modulation (FM) band spans the frequency range from 88 to 108 MHz with multiple 200 kHz channels. Similarly, cable TV modems select, down-convert, and demodulate 5 MHz symbol rate quadrature amplitude modulation (QAM) signals spanning from 54 to 216 MHz with 6 MHz channels interspersed with legacy frequency gaps. The traditional architecture of a radio receiver that performs this task is shown in Figure 14.1. This standard architecture performs two frequency translations and is called a dual-conversion receiver. The receiver passes the input signal through an image reject filter, amplifies, and then down-converts a selected RF channel to an intermediate frequency (IF) filter that performs initial bandwidth limiting. The output of the IF filter is again down-converted to baseband by matched quadrature mixers that are followed by matched analog baseband filters that perform final bandwidth control. Each of the quadrature down-converted signals is then converted to their digital representation by a pair of matched analog-to-digital converters (ADC). The output of the ADCs is processed by digital signal processing (DSP) engines that perform the required synchronization, equalization, demodulation, detection, and channel decoding.

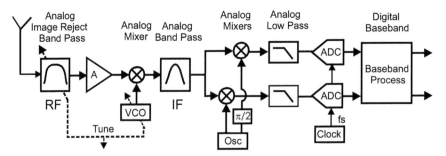

Figure 14.1 Standard radio receiver architecture.

Gain and phase imbalance between the two paths containing the quadrature mixers, the analog baseband filters, and the ADCs in the receiver is the cause of cross-talk between the in-phase and quadrature (I/Q) components. In addition, the ADCs inject a DC term at the center of the baseband signal and the analog filters introduce group delay distortion. Adaptive algorithms can remove the imbalances, the DC terms, and the phase distortion as background processing tasks in the DSP segment of the receiver. Rather than repairing the analog defects in the DSP domain, we have high motivation to avoid these distortion effects entirely by performing the entire baseband processing task in the DSP domain. Besides the advantages of avoiding performance degradation due to path imbalance due to analog component tolerance, avoiding performance degradation due to component parameter drift with time and temperature, avoiding the cost of quadrature mixers, and avoiding group delay distortion associated with analog filters, DSP insertion also offers the attraction of flexibility related to filters with programmable bandwidth and sample rates.

Figure 14.2 presents the block diagram of a second-generation receiver in which the conversion from analog to digital occurs at IF rather than at baseband. Examining the receiver, we note two significant changes in the processing stream. First, due to the higher center frequency of the IF signal, the ADC must operate at a higher sample rate than that in the baseband version. Second, we see that the down-conversion of the selected channel is performed by a digital down converter and digital low-pass filter which because of the higher sample rate now includes a resampling operation. We are willing to accommodate this extra processing burden to gain the advantage that the DSP-based down-conversion is free of imbalance related distortion terms. A second advantage of digital translation process

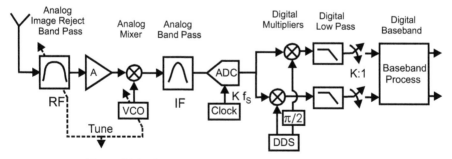

Figure 14.2 Second-generation radio receiver architecture.

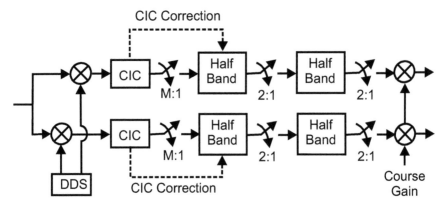

Figure 14.3   Standard digital down converter with CIC and half-band filters.

is that the digital filters in the process are designed to have linear phase characteristics, a characteristic trivially simple to realize in digital non-recursive filters. Another option we will examine shortly is the ability to embed the down-conversion in the resampling process.

To control the computational workload, the filtering and down-sampling is usually performed in two stages, a four- or five-stage cascaded integrator comb (CIC) filter that performs filtering without multiplications while performing an internal M-to-1 down-sampling to an output rate of $4 f_s$. The CIC is followed by two or more half-band filters that correct for the main lobe gain of the CIC while rejecting the main lobe spectral region containing significant aliased energydue to the M-to-1 down-sampling. Figure 14.3 shows a possible realization of the standard second-generation receiver front end implemented by a digital down converter. The converter performs a cascade of simple operations consisting of a quadrature mixer driven by a direct digital synthesizer (DDS), a K-stage CIC filter capable of large integer resampling of M-to-1, a 2-to-1 half-band filter with CIC compensation, a second 2-to-1 half band to finish the spectral control and a course gain control.

Figure 14.4 shows the log magnitude frequency response of a five-stage CIC filter operating as a 10-to-1 resampling filter. The spectrum is shown at the input and output sample rates. The spectrum at the output rate shows the aliased main lobe folding back into the main lobe spectral interval due to the 10-to-1 down-sampling. The alias free region, approximately one-fourth of the output sample rate, is extracted from the main lobe by the half-band filters following the CIC. These filters not only extract the desired bandwidth, they correct the droop in pass-band gain due to the curvature of the CIC main lobe

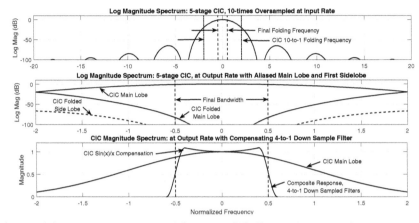

Figure 14.4 Frequency response of five-stage CIC filter, at input sample rate, at output sample rate illustrating main lobe folding due to 10-to-1 resampling, and main lobe response with overlaid compensating 4-to-1 down-sample filter.

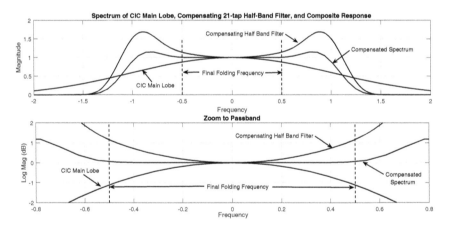

Figure 14.5 Spectra of CIC main lobe, of 21-tap compensating half-band filter, and the composite response.

response. Shown as an overlay on the main lobe is the desired response of the half-band filters following the CIC filter.

Figure 14.5 shows the frequency response of the CIC filter main lobe along with the frequency response of the compensating first half-band filter and their composite response. Note the spectral peaking of the compensated spectrum beyond the pass-band edges. This peaking is suppressed by the frequency response of the second half-band filter following the compensating half-band filter. The compensating half-band filter was designed using the Remez algorithm with the minor modification that the pass-band gain is the reciprocal of the CIC gain over the normalized frequency interval of $0-0.5$. The gains were set over a grid spanning the normalized frequency interval, and the Remez algorithm does linear interpolation between the grid points. Rather than measuring the CIC gain, we computed it at the desired grid points using the first three terms of its Taylor series as indicated in the following equation:

$$gg = (1 - \frac{1}{6}\theta^2 + \frac{1}{129}\theta^4 - \frac{1}{5040}\theta^6)^5. \tag{14.1}$$

The MATLAB call to the Remez routine using this expression for the desired pass-band gain is shown here.

```
phi=[0.00  0.01  0.011  0.02  0.021  0.03  0.031...
            0.04  0.041  0.05  0.051  0.06  0.061...
            0.07  0.071  0.08  0.081  0.09  0.091      0.10];
phi=phi*pi;
tt=((1-(phi.^2)/6+(phi.^4)/120-(phi.^6)/1540)).^5);
hh=remez(20,[phi 0.4 0.5]/0.5,[1./tt 0 0]);
```

MATLAB Call for CIC Compensating Half-Band Filter

14.2 Aliasing Digital Down Converters

In the conventional sampled data collection process, the sample rate is selected to satisfy Equation (14.2) where $2f_{BW}$ is the two-sided bandwidth of the analog anti-alias filter and f_Δ is the transition bandwidth of the same filter

$$f_S = 2f_{BW} + f_\Delta. \tag{14.2}$$

Signals collected from the output of the anti-alias filter at this data rate are said to satisfy the Nyquist sampling criterion. The Nyquist criterion is sometimes stated as: "The sample rate must be greater than twice the highest frequency of the input signal." This is an over-restrictive or a narrow interpretation of the Nyquist criterion. The less restrictive interpretation is that the sample rate must exceed the two-sided bandwidth of the signal. This second interpretation is important when we have a narrow bandwidth signal centered on a high frequency carrier. In a digital receiver, the signal processing following the data collection process removes the carrier to extract the complex envelope of the narrowband signal on the carrier. A digital down-conversion process normally performs this task. Since the carrier frequency is discarded as part of the signal extraction, there is no need to preserve it during the data sampling process. We are thus free to violate the Nyquist criterion for the carrier frequency as long as we satisfy the criterion for the bandwidth of its complex envelope.

This narrowband interpretation of the Nyquist criterion leads to an alternate data collection process known as subsampling or IF sampling. In this process, the sample rate is selected to be less than the signal's center frequency to intentionally alias the center frequency to a lower frequency less than the sample rate. Since we are intentionally violating the Nyquist sampling criterion, we must condition the analog signal to prevent multiple frequency intervals from aliasing to the same frequency location to which our desired signal component will alias. To minimize the cost of the analog signal-conditioning filter, we arrange for the signal band of interest to alias to one-fourth of the selected sample rate. Aliasing to the quarter sample rate maximizes the separation between the positive frequency alias and the negative frequency alias, which permits the maximum transition bandwidth of the analog band-pass filter. Aliasing the center frequency f_C to the quarter sample rate during the sampling process is assured if the sample rate satisfies Equation (14.3). The $k + 1/4$ option aliases the signal to the positive quarter sample rate while the $k - 1/4$ option aliases the signal to the negative quarter sample rate. Use of the two options simply makes available a larger set of possible sample rates with either option equally acceptable.

$$f_C = k\, f_S \pm \frac{1}{4} f_S. \qquad (14.3)$$

To better understand the process, we now present an example of IF subsampling with aliasing to the quarter sample rate.

14.2.1 IF Subsampling Example

Signal Two-Sided Bandwidth	10 kHz
Center Frequency	450 kHz
Signal Dynamic Range	80 dB
Output Sample Rate	20 kHz

First Option: Our first option is to sample the input signal at 2.0 MHz, placing the band of interest near the quarter sample rate, and then using a

Table 14.1   List of possible sample rates that alias 450 kHz to quarter sample rate.

Integer k	f_s for $450 = (k + 1/4)$	f_s for $450 = (k - 1/4)$
	f_s	f_s
0	1800	1800
1	360	600
2	200	257.143...
3	138.46...	163.37...
4	105.88...	120

standard digital down converter, we would perform a complex translation to baseband, filter and down-sample 25-to-1 with a CIC, and then filter and down-sample 4-to-1 with a pair of half-band filters.

Second Option: Our second option, the focus of this example, is to perform IF sampling at a sample rate satisfying Equation (14.3). A list of possible sample rates that satisfy Equation (14.3) for the center frequency of 450 kHz is presented in Table 14.1.

We will select the 200 kHz sample for this example and suggest that the reader consider how the solution would be different had we selected 120 kHz as the sample rate. The aliasing caused by sampling the signal centered at 450 kHz at a 200 kHz sample rate is illustrated in Figure 14.6. Also note that in this figure, the transition bandwidth of the analog IF filter is maximized when the alias is to the quarter sample rate. Figure 14.7 presents the same spectrum on a frequency scaled periodic circle. Here we can start at 0 and travel in the positive direction around the 200 kHz circle passing the listed frequencies. After two encirclements, we see that frequency 450 kHz is located at the same location as 50 kHz. This is the desired result that we expected from the IF sampling process.

At this point, we could return to the standard digital down converter that can translate the spectral region to baseband with a complex heterodyne followed by a 10-to-1 resampling filter. This filter can be a 5-to-1 CIC followed by a 2-to-1 compensating half-band filter. The complex heterodyne required to move the spectrum from the quarter sample rate is of the form shown in Equation (14.4). The values of the cosine and sine terms in Equation (14.4) are trivially the +1 and –1 and zero values shown in Equation (14.5) so that, in fact, the heterodyne is trivial.

$$\exp(-j\frac{\pi}{2}n) = \cos(j\frac{\pi}{2}n) - j\sin(j\frac{\pi}{2}n) \tag{14.4}$$

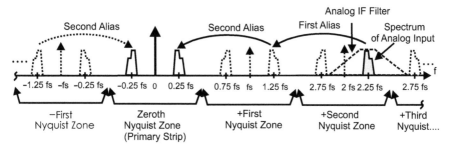

Figure 14.6 Aliasing spectra at $2.25\,f_s$ in second Nyquist zone to $0.25\,f_s$ in zeroth Nyquist zone by IF sampling at f_s.

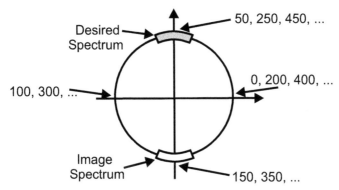

Figure 14.7 Periodic spectra on scaled circle showing how 450 kHz from the second Nyquist zone folds through 250 kHz in the first Nyquist zone to 50 kHz in zeroth Nyquist zone.

$$\exp\left(-j\frac{\pi}{2}n\right) = \{+1,\ 0,-1,\ 0,\ldots\} + j\{0,-1,\ 0,+1,\ldots\}. \qquad (14.5)$$

We also note that due to the zero-valued samples of the heterodyne, half the data samples in the cosine product are set to zero, and the complementary set of the data samples in the sine heterodyne are also set to zero. These zero-valued terms cannot contribute to the outputs of the filter following the heterodyne. Since we know the location of these zeros, we can discard them without error as long as we account for their effect in shifting the non-zero data samples through the filter. It would be a shame to waste the unique attributes of this heterodyne and data variation on a standard digital down converter. Instead, we continue this example with a number of techniques that take advantage of the signal being located at the quarter sample rate.

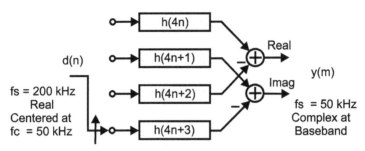

Figure 14.8   Four-path polyphase filter: simultaneously translates frequency band from quarter sample rate to baseband, down-samples 4-to-1, and converts real input to complex output.

The most obvious option to apply at this point is a four-path polyphase filter to down-sample 4-to-1 and thus alias the quarter sample rate to DC. A phase rotator at the output of the four-stage filter would unwrap the aliases and extract samples corresponding to the band centered at the quarter sample rate. By setting the center frequency of the sines and cosines at the quarter sample rate, the phase rotator sequence $\exp[j\,(\pi/2)r]$, for $r = 0, 1, 2$, and 3, is a particularly simple sequence of $1, j, -1$, and $-j$. Merging the phase-rotated outputs of the polyphase filter stages proves to be a particularly simple task of identifying real and imaginary output sample and forming the difference of the two real paths and of the two imaginary paths. The four-path polyphase filter using these rotators is shown in Figure 14.8.

The remaining operation is to down-sample the output of the four-path polyphase filter from 50 kHz to the desired 20 kHz. This down-sampling task requires a 2.5-to-1 change in sample rate, which is accomplished as a 1-to-2 up-sampling followed by a 5-to-1 down-sampling. The 1-to-2 up-sampling is performed conceptually by zero packing the input series 1-to-2. Of course, the alternate zero-valued samples resulting from the zero packing cannot contribute to the filter output and, as observed earlier, can be discarded, provided we account for their data shifting function in the five-stage polyphase filter. Two successive cycles that deliver five zero-packed data samples to the five-stage filter are shown in Figure 14.9.

Note that in the first cycle, we deliver three non-zero samples $d(n)$, $d(n + 1)$, and $d(n + 2)$ and then compute an output. In the second cycle, we deliver two non-zero samples $d(n + 3)$ and $d(n + 4)$ and then compute the next output. Over two successive input cycles, we deliver 5 inputs and compute 2 outputs for a resampling ratio of 5-to-2 or 2.5-to-1. The inserted zero-valued samples simply move the data samples in their paths one data sample to the right. We

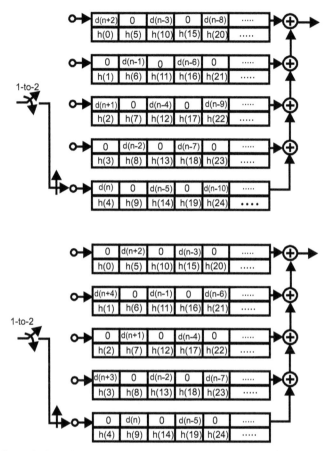

Figure 14.9 Indexing of successive zero-packed data samples to five-path polyphase filter.

can omit the inserted zeros by following the two-cycle state machine rules listed in Table 14.2.

Figure 14.10 shows the block diagram of the down converter using under-sampling to alias the narrow band spectrum to the quarter sample rate and then a four-path polyphase filter to alias it again to baseband for a final down-sample of 5-to-2 with a five-path polyphase filter. The first filter requires 16 coefficients, which means that each of the four paths requires inner products with 4 coefficients, while the second filter requires 40 coefficients that are split evenly between the two cycles of the state machine so that each of the five paths also requires inner products with 4 coefficients.

Table 14.2   Two-state state machine input and inner product schedule for five-path 2.5-to-1 resampling filter.

Cycle-0	Cycle-1
Input data:	Input data:
$d(n)$ in path-4	$d(n + 3)$ in path-3
$d(n + 1)$ in path-2	$d(n + 4)$ in path-1
$d(n + 2)$ in path-0	
Inner products	Inner products
path-4, data with $h(10n + 4)$	path-4, data with $h(10n + 9)$
path-3, data with $h(10n + 8)$	path-3, data with $h(10n + 3)$
path-2, data with $h(10n + 2)$	path-2, data with $h(10n + 7)$
path-1, data with $h(10n + 6)$	path-1, data with $h(10n + 1)$
path-0, data with $h(10n)$	path-0, data with $h(10n + 5)$

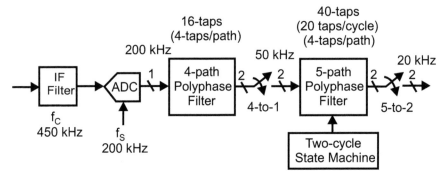

Figure 14.10   Signal processing structure of down converter using aliasing to and from quarter sample rate by subsampling and polyphase filtering.

Third Option: In the example we are using to demonstrate subsampling, we selected a sample rate of 200 kHz to translate the 450 kHz IF signal to the quarter sample rate of 50 kHz. We selected the 200 kHz for the ease with which we could perform the remainder of the processing task, filtering, and down-sampling by a factor of 10 from 200 to 20 kHz. We accomplished the 10-to-1 down-sampling in two steps: first the 4-to-1 in the four-path polyphase filter and then a 5-to-2 in the five-path polyphase filter. We were guided to the 4-to-1 down-sampler because it was natural to alias the quarter sample rate to baseband in this stage. A signal at any center frequency can be aliased to other center frequencies besides the baseband under the resampling operation. The quarter sample rate is quite remarkable in that it is only aliased to the four cardinal directions by other resample ratios besides the obvious 4-to-1. We now consider another option for this example in which we use a 5-to-1 first stage filter followed by a 2-to-1 second stage filter.

We can design a prototype low-pass filter that will support a 5-to-1 resampling operation in a five-path polyphase filter. In the standard way we would apply this filter, we would permit multiples of the output sample rate to alias to baseband where phase rotators would unwrap the desired alias. The problem here is that the signal of interest resides at 50 kHz, which is not one of the multiples of 40 kHz, so the standard polyphase structure appears to be inappropriate for this problem. By a slight modification of how we use the polyphase filter, we are able to take advantage of the unique attributes of the quarter sample rate center frequency. In particular, we translate the prototype low-pass filter to the quarter sample rate prior to the five-path polyphase partition. We accomplish this translation by the trivial heterodyne of the filter coefficients with the terms $\exp[j(\pi/2)n]$. These heterodyne terms, the periodic sequence $\{1, j, -1, -j\}$ merely guide the output samples to the real or imaginary output and at most effect a sign change in the weighted sum performed in the filtering operation. This form of the filter is a close cousin of the Hilbert transform filter.

When the filter is used in this manner, as a complex polyphase band-pass filter, the resampling operation exhibits an interesting property. This property is that a signal at the quarter sample rate will always alias to a multiple of the quarter sample rate under any integer resampling operation. In particular, for our example, we have a signal centered at 50 kHz and sampled at 200 kHz so that it is located at the quarter sample rate when it enters the five-path polyphase filter. Here the signal is resampled 5-to-1 to 40 kHz which aliases the 50 kHz centered signal to 10 kHz, the quarter sample rate at the output of the filter. Figure 14.11 shows the spectrum of a 25-tap baseband filter with the frequencies that alias to baseband under a 5-to-1 resampling as well as translated to the input quarter sample rate filter with the frequencies that alias to the output quarter sample rate when resampled 5-to-1.

We note that the sampled data sequence that enters the five-path filter is real, while the sampled data sequence that leaves the filter is complex and centered at the quarter sample rate. This output sequence is heterodyned to zero by the trivial heterodyne $(-j)^n$ prior to final processing in the half-band filters. Figure 14.12 presents the block diagram of the signal processing performed by this option.

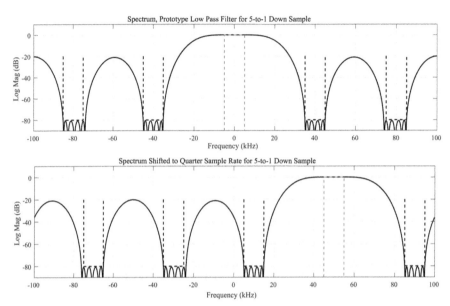

Figure 14.11   Spectrum of 25 tap prototype low-pass filter with frequency bands that alias to baseband when resampled 5-to-1 (upper figure) and spectrum of same filter translated to quarter sample rate with frequency bands that alias to quarter sample rate when resampled 5-to-1 (lower figure).

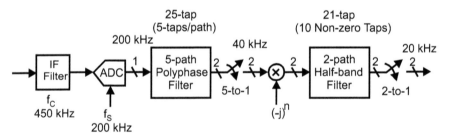

Figure 14.12   Signal processing structure of down converter using aliasing to quarter sample rate by subsampling and again by five-path polyphase filter.

14.3 Timing Recovery in a Digital Demodulator

14.3.1 Background

The modulator of a communication system transmits a sequence of scaled and time translated band-limited waveforms of the form shown in Equation (14.6). The scale factors a_n are selected from a finite list of permitted amplitudes in response to a sequence of binary words delivered at a periodic rate to the modulator. The alphabet may, for instance, contain two elements such as +1 and –1 selected by a 1-bit input or it may contain four elements such as +3, +1, −1, and –3 selected by a 2-bit input and so on.

$$d(t) = \sum_n a_n h(t - nT).\qquad(14.6)$$

In an ideal communication system, the demodulator receives a version of the transmitted waveform that has been corrupted by additive white Gaussian noise (AWGN) as shown in the following equation:

$$d(t) = \sum_n a_n h(t - nT) + N(t).\qquad(14.7)$$

To minimize the effects of the received noise in the ensuing decision process, the received noisy signal is passed through a matched filter. The impulse response of the matched filter is, as shown in Equation (14.8), a time reversed and delayed version of the transmitted waveform $h(t)$. Here we assume that the wave shape extends over k symbol durations.

$$g(t) = h(kT - t).\qquad(14.8)$$

The convolution process performs a running weighted average with the filter's time-reversed impulse response. We purposely time-reversed the filter impulse response in anticipation of the time reversal that will occur in the convolution process. The reversal in the convolution process undoes the initial reversal so that the running weighted average is performed with a replica of the transmitted waveform and the convolver performs a replica correlation. The peak of the correlation function corresponds to the projection of the noisy signal on a noiseless replica of the signal and this projection exhibits the maximum signal-to-noise ratio (SNR) that can be obtained by any processing of the received signal. In order for the receiver to access these peak values, it invokes a timing recovery mechanism to align the frequency and phase of its sampling clock with the epochs of the correlation peaks. The signal flow of the entire process is shown in Figure 14.13.

In modern receivers, the matched filter operation is performed in the sampled data domain. The receiver structure changes slightly to accommodate the anti-aliasing filter, the sample and hold (S&H), and the ADC. A first-generation digital demodulator structure is shown in Figure 14.14. Note here that the timing recovery is now controlling the sampling clock to the digital matched filter rather than the sampling at the

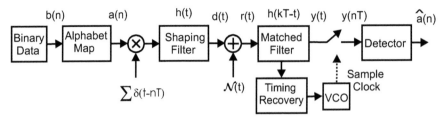

Figure 14.13   Signal flow in modulator and demodulator for communicating through band-limited AWGN channel.

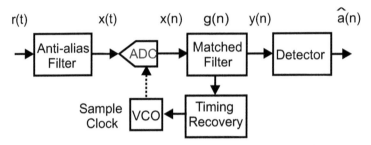

Figure 14.14   Signal flow in first-generation digital demodulator.

output of the analog matched filter. The sampling performed at the input to the digital filter must satisfy the Nyquist criterion for the collected wave shape and in many systems, the sample rate is two samples per symbol. The sampling performed at the output of the analog matched filter to supply samples to the detector occurs at the symbol rate or one sample per symbol.

The timing recovery process must ascertain if the clock sample position is at the correct position or should be advanced or retarded relative to the input time waveform. It does this by posing the question, "How do I know that I am at the local peak of the correlation function?" It knows that at the peak, the correlation function has a zero derivative, so it poses the nearly equivalent question, "What is the derivative of the correlation function zero at this sample time?" Traditionally auxiliary matched filter outputs, called early and late gates, that are time advanced and time delayed relative to the sample test point supply an answer to this question as the average difference between the early and late gate output values. This can be visualized in Figure 14.15.

The derivative alone contains insufficient information to determine if the timing should be time advanced, held, or retarded. A conditional piece of information is missing, and this information is the answer to the question "What is the sign of this sample of the correlation function?" As seen in

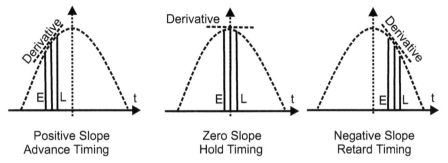

Positive Slope
Advance Timing

Zero Slope
Hold Timing

Negative Slope
Retard Timing

Figure 14.15 Three samples, early, punctual, and late samples, on correlation function for three operating conditions. (1) Positive Slope: peak is ahead; (2) zero slope: at peak; (3) negative slope: peak is behind.

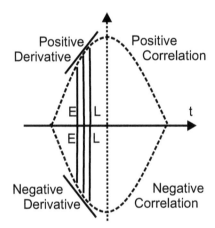

Figure 14.16 Three samples, early, punctual, and late samples, on positive and on negative correlation.

Figure 14.16, a sample set formed prior to the peak will generate a positive slope if the correlation values are positive but will generate a negative slope if the correlation values are negative.

The conditional information is folded into the observable error parameter in the timing recovery process as the products shown in Equation (14.9). When the SNR is high, the sign of the data sample $y(t)$ is reliable and the sign$[y(t)]$ is a sufficient modifier. When the SNR is small, the modifier is the data itself rather than the sign of the data. The maximum likelihood timing recovery process minimizes the product of the derivative matched filter output $\dot{y}(t)$ with SNR conditioned selection of the matched filter output $y(t)$ or sign of the matched filter output sign[$y(t)$]. Many systems do not bother switching modes and simply use $y(t)\,\dot{y}(t)$ for the full range of SNR.

$$e(t) = \dot{y}(t) \cdot \text{sign}[y(t)] \cong [y(t + \Delta t) - y(t - \Delta t)] \cdot \text{sign}[y(t)], \quad \text{High SNR}$$
$$e(t) = \dot{y}(t) \cdot y(t) \cong [y(t + \Delta t) - y(t - \Delta t)] \cdot y(t), \quad \text{Low SNR.}$$
$$(14.9)$$

14.3.2 Modern Timing Recovery

Examining Figure 14.14, we note that the timing recovery process controls the phase and frequency of the Voltage Controlled Oscillator (VCO) that is supplying the sampling clock to the ADC. It is likely that the error detector, the $y(t)\dot{y}(t)$ of Equation (14.9) is performed in the sampled data domain as $y(n)\dot{y}(n)$ and that the loop filter that controls the transient and steady-state behavior of the timing recovery system is also performed in the sampled data domain. The control signal formed to operate the VCO resides, as a set of numbers, in the sampled data domain while the control signal required to operate the VCO is an analog voltage level. To convert the digital control signal to an analog control signal requires a high precision digital-to-analog convertor (DAC), a low bandwidth analog filter, and a bus and control mechanism to transfer the control words to the DACs internal register. Rather than incur the overhead of changing between the sampled data and continuous domains, we can perform the entire timing recovery process in the sampled data domain. We can accomplish this in two different ways. We can either move the data samples to be aligned with the filter coefficients or we can move the filter coefficients to be aligned with the data samples. In the first method, we use an interpolator to raise the input sample rate by a factor of M, say 32, and then down-sample back to the same input rate with a fixed time offset to the sample locations required to be aligned with the impulse response of the matched filter. Figure 14.17 presents the signal flow for this demodulator architecture. In the second method, we increase the sample rate of the matched filter and then resample the filter response to the original sample rate with successive time offsets of $1/M$, $2/M$, $3/M$, etc., to form a set of M filters matched to different time offsets between the input sample location and the envelope of the received waveform. Figure 14.18 presents the signal flow for this demodulator structure. In both cases, we are required to operate only one filter path out of the M possible paths. In this mode, the interpolator is used as a 1-to-M up-sampler followed by an M-to-1 down-sampler with the desired time offset between input and output sample

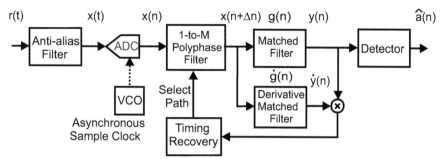

Figure 14.17 Signal flow for timing recovery with polyphase interpolator processing and shifting asynchronous samples to desired time locations.

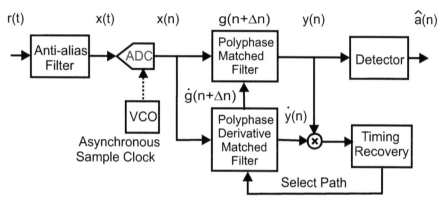

Figure 14.18 Signal flow for timing recovery with polyphase matched filter. timing recovery selects matched filter path aligned with input sample position.

locations. The timing recovery process determines which filter path is the one required to align the filter with the signal time offset.

In Figure 14.15, we used the analog perspective that used an early and a late gate to determine the derivative at the output of the matched filter. Doing so requires the operation of three filters, early, punctual, and late. Note that the filters corresponding to the early and late gate filters are trivially the polyphase segments $(k - 1)$ and $(k + 1)$ when testing polyphase segment (k). This is seen in Figure 14.19. As shown in Equation (14.10), the same data is convolved with filters $(k - 1)$ and $(k + 1)$ to form corresponding outputs, which we then subtract to form the derivative estimate. Factoring out the common data, we are left with a filter with coefficients obtained as the difference of the two adjacent filters. We recognize this single filter

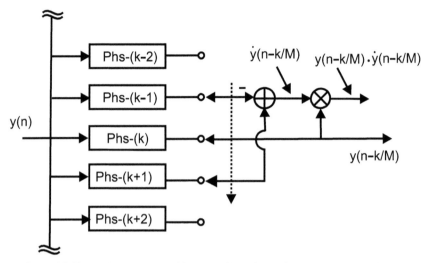

Figure 14.19   Early, prompt, and late gate filters for timing recovery control signals.

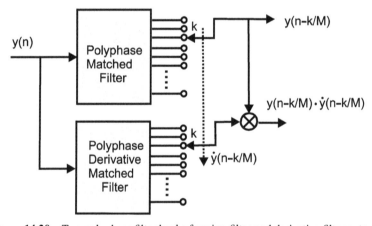

Figure 14.20   Two polyphase filter banks forming filter and derivative filter outputs.

as one that directly computes the desired derivative. We can now replace the early and late gate filters with the derivative filter and thus reduce the workload required to implement control data for the timing recovery process. Figure 14.20 shows a process using two banks of polyphase filters, one for the signal, and one for the derivative, while Figure 14.21 shows a simple structure with two simple filters with a coefficient selection process. It is this later structure that is embedded in the block diagrams shown in Figures 14.17 and 14.18.

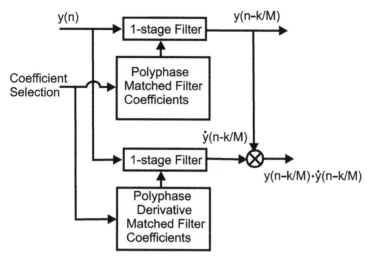

Figure 14.21 Two single-stage filters with selection of coefficient sets from polyphase filter bank for matched filter and derivative matched filter.

$$\dot{y}(n + k/M) = y(n) * h_{k+1}(n) - y(n) * h_{k-1}(n)$$
$$= y(n) * [h_{k+1}(n) - h_{k-1}(n)] \qquad (14.10)$$
$$= y(n) * \dot{h}_k(n).$$

Figure 14.22 shows the time history of the address pointer in a modem using a 40-path polyphase-matched filter for timing recovery. Since the input signal is collected at two samples per symbol, one sample, the one with an even index for instance, is chosen as a data sample and the other is the non-data sample, sometimes called the timing sample. The data sample is the one delivered to the detector. If the address pointer tries to cross the address boundaries, an address lower than 1 or greater than 40, the address wraps circularly and the input sample identified as the data sample is switched to the odd index. When this happens, the pointer switches from, say, a forward progression to a large forward jump and a backward progression. You can see this pointer direction reversal (or reflection) in the lower half of Figure 14.22. Beyond the scope of this discussion is the observation that there may also be data sample insertion (data stuff) or deletion (data skip) at the boundary crossing depending on the direction of the crossing and whether the data sample is an even or odd index input sample. Both operating conditions are presented in Figure 14.22.

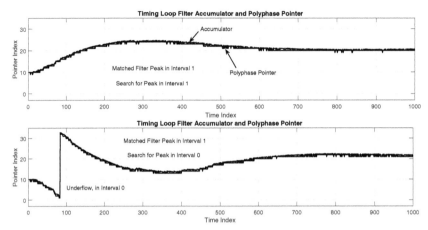

Figure 14.22 Polyphase address pointer for timing recovery loop with 40-path filter and two samples per symbol. In the upper figure, the pointer does not cross the address pointer boundary, and in the lower figure, the pointer crosses the address pointer boundary and converts Address 0 to Address 40 and switches from even-indexed to odd-indexed data sample.

14.4 Modem Carrier Recovery

Many communication systems use a complex heterodyne to translate a complex baseband communication signal to a center frequency in the band allocated to the particular radio service. We perform spectral translation to different center frequencies for a number of reasons. One reason is the sharing of the radio spectrum by multiple users through use of frequency division multiplexing. Another reason is the relative ease of designing analog circuits when the signal bandwidth occupies a small fraction of the center frequency. Yet, another reason is access to smaller or more efficient antennae at higher frequencies as well as access to spectral bands with specific propagation characteristics.

The transmitter in the communication system performs the spectral translation of the baseband signal to the selected carrier center frequency by an up-conversion process. To cull the desired signal from the numerous channels sharing the spectral region, the receiver must invert the spectral translation process. In order to accomplish this task, the receiver requires a phase coherent replica of the unmodulated carrier delivering the signal. In early radio systems, a copy of the carrier was embedded in the modulated signal to enable the down-conversion process. In modern radios, the receiver forms its own copy of the carrier from side information residing in the modulated signal. This process is called carrier recovery.

14.4.1 Background

At the simplest level, modulation is simply the translation of a complex baseband modulation envelope $m(t)$ with baseband spectral characteristic $M(\omega)$ to an arbitrary center frequency called the carrier. The relationship between the time and frequency descriptions of the baseband and translated signal is shown in the following equation:

$$m(t) \Leftrightarrow M(\omega)$$
$$m(t) \exp(j\omega_C t) \Leftrightarrow M(\omega - \omega_C). \quad (14.11)$$

Product modulators that modify the amplitude and phase of the carrier perform this translation. To form the signal required for the actual transmission, the conjugate of the complex waveform is added to the expression shown in Equation (14.11) to make the signal real. As shown in Equation (14.12), the real signal is formed as the Cosine heterodyne of the real part of the complex envelope minus the sine heterodyne of the imaginary part of the complex envelope. The real and imaginary parts of the complex waveform are normally called the in-phase and quadrature phase or the I and Q components. These I and Q designations are inherited from the phase of their respective carriers.

$$s(t) = m(t) \exp(j\omega_C t) + m^*(t) \exp(-j\omega_C t)$$
$$= 2RL[m(t)] \cos(\omega_C t) - 2IM[m(t)] \sin(\omega_C t) \quad (14.12)$$
$$= I(t) \cos(\omega_C t) - Q(t) \sin(\omega_C t).$$

The narrowband signal is launched through the channel by the transmitter and is delivered through a noisy channel at the receiver. The receiver must remove the complex envelope from the carrier and present the noisy versions $I(t)$ and $Q(t)$ to the demodulator. To do so, it must align the frequency and phase of its local oscillator to match the frequency and phase of the carrier in the received signal. The oscillator must do this in spite of temperature variation of its frequency-dependent components in spite of manufacturing tolerance spread, Doppler-related frequency offsets due to a velocity component between platforms, and the fact that there is no spectral line at the carrier frequency in the received signal. It accomplishes this feat with a carrier recovery loop formed around a phase locked loop (PLL). The loop is composed of a phase detector operating on demodulated data, a loop filter, and a controlled oscillator. The high-level block diagram of the up-conversion and down-conversion performed at the two ends of the channel is shown in Figure 14.23.

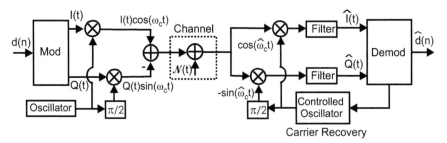

Figure 14.23   Block diagram of quadrature up converter at transmitter and quadrature down converter at receiver.

The PLL has two distinct modes of operation, acquisition and tracking. The tracking bandwidth of a PLL, the range of frequencies over which it can follow frequency offsets is limited by the control range of the oscillator. The acquisition bandwidth is limited by the pass-band bandwidth of the loop filter. The acquisition range is considerably smaller than the tracking range. A receiver is directed to acquire an RF signal by being tuned to the expected center frequency of the signal.

Previous generation, or legacy design, receivers operate in the following manner: if the frequency offset between the local oscillator and the received center frequency is less than the acquisition bandwidth, the carrier recovery loop is able to generate a control signal to shift the local oscillator in the direction to servo the offset to zero, thus accomplishing the acquisition task. If the frequency offset is greater than the acquisition bandwidth, a lock detection signal will not be generated in the time-out interval that monitors the operating state of the receiver. The receiver enters a *failure to acquire* state in which it invokes an acquisition aid. The aid conducts a search for the signal by directing the controlled oscillator to perform a frequency sweep through the expected range of frequency offsets. When the frequency difference between the local oscillator and the errant signal becomes smaller than the acquisition bandwidth, the loop successfully acquires the signal, generates the lock detection signal, and disables the acquisition aid.

14.4.2 Modern Carrier Recovery

A modern receiver will use a maximum likelihood frequency estimator as an aid to pull the local oscillator close to the center frequency of an offset signal and thus bring the offset signal within the acquisition range of the PLL. The derivation of the maximum likelihood frequency estimator is beyond

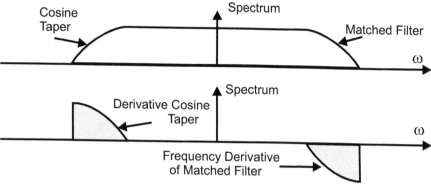

Figure 14.24 Spectra of matched filter for square-root Nyquist filter and frequency derivative matched filter.

the scope of this text, so we will simply describe its operation and then its implementation. In a manner similar to the maximum likelihood timing recovery process, we form a filter that is the derivative of the matched filter. In the timing recovery process, we took the time derivative of the matched filter and then drove the product of the matched filter and the time derivative matched filter outputs to zero. In the frequency recovery process, we take the frequency derivative of the matched filter and, in a similar manner, we can drive the product of the matched filter and the frequency derivative matched filter outputs to zero. The frequency derivative matched filter is called a band-edge filter, and we now describe why it is so called and describe a common variant of its operation.

Figure 14.24 presents the power spectrum of the matched filter for a square-root (SQRT) Nyquist filter as well as the frequency derivative of the same filter. Note that the derivative is zero everywhere except in the transition band of the filter. Seeing that the non-zero spectral response of the derivative matched filter spans the band edges of the matched filter, we can readily understand why it is called a band-edge filter.

Figure 14.25 illustrates how the band-edge filter responds to input signals with significant frequency offsets. The top two figures present the spectrum of an input signal without a frequency offset and the response of the two spectral segments in the band-edge filter. We note that the two band-edge segments contain the same spectral energy. By comparison, the bottom two figures present the spectrum of an input signal with a small frequency offset and the response of the two spectral segments in the band-edge filter. Here, we note that as a result of the spectral shift to the right, the energy contained

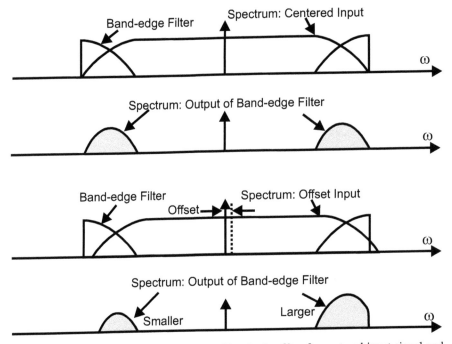

Figure 14.25   Spectra at input and output of band-edge filter for centered input signal and for input signal with spectral offset.

in the filter segment on the right increased while the energy contained in the filter segment on the left decreased. Energy difference in the two band-edge segments is a simple and accessible observable that can be used to drive the input spectrum to be centered at baseband. The modification to the receive that incorporates the band-edge filter is shown in Figure 14.26. The polyphase filter following the band-edge filter separates the two band-edge spectral regions and their outputs become the frequency error detector as the difference of their squared magnitudes.

14.4.2.1 Design and Partition of Band-Edge Filter
We now address the task of designing the band edge-filter and the polyphase filter that separates the positive and negative frequency spectral segments. The first task of designing and implementing the band-edge filter is a three-step process shown in the subplots of Figure 14.27. The top subplot shows the impulse response and spectrum of the SQRT Nyquist shaping filter. The second subplot shows the frequency derivative of the shaping filter. The

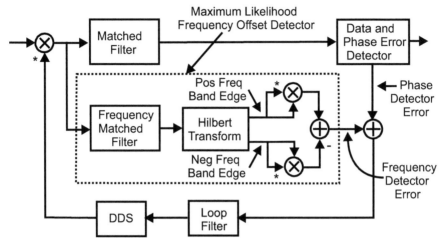

Figure 14.26 Band-edge filter and frequency error detector appended to carrier recovery process.

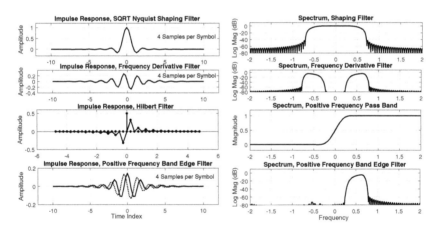

Figure 14.27 Time and frequency domain development of band edge filter. Top subplot: shaping filter; second subplot: frequency derivative matched filter; third subplot: positive frequency Hilbert filter, band edge filter, and Hilbert filtered frequency derivative filter.

derivative band edge filters are shown in the Log Mag form which masks the negative amplitude values of the negative frequency band edge. The frequency derivative filter impulse response, guided by the time frequency relationships shown in Equation (14.13), is shown in Equation (14.14). Here we see the sampled data time ramp reflecting the frequency derivative and the Kaiser window to suppress the boundary discontinuity and thus avoid Gibbs phenomena ringing. Note that the derivative inherits the same scale factor used by the shaping filter and should be rescaled for the demodulator matched filter.

$$h(t) \Leftrightarrow H(\omega)$$

$$\frac{d}{dt}h(t) \Leftrightarrow j\omega \cdot H(\omega) \tag{14.13}$$

$$\frac{d}{d\omega}H(\omega) \Leftrightarrow -jt \cdot h(t)$$

```
h1   = rcosine(1,4,'sqrt',0.4,10);          % shaping filter :
h1   = h1/max(h1);                          % scaling
dh1  = -j(-10:0.25:10).*h1.*kaiser(41,5)';  % frequency derivative
```
(14.14)

The third subplot shows the complex impulse response and spectrum of a Hilbert filter with a positive frequency pass band. The fourth and last subplot shows the positive frequency band edge filter separated from the band edge pair by the Hilbert filter. The negative frequency band edge filter has an impulse response which is the conjugate of the positive frequency band edge filter. The pair of band edge filters can be used as shown in Figure 14.26.

An alternate, and more interesting, option to the development illustrated in Figure 14.27 but using the same resources is shown in Figure 14.28. In this figure, the spectrum is not shown in Log Mag but rather as real or imaginary parts of the spectrum. The top subplot shows the impulse response and spectrum of the SQRT Nyquist shaping filter. The second subplot shows the frequency derivative of the shaping filter. Note that the time series and the spectrum are odd symmetric, and we can see the negative amplitude values of the negative frequency band edge. The third subplot shows the imaginary impulse response and spectrum of a Hilbert filter which we recognize as a wide band $\pi/2$ phase shifter, $\theta(f) = -j \, sgn(f)$. The fourth and last subplot shows the Hilbert filtered band edges with both time and spectrum being even symmetric functions. Interestingly, the even and odd time series developed in Figure 14.28 are simply half the real and imaginary parts of the positive and negative band edge filters developed in Figure 14.27. This is quite reminiscent of Euler's identity for cosines and sines.

The question now is: what can we do differently with the complex time signals with two sided even and odd symmetric spectra? Figure 14.29 shows a possible asymmetric spectral response in the band edge pairs due to a frequency shift of the input spectrum toward high positive frequencies to the

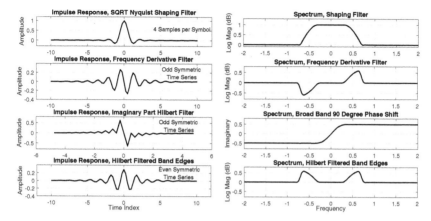

Figure 14.28 Time and frequency domain development of band edge filter. Top subplot: shaping filter; second subplot: odd symmetric frequency derivative matched filter; third subplot: imaginary part of Hilbert filter, band edge filter, and even symmetric Hilbert filtered frequency derivative filter.

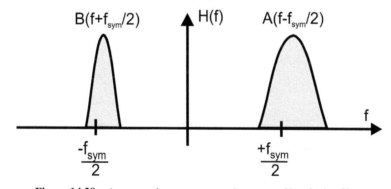

Figure 14.29 Asymmetric power spectra in output of band edge filter.

right. We can identify the complex time signals $a(t)$ and $b(t)$ with baseband spectra $A(f)$ and $B(f)$ translated to their respective center frequencies at $f_{sym}/2$ and at $-f_{sym}/2$ corresponding to the pair of band edge filter bands. We can combine them by a sum for the signal $c(t)$ in the even symmetric filter and by a difference $s(t)$ for the odd symmetric filter. Here the $c(t)$ and $s(t)$ are named for their related symmetries in the cosine and sine functions. Equation (14.19) shows this definition as frequency translated versions of their complex baseband counterparts. We then form the conjugate product of the two signals and collect and combine parts as shown in Equation (14.20).

$$c(t) = \{a(t)\exp(+j0.5f_{\text{sym}}\,t) + b(t)\exp(-j0.5f_{\text{sym}}\,t)\}$$
$$s(t) = \{a(t)\exp(+j0.5f_{\text{sym}}\,t) - b(t)\exp(-j0.5f_{\text{sym}}\,t)\} \tag{14.15}$$

$$
\begin{aligned}
d(t) &= c(t)\cdot\text{conj}(s(t))\\
&= [|a(t)|^2 - |b(t)|^2]\\
&\quad -[a(t)b^*(t)\,\exp(j\,f_{\text{sym}}\,t) - a^*(t)b(t)\,\exp(-j\,f_{\text{sym}}\,t)]\\
&= [|a(t)|^2 - |b(t)|^2] - \text{Imag}\{r(t)\exp(j\,[f_{\text{sym}}\,t + \phi(t)])\}\\
&= [|a(t)|^2 - |b(t)|^2] - j\,r(t)\sin(f_{\text{sym}}\,t + \phi(t)).
\end{aligned}
\tag{14.16}
$$

Note that the real part of this conjugate product contains the energy difference between the signals in the upper and lower frequency band edge filters which we know is proportional to offset frequency of the input signal. The surprise is that the imaginary part is a sine wave at the symbol rate with a phase we can show is the phase offset of the symbol clock to the modulation waveform in the received signal. We started this trip of discovery seeking a frequency offset detector and found, serendipitously, a timing recovery process as well. A receiver structure that uses this band edge process to accomplish the two tasks of frequency acquisition and timing acquisition is shown in Figure 14.30. The outer loop acquires frequency lock by shifting the input spectra to the mid-point of the band edge filter locations. The inner loop acquires modulation timing by selecting the appropriate time offset

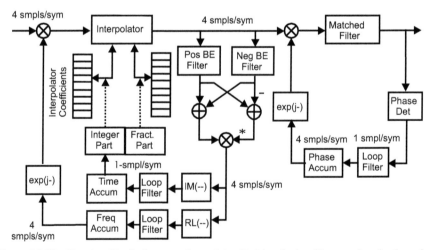

Figure 14.30 Receiver block diagram of non-data aided band edge filter synchronization of carrier frequency and modulation timing.

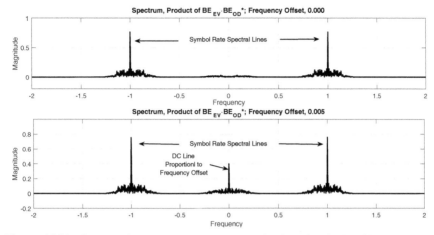

Figure 14.31 Spectra of conjugate product time series from band edge filters. A pair of symbol rate spectral lines aligned with frequency and phase of modulation waveform and DC spectral line proportional to frequency offset of modulation spectrum.

coefficients from a set of M interpolator weights. The band edge process in this configuration runs at four samples per symbol. This higher sample rate is required because the conjugate product of the two band edges double the band edge frequency from $f_{sym}/2$ to f_{sym}. If the loop ran at two samples per symbol, the $f_{sym}/2$ would reside at the quarter sample rate and the doubling would place lines at f_{sym} which is the half sample rate. The positive and negative spectral lines would meet at the half sample rate and destructively cancel their phase information. There is a version of this process running the band edge filters at two samples per symbol with a 1-to-2 interpolator doubling sample rates at the BE filter outputs prior to forming their conjugate product. See the paper listed in the reference list at the end of this chapter for more details. Figure 14.31 shows the spectra of the band edge filter product without and with a frequency offset.

14.5 Digitally Controlled Sampled Data Delay

A digitally controlled sampled data delay line described here is implemented with recursive all-pass filter sections. This is in marked contrast to the standard implementation of programmable time delays that use fixed non-recursive polyphase stages or adjustable Farrow finite impulse response (FIR) filters. The recursive filter exhibits an equal-ripple approximation to constant group delay. The phase slope is programmable to present a continuously variable time delay network. Applications of a continuously adjustable, linear-phase time delay structure offers unique signal processing options to address various communication system tasks. These include timing recovery in DSP-based receivers, adaptive beam-forming and steering, communication systems channel modeling, and reverberation modeling in acoustic chambers and instruments.

14.5.1 Recursive All-Pass Filter Delay Lines

A polyphase M-path filter can be formed with recursive as well as with non-recursive filter segments. The recursive system is composed of all-pass filters with numerator and denominator formed with reciprocal polynomials in Z^M. For ease of design, and without loss of generality, the polynomials are formed as a cascade of first- and second-order all-pass filter stages. The first-order filter in Z^M forms M-poles and M-zeros with a single multiply, while the second-order filter in Z^M forms $2M$-poles and $2M$-zeros with two multiplies. These filters offer particularly efficient implementations of all-pass transfer functions. Figure 14.32 presents the standard structure for an M-path filter, which, because of the Z^M polynomial structure, can be used as an M-to-1 down-sampler or as a 1-to-M up-sampler. When the zero-indexed path in the M-path filter is selected to be pure delay, a particularly simple all-pass filter, every path in the M-path filter becomes an equal-ripple approximation to that path delay. Figure 14.33 presents the structure of a sixth-order recursive all-pass filter that make up each path in the M-path polyphase filter. Each path contains two first-order polynomials and two second-order polynomials. Each path requires six multiplies per output data point.

Figure 14.34 presents the phase shift of the 10 paths in the 10-path recursive polyphase partition. Note the linear phase as a function of normalized frequency. Figure 14.35 presents the phase slopes for the same 10 paths of the 10-path filter formed with six coefficients per path. Note that the 10 paths present almost linear time delay over $\pm$ 35% of sample rate. If a signal is sampled at twice its nominal bandwidth, a common practice in many digital receivers, it will experience linear time delay over its whole bandwidth when processed by the 10 stages shown here. For comparison, a 10-stage polyphase FIR filter with the same linear phase shift over the same fractional bandwidth would require 12 taps per path. Figures 14.36 and 14.37 present the phase and phase slopes of the equivalent 10-path FIR filter.

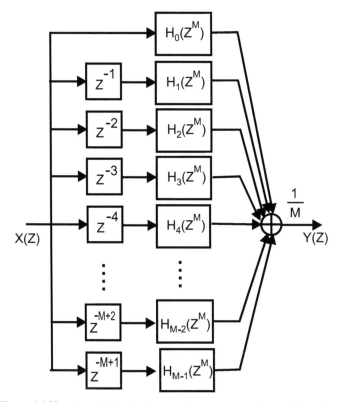

Figure 14.32 *M*-path filter implemented as recursive all-pass filters in Z^M.

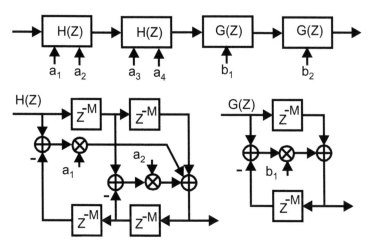

Figure 14.33 Structure of each path in *M*-path recursive all-pass polyphase filter.

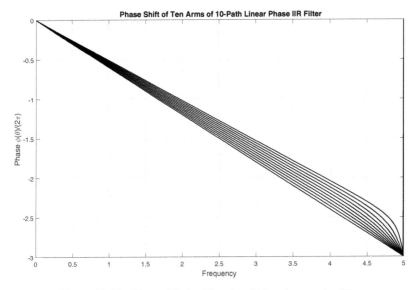

Figure 14.34   Phase shift for 10 paths of 10-path recursive filter.

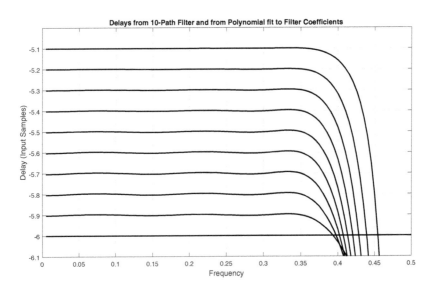

Figure 14.35   Phase slopes (group-delay) for 10 paths of 10-path recursive filter.

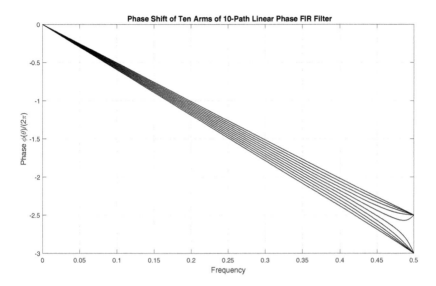

Figure 14.36 Phase shift for 10 paths of 10-path non-recursive filter.

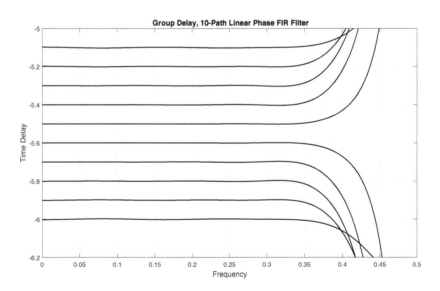

Figure 14.37 Phase slopes (group-delay) for 10 paths of 10-path non-recursive filter.

We can imagine that the six coefficients of the all-pass filter on each of the nine non-trivial paths are samples of a smooth continuous curve which, when sampled at increments of 0.1, present the values that define the nine paths. If we had the smooth continuous functions, we could sample it at any location (besides the multiples of 0.1) to determine the weights required to obtain arbitrary time delay corresponding to the desired sample point. Figure 14.38 presents the coefficients of the nine path filters and a fifth-order polynomial fitted with equiripple error to the sample values. As seen, the polynomials offer the smooth continuous function that relates weight values to delay. These polynomials can then be sampled at any location to obtain the weights that will present the delay at that sample position. The separate curves are labeled with their parameter values as shown in Figure 14.31.

These polynomials are embedded in the structure of an arbitrary time delay network as shown in Figure 14.39. Figure 14.40 presents the original delays plus those obtained by evaluating the polynomials at increments of 0.1 starting from an initial offset of 0.05. These samples correspond to delays midway between those obtained from the original 10-path filter bank. As can be seen, the delays from the coefficients obtained by sampling the polynomials are also linear with frequency.

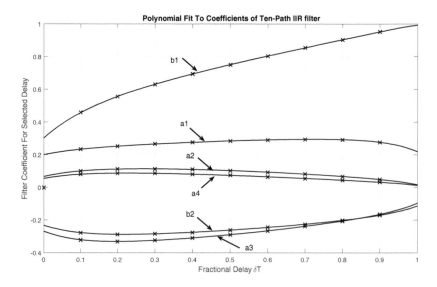

Figure 14.38   Coefficient values for 10-path filter and fifth-order polynomial fitted to values.

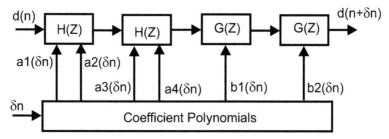

Figure 14.39   Programmable time delay network with associated coefficient generator.

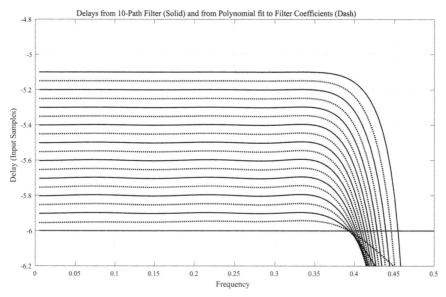

Figure 14.40   Original delays plus mid-value delays obtained by evaluating coefficient polynomials defined by original 10-path filter.

The digital delay line described here has been used in a timing recovery loop as shown in Figure 14.41. Here the input data is collected at two samples per symbol, the envelope is time shifted by the variable digital time delay line, and is presented to the matched filter and derivative matched filter. Figure 14.42 illustrates time responses of the receiver structure of Figure 14.41 that uses the variable time delay network in its timing recovery loop. The timing phase accumulator is the variable presented to the coefficient generator shown in Figure 14.39. The inserted time delay varies continuously

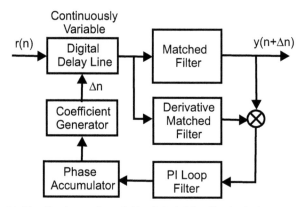

Figure 14.41   Continuously variable digital delay line in timing recovery loop.

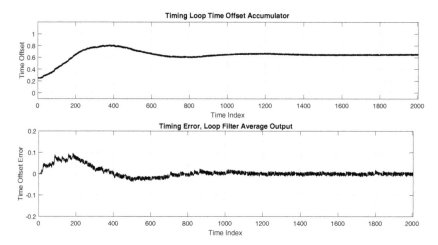

Figure 14.42   Timing phase accumulator that controls timing delay of digital delay line and timing error, and the loop filter output input to phase accumulator.

with the timing control. The average time delay error is shown in the lower subplot of Figure 14.42.

We have demonstrated here that a single path of a standard *M*-path recursive polyphase filter can be used as a programmable linear phase time delay network in a manner similar to the FIR Farrow filter. The example we used to illustrate the performance of this concept used two second-order polynomials in Z^2 and two first-order polynomials in Z^2, for a total of six multiplies per filter path to obtain the variable time delay

Table 14.3 Coefficients of all-pass filters of Paths 1−9 of 10-path polyphase filter.

Path #	B1	B2	A1	A2	A3	A4
1	0.458758	−0.276820	0.233603	0.100849	−0.320550	0.080736
2	0.556790	−0.287255	0.252152	0.112364	−0.330782	0.087489
3	0.630817	−0.284168	0.265170	0.113768	−0.324013	0.086214
4	0.693992	−0.274731	0.275704	0.110127	−0.309201	0.081185
5	0.751025	−0.261264	0.284359	0.103237	−0.289275	0.073988
6	0.804237	−0.244646	0.290925	0.093857	−0.265387	0.065386
7	0.854981	−0.224961	0.294424	0.082187	−0.237745	0.055720
8	0.904135	−0.201338	0.292323	0.067839	−0.205482	0.044960
9	0.952320	−0.170104	0.276523	0.048997	−0.164700	0.032264

process. Time delay filters can be formed with fewer stages per arm, and we have tested the performance of systems with one, two, or three multiplies. These filters, with smaller number of computations, perform very well. They simply exhibit larger ripple in the time delay versus frequency characteristic or exhibit linear group delay over a smaller fraction of the input sample rate due to larger transition bandwidth. Interestingly, when coupled with a decision feedback equalizer, the performance of the time delay network with a small number of coefficients is enhanced since the equalizer attributes residual group delay distortion to the channel and compensates appropriately. For those interested in playing with the structure we describe here, Table 14.3 lists the six coefficients of the nine non-trivial paths of a linear-phase 10-path recursive polyphase filter. The MATLAB script file *time_10* that contains the same coefficients and generated the phase portraits is also available in the MATLAB software available for this text. https://www.riverpublishers.com/book_details.php?book_id=788

14.6 Interpolated Shaping Filter

Throughout this book, we have examined a number of techniques to implement resampling filters. In this section, we select a specific task of up-sampling by a factor of 8, the impulse response of shaping filter used in a modulator. The particular shaping filter we examine is a square-root cosine tapered Nyquist filter used in a third-generation cellular phone system. A friend supplied the spectral mask that the system had to satisfy and asked us to design the filter and the interpolator to perform the up-sampling task. The question then arose as to what would be an efficient up-sampler and what signal degradation effects would be introduced by the up-sampling options? The distortion terms are related to the pre- and post-echoes that the interpolating filter contributes to the time response due to ripple in spectral

Table 14.4   Interpolator options and their comparative measures.

Interpolator type	Length or number of coefficients	Operations per Output Sample	Peak and RMS ISI
Shaping filter No interpolation	41 Taps	41 M, 41 A	PK; 0.0101 RMS; 0.0036
Eight-path Polyphase FIR	72 taps, 9 taps/path	9 M, 9 A	PK; 0.0158 RMS; 0.0040
Three-stage half-band FIR filter set	First stage; 21 taps, second stage; 13 taps third stage; 9 taps	4.8 M, 4.8 A	PK; 0.0109 RMS; 0.0036
Eight-path linear-phase polyphase Infinite Duration Impulse Response (IIR filter)	21 Coef. 3 Coef./path	2.7 M, 5.3 A	PK; 0.0128 RMS; 0.0040
Three-stage linear-phase IIR half-band filter set	First stage; 4-Coef. second stage; 2-Coef. third stage; 1-Coef.	1.5 M, 3.0 A	PK; 0.011 RMS; 0.0036
Eight-path non-linear phase polyphase IIR filter	14-Coef. 2-Coef./path	1.8 M, 3.6 A	PK; 0.0817 RMS; 0.0259
Three-stage non-linear phase IIR half-band filter set	First stage; 3-Coef. second stage; 2-Coef. third stage; 1-Coef.	1.4 M, 2.8 A	PK; 0.0808 RMS; 0.0284

amplitude and spectral phase. We also chose to examine distortion levels related to non-uniform phase characteristics of efficient recursive filters. The measure of distortion that is easy to access is the peak and RMS level of ISI at the output of the matched filter. The types of interpolators we examined are listed in Table 14.4 along with an indication of their relative computational complexity and their ISI levels. Also included in this table is the level of ISI contributed by the original shaping filter operating at its design frequency of two samples per symbol and then down-sampling to symbol rate. The remaining table entries were obtained by operating the filters at 16 samples per symbol and then down-sampling the filter output to symbol rate.

Figure 14.43 shows the impulse response and the frequency response of the shaping filter designed to meet the spectral mask indicated on the spectral plot. Also presented in the spectral plot is an inset figure zoomed to show the in-band ripple levels of the shaping filter. The in-band ripple level was specified to be less than 0.1 dB, and here it is seen to be 0.022 dB. We see here that the frequency of the ripple is 10 cycles per interval of symbol bandwidth

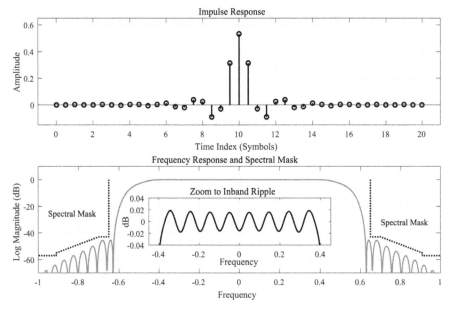

Figure 14.43 Impulse response and spectra of shaping filter.

that has been normalized to 1.0 in this figure. The ripple frequency tells us that the matched filter output will exhibit pre- and post-echoes at positions ±10 symbols from the location of its peak value. Figure 14.44 presents output samples of the matched filter taken at two samples per symbol with circles marking the one-sample per symbol time marks aligned with the peak output. The subplot accompanying the matched filter is a zoom to the low ISI levels where we see, as expected, the pre- and post-echoes at ±10 symbols from the peak position. The maximum ISI level that the matched filter can exhibit, computed as the sum of the absolute values, was found to be 0.0101, or approximately 1% of the peak output level. The RMS value of the ISI cloud around any constellation point was found to be 0.0036 approximately 1/3 of the peak ISI.

The first contender that comes to mind for use as the 1-to-8 up-sampler is an eight-path polyphase FIR filter. A prototype FIR filter of length 72 taps satisfied the filtering requirements dictated by the transition bandwidth and out-of-band attenuation levels. This filter is partitioned into an eight-path polyphase structure that is fed by the shaping filter. The polyphase filter is likely implemented in the form shown in Figure 7.13 rather than in the expanded form suggested by Figure 14.45. We see here that the

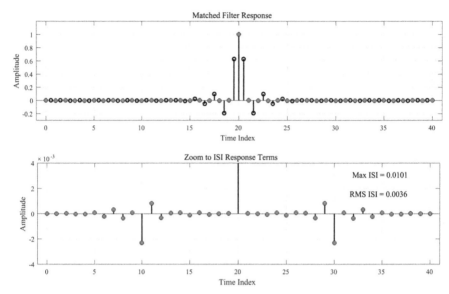

Figure 14.44   Matched filter response and details showing ISI levels.

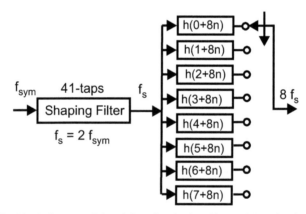

Figure 14.45   Block diagram of signal flow for shaping filter and 1-to-8 polyphase FIR filter interpolator.

polyphase form of the filter requires nine multiplies and adds per interpolated output sample point. We use this workload as the reference against which we compare the other interpolator options. The top subplot of Figure 14.46 shows the phase response of the eight paths of the polyphase partition of this prototype filter along with its log magnitude frequency response in the bottom subplot. The same subplot also includes a zoom to the interpolator's

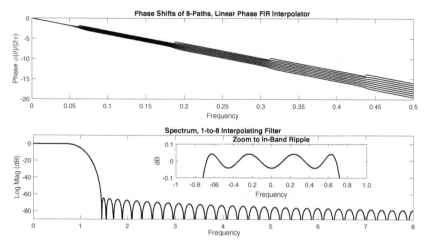

Figure 14.46 Path phase responses of eight-path polyphase filter and frequency response of prototype filter.

in-band ripple. This ripple is seen to be approximately four cycles per signal bandwidth and is approximately the same amplitude of the shaping filter ripple. We would expect an increase in ISI level due to the composite filtering action of the two filters. In fact, the ISI exhibited by the time series formed by the eight-path interpolator increased from peak levels of 1% to 1.6% and from RMS levels of 0.36%–0.4 %.

The phase response imparts little value here except that it does show the constant phase offsets present in the different Nyquist zones that can be used in a polyphase channelizers. The reason we show the phase here is that we will shortly compare this phase profile with corresponding profiles of recursive counterparts of the eight-path filter. Figure 14.47 shows the spectrum of the interpolating filter overlaid on the periodic extension of the shaping filter's spectrum and the spectrum of the composite impulse response of the shaping filter and the 1-to-8 interpolating filter.

The next contender for use as the 1-to-8 up-sampler is a cascade of three stages of half-band FIR filters. The half-band filters offer two primary advantages over a single polyphase filter. The first is that by ratcheting the sample rate up in increments of 2; the successive filters in the cascade, while operating at higher sample rates, do so with shorter filters due to the increased separation between the spectral replicates at the higher sample rates. The second advantage is the nearly 50% reduction in workload due to the zero-valued weights in the half-band filter. A prototype FIR filter of lengths 21,

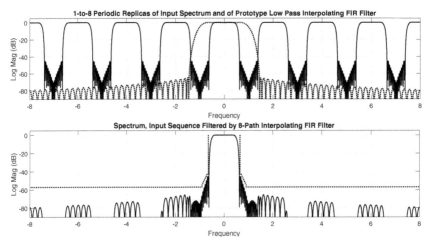

Figure 14.47   Spectra of up-sampled shaping filter with spectra of interpolating filter and spectra of interpolated shaping filter response.

13, and 9 taps respectively satisfied the filtering requirements dictated by the successive transition bandwidths. These filters are partitioned into two-path polyphase structure in which the upper path of each defaults to a simple delay line. The filter cascade is shown in Figure 14.48 where we see the delays in the upper arm and the non-trivial weights in the lower arm. The lower arm is tagged with its workload per output sample. Note that the first stage is used once per input to output two samples with a stage workload of 10 operations. These samples in turn are presented to the next stage, which must operate once per input it sees and thus operates twice to output four samples for a stage workload of 12 operations. Finally, the last stage sees four inputs and thus operates four times to output eight data samples with a stage workload of 16 operations. The total workload for the three stages is [10 + 12 + 16] or 38 operations which when amortized over the eight output samples leads to 4.75 operations per output. This workload is approximately half that of the eight-stage polyphase filter. We note that the half-band filters are too short to exhibit ripple in the signal bandwidth they are interpolating. Thus, a second benefit of the half-band filter is that it contributes an insignificant increase to the composite ISI, with the peak ISI changing from 1.0% to 1.1% and the RMS ISI not changing within the measurement resolution of 0.01%. Figures 14.49–14.51 present the spectra of the successive 1-to-2 up-sampled input data with the overlaid half-band filter responses as well as the spectra of the up-sampled and filtered time responses. These spectra are presented to

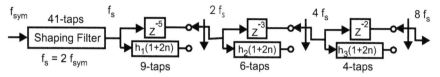

Figure 14.48 Block diagram of signal flow for shaping filter and cascade of three levels of half-band 1-to-2 up-sampling FIR interpolating filter.

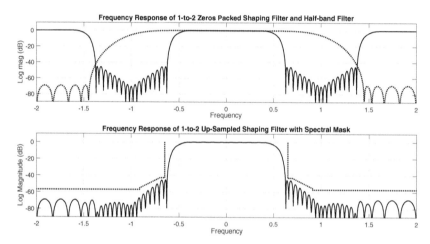

Figure 14.49 Spectra of up-sampled shaping filter with spectra of first half-band FIR interpolating filter and spectra of 1-to-2 interpolated filter response.

illustrate the ratcheting process that enables the higher speed processing tasks to be implemented with simpler, shorter filters.

At bandwidths, hence sample rates, for which we can use recursive filters for the interpolating function, we can bring new contenders to the table. These contenders are the resampling IIR filters described in detail in Chapter 11. There are recursive counterparts to the two non-recursive options we have just examined. These are an eight-path polyphase filter and a three-stage cascade of half-band filters. Figure 14.52 presents the approximately linear phase IIR polyphase filter structure. This filter approximates the linear phase characteristics by setting its top arm to a delay line with linear phase to which the remaining arms approximate with equal-ripple error. The non-trivial arms each have two filters in cascade, one filter requiring a single multiply and two adds and the second filter requiring two multiplies and four adds. The structural details of these filters can be found in Figures 11.11 and 11.35.

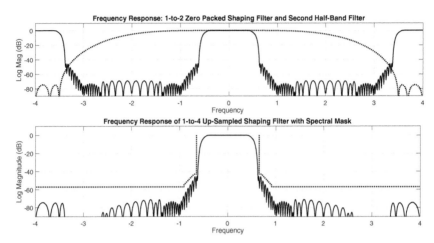

Figure 14.50   Spectra of 1-to-4 up-sampled shaping filter with spectra of second half-band FIR interpolating filter and spectra of 1-to-4 interpolated filter response.

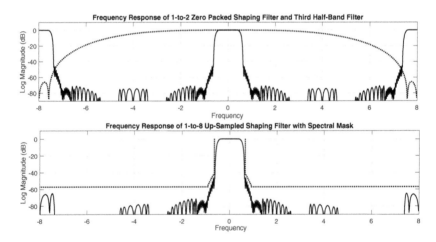

Figure 14.51   Spectra of 1-to-8 up-sampled shaping filter with spectra of third half-band FIR interpolating filter and spectra of 1-to-8 interpolated filter response.

The workload for this polyphase filter is seen to be less than three operations per output sample. Figure 14.53 shows the group delay response and the in-band ripple response of the prototype eight-path IIR filter. We first note the remarkable low level of amplitude ripple, the peak ripple being approximately 2 μ-dB presenting an inconsequential level of ISI. The group delay ripple on

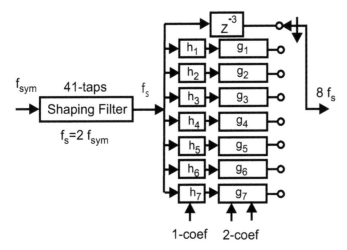

Figure 14.52 Block diagram of signal flow for shaping filter and 1-to-8 polyphase, approximately linear, recursive filter interpolator.

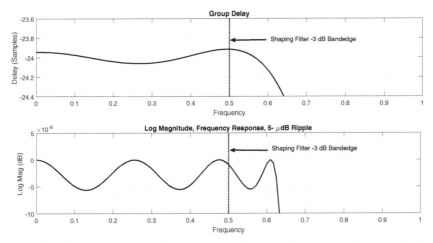

Figure 14.53 Zoom to group delay ripple and magnitude ripple of 1-to-8 polyphase recursive, approximately linear, recursive filter interpolator.

the other hand has amplitude of 0.05 and a frequency of approximately 2 cycles per signal bandwidth or 32 cycles per sample rate. We would expect this filter to contribute odd symmetric echoes of amplitude 0.05/32 at a two-sample offset from the maximum filter response. Figure 14.54 presents

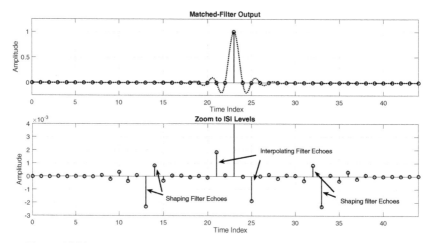

Figure 14.54   Interpolated matched filter response and details showing ISI levels.

the matched filter response and a zoom to the ISI levels to show the ISI contributors. The echoes are at the location and at the levels we expected.

Figure 14.55 shows the phase response of the eight paths of the approximately linear phase IIR polyphase filter along with its log magnitude frequency response. The phase response shows the phase of the delay-only reference path and the phase response of the other paths as they approximate the reference slope. Note that at the Nyquist boundaries, the phase difference expands to the next multiple of $2\pi/8$ and that this expansion occurs by inserting additional phase shift. Compare this behavior to the related phase expansion of the eight-path FIR filter in Figure 14.46. For the IIR, the spectral intervals corresponding to the phase transitions between Nyquist zones, the phase terms do not destructively cancel and the magnitude response exhibits a transition bandwidth between stop-band intervals in successive Nyquist zones. We can clearly see the transition bandwidths in the associated log magnitude spectrum. Our first reaction to these spectral bumps is horror; possibly "Good grief!" These transition intervals do not bother us; they can be thought of as do-not-care bands matching the spectral intervals in which we know there is no input energy by virtue of the input signal being oversampled. Figure 14.56 shows the spectrum of the IIR interpolating filter overlaid on the periodic extension of the shaping filter's spectrum. As expected, the transition regions of the polyphase IIR filter match the stop bands already present in the input spectrum. The spectrum of the up-sampled and filtered

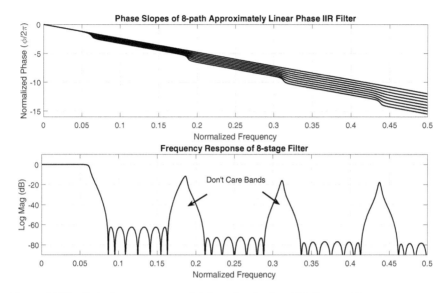

Figure 14.55 Path phase responses of eight-path polyphase IIR filter and frequency response of prototype filter.

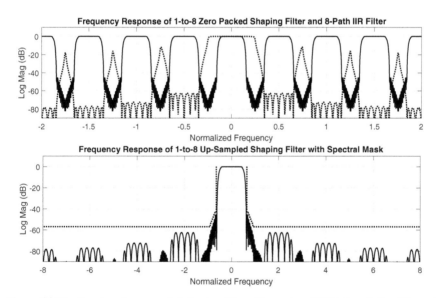

Figure 14.56 Spectra of up-sampled shaping filter with spectra of IIR interpolating filter and spectra of interpolated shaping filter response.

composite impulse response is seen to meet the spectral mask requirements of the filtering task.

The final interpolator option we examine in this discussion is the use of the nearly linear phase recursive half-band filters. These filters have the same property as that of the FIR half-band filters, in that the upper arm of the polyphase half band is delay only. The IIR version of a three-stage cascade that meets the filtering requirements of the interpolator task is shown in Figure 14.57. Here we see the delays in the upper arm and the all-pass filter structures in the lower arm. The workload for the filters in the lower arm matches the number of input coefficients that represent a small workload to perform its phase shift task. In this cascade, the first stage is used once per input to output two samples with a stage workload of four operations. These samples in turn are presented to the next stage, which must operate once per input it sees and thus operates twice to output four samples for a stage workload of four operations. Finally, the last stage sees four inputs and thus operates four times to output eight data samples with a stage workload of four operations. The total workload for the three stages is [4 + 4 + 4] or 12 operations which, when amortized over the eight output samples, leads to 1.5 operations per output. This workload is approximately half that of the eight-stage polyphase IIR filter and one-third of the workload for the FIR half-band cascade. We note that the IIR half-band filters do not exhibit amplitude ripple and that their phase ripple has a long period relative to the bandwidth being processed during the interpolation process. Thus, a second benefit of the IIR half-band filter is that it contributes an insignificant increase to the composite ISI, with the peak ISI changing from 1.0% to 1.1% and the RMS ISI not changing within the measurement resolution of 0.01%. Figures 14.58–14.60 present the spectra of the successive 1-to-2 up-sampled input data with the overlaid half-band IIR filter responses as well as the spectra of the up-sampled and filtered time responses. These spectra are presented to allow comparison between FIR and IIR filters performing the same ratcheting process to permit shorter filters at higher sample rate. Be sure to compare these figures with

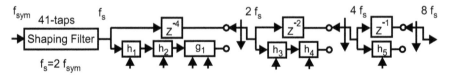

Figure 14.57 Block diagram of signal flow for shaping filter and cascade of three levels of half-band 1-to-2 up-sampling IIR interpolating filter.

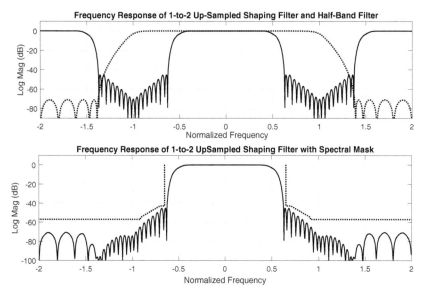

Figure 14.58 Spectra of up-sampled shaping filter with spectra of first half-band IIR interpolating filter and spectra of 1-to-2 interpolated filter response.

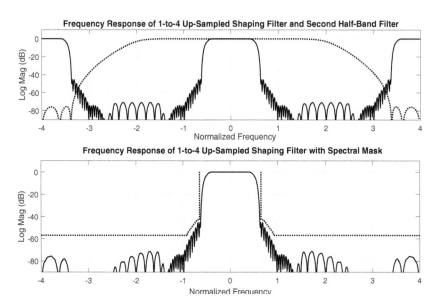

Figure 14.59 Spectra of 1-to-4 up-sampled shaping filter with spectra of second half-band IIR interpolating filter and spectra of 1-to-4 interpolated filter response.

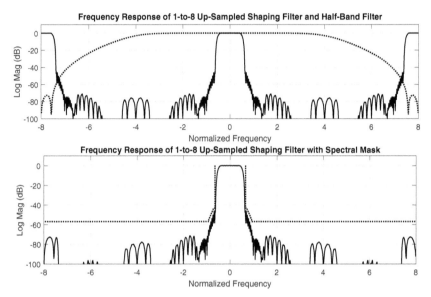

Figure 14.60 Spectra of 1-to-8 up-sampled shaping filter with spectra of third half-band IIR interpolating filter and spectra of 1-to-8 interpolated filter response.

Figures 14.49–14.51 keeping in mind the relative workloads expended while performing the interpolation process.

A final note in this area is that the eight-path polyphase filter and the three-stage half-band filter can also be implemented with recursive filters that exhibit non-uniform phase responses. The workload required of these filters is again reduced relative to that of the linear phase recursive filter set. These filters can be mixed and matched, linear filters at the low sample rate and non-linear phase filters at the higher sample rate. At the higher rates, the fractional bandwidth is sufficiently small that the non-uniform phase IIR filters are approximately linear phase filters. Another, tongue in cheek, comment is that group delay distortion introduced in the shaping filter and interpolator at the modulator will be attributed to channel effects, including analog filters in the signal flow path and that these small effects will be suppressed by an equalizer in the demodulator. A less onerous comment is that the group delay distortion terms generated by the low-cost, non-uniform phase recursive interpolating option are known *a priori* and can be pre-equalized by inserting their conjugate phase in the initial shaping filter.

14.7 Sigma-Delta Decimating Filter

Multirate filters find great application in source coding applications. One ubiquitous application is in the filtering and sample rate change associated with sigma-delta modulators. The sigma-delta modulation process is a source-coding technique that uses memory to represent samples of data at a given level of fidelity with a smaller number of bits than normally required to achieve that fidelity level when used without memory. The source coding performed by an A-to-D converter is that of representing a sampled data process with a finite set of amplitudes in such a way that the error between the two representations is made acceptably small. Many A-to-D and D-to-A converters perform their conversion process on a sample-to-sample basis without regard to past or future conversion steps. These are often called memory-less converters or, as a friend once observed, converters with amnesia. The memory required for this conversion process resides in the correlation between data samples. If data samples are highly correlated, and if we have the current and recent past samples of the data, there is little uncertainty of the value of the next sample. Predictive encoders operate on this basis. We predict the next sample and use the converter to measure or resolve the error in the prediction process. Similarly, if the data is highly correlated, the error generated by the conversion process is also highly correlated and hence predictable. If we are able to predict the error made by the quantizer, we can subtract the estimate prior to the conversion process. Converters that predict and cancel errors prior to their occurrence are called noise feedback or noise shaping converters. The classic sigma-delta converter is a member of this class of systems.

To assure high levels of correlation between data samples, we significantly oversample the process being encoded. Typical oversample ratios are 16 to 128 times the Nyquist rate of the process. Figure 14.61 presents the block diagram of a simple noise feedback quantizer. In this figure, the quantizer is modeled as an additive noise source. The difference between input and output of the quantizer is the quantizer error. We measure this error as the difference between the quantizer input and output ports and present the error to the predicting filter with Z-transform $P(Z)$. This filter, based on previous samples of the quantizer error, predicts a value for the current error and subtracts it from the input signal prior to delivering it to the quantizer. The simplest predicting filter uses the previous error as an estimate of the current error and is, in fact, merely a one-sample delay register, that is $P(Z) = Z^{-1}$. This substitution for the simple predicting filter

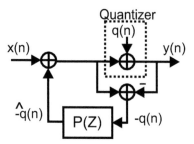

Figure 14.61   Block diagram of a noise feedback quantizer with predicting filter P(Z).

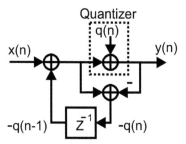

Figure 14.62   Block diagram of a noise feedback quantizer with delay line as predicting filter.

is shown in Figure 14.62. In this model, we see clearly that the noise is being fed back to the input of the system. The block diagram shown in Figure 14.62 is often redrawn as a two-loop feedback system. The first loop starts at the input goes to the quantizer, but not through it, and returns to the input through the delay line. This loop describes a digital integrator. The second loop starts at the input, goes through the quantizer to the output port, and returns to the input through a delay and negative sign. This alternate description of the noise feedback process is shown in Figure 14.63. Finally, the explicit feedback around the delay line in the first loop forming the digital integrator is suppressed by replacing the loop by its transfer function as shown in Figure 14.64. This is the most common representation of a sigma-delta modulator, one or more integrators, and a quantizer in a unity gain feedback loop.

If we model the sigma-delta modulator as a two-input, one-output system, we can form the output as the sum of two inputs through their corresponding transfer functions. This is shown in Equation (14.13). When we replace the predicting filter with a delay line Z^{-1}, we obtain the relationship shown in

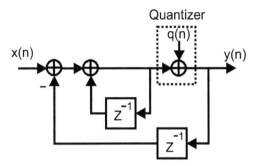

Figure 14.63 Alternate block diagram of a noise feedback quantizer showing digital integrator in feedback loop.

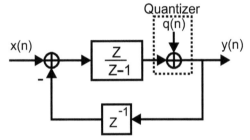

Figure 14.64 Block diagram of a noise feedback quantizer with block diagram of digital integrator replaced by its transfer function.

Equation (14.22) where we see that the noise transfer function (NTF) is a simple differentiator, $[1 - Z^{-1}]$.

$$Y(Z) = X(Z) + Q(Z)[1 - P(Z)] \qquad (14.17)$$

$$Y(Z) = X(Z) + Q(Z)[1 - Z^{-1}] = X(Z) + Q(Z) \left[\frac{Z-1}{Z} \right]. \qquad (14.18)$$

The single zero of the NTF resides at $Z = 1$, the zero frequency position for sampled data spectra. A stylized power spectrum of the output signal is shown in Figure 14.65. Here we see that the noise has been shaped by the NTF, which has placed a double zero at DC. The double zero suppresses noise in the vicinity of DC but permits noise away from the DC area. Since the input signal has been oversampled by, say a factor of 64, the spectrum of the input is confined to a small neighborhood around DC, the bottom ±0.8 % of the sample rate. Thus, the sigma-delta loop has arranged to keep noise away from the low-pass spectral region that contains input signal and

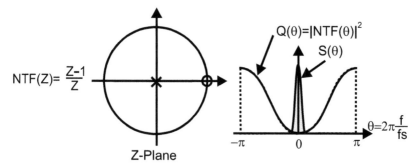

Figure 14.65   Roots of NTF and power spectrum of composite output signal of a noise feedback quantizer.

place it in a spectral region that contains no input signal. Since the zero of the NTF suppresses the noise, the level of noise injected by the quantization process is not important since the product of any finite noise power spectral density and the zero of the NTF is always zero. Consequently, the quantizer is often implemented as a 1-bit decision device. While a number of sigma-delta systems are implemented with as many as four bits performing the quantization process, a 1-bit quantizer with multiple integrators in the loop is the most common structure. A low-pass filter can reject the out-of-band quantization noise and the sample rate of the filtered data can be reduced in concert with the bandwidth reduction. This certainly sounds like a task for multirate filters.

14.7.1 Sigma-Delta Filter

The filtering task we address here is that of reducing the bandwidth and sample rate of an input data stream formed by a 1-bit, two-loop sigma delta converter operating at 64 times the signal's Nyquist rate. For this example, the signal two-sided bandwidth is 30 kHz and we require an output sample rate of 60 kHz. Additional post processing of the oversampled output series is not of interest to us here. The sigma-delta converter input data rate is 3.840 MHz. The dynamic range of the two-loop modulator is 90 dB, so the filtering to be accomplished is equivalent to a single low-pass filter with a two-sided bandwidth of 30 kHz, a transition bandwidth of 15 kHz, and a dynamic range of 90 dB with 0.1 dB in-band ripple. The 90 dB is defined by the performance of the two-loop sigma-delta, which improves quantizing SNR at the rate of 15 dB per doubling of sample rate. The filter following the modulator will

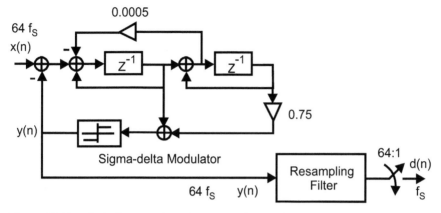

Figure 14.66   Block diagram of two-loop, 1-bit sigma-delta modulator, and its companion resampling filter.

of course reduce the sample rate in proportion to the bandwidth reduction. Figure 14.66 is a block diagram of the sigma-delta loop and Figure 14.67 presents a plot of the input and output time series formed by the modulator as well as the spectrum and a zoomed baseband detail of that spectrum of the output time series.

The traditional approach to the filtering and resampling task following the sigma-delta modulator is a two-step process comprising a 16-to-1 CIC filter followed by a four-path polyphase filter or a pair of half-band filters. This approach is similar to the filtering used in digital down converters. A fourth order CIC is required to meet the 90 dB suppression requirement. The CIC is operated as a Hogenauer filter, simultaneously performing filtering and 16-to-1 down-sampling. Figure 14.68 presents the time series formed at the output of the CIC as well as the spectrum of the output and with an overlaid frequency response of the four-stage CIC filter. Note that the input spectrum contains multiple equal amplitude non-harmonically related sinusoids. This signal is used to probe the frequency response of the process. Figure 14.69 presents the same information as did Figure 14.66 except that, here, the time series and the spectra are formed after the time series has been down-sampled 16-to-1. Here is where we would observe undesired aliasing into the baseband spectral region. None is seen!

The 4-to-1 down-sampling filter following the CIC performs spectral housekeeping, which entails spectral suppression of out-of-band quantizing noise, and in-band spectral equalization if required, as well as the sample

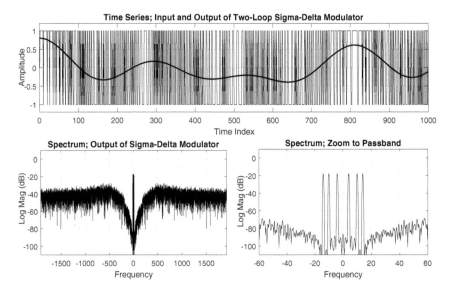

Figure 14.67   Time series of input and output of sigma-delta modulator and spectrum with zoomed detail of output spectrum from modulator.

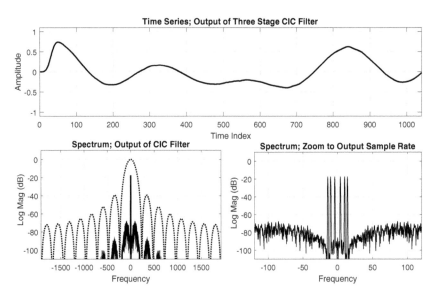

Figure 14.68   Time Series and spectrum of CIC filtered output of sigma-delta modulator with overlaid filter response and zoomed detail of output spectrum.

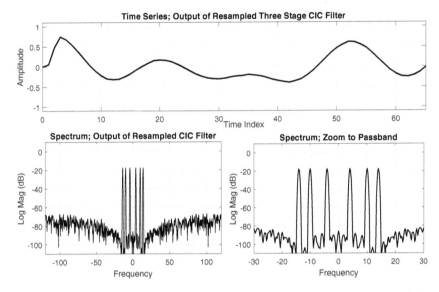

Figure 14.69 Time series and spectrum of CIC filtered and down-sampled output of sigma-delta modulator and zoomed detail of output spectrum.

rate change. The amount of spectral suppression required of this filter is surprisingly small. The required level of additional attenuation can be estimated from the subplot in the lower left side of Figure 14.69. Here we see that we only require another 20 dB or so to achieve the desired 90 dB attenuation levels. Figure 14.70 presents the time response and spectrum of the output from the final 4-to-1 down-sampling filter. Also overlaid on the output spectrum is the frequency response of this 24-tap filter. We also see the spectrum of the final down-sampled time series and note that the spectral terms are 90 dB below full-scale response. What we have illustrated here is that the combination of noise shaping with the two-loop sigma-delta modulator along with the spectral suppression performed by the cascade of two resampling filters has been able to convert a highly oversampled, hence highly correlated, signal with a 1-bit noise shaping converter to 15-bits of uncorrelated data samples.

It may have occurred to you that the filtering performed by the cascade of the CIC and the four-path polyphase filter did not take advantage of the fact that the data formed by the sigma-delta modulator is bipolar and of fixed amplitude, in fact only the sign of the modulated process. Processing such data in a FIR filter can be accomplished without the need for multiplications.

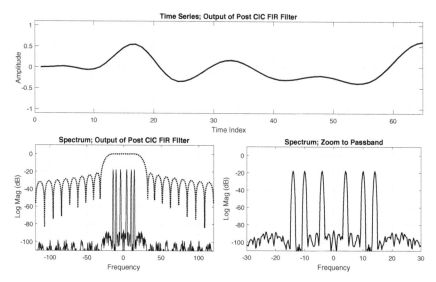

Figure 14.70   Time series and spectrum of second FIR filtered and down-sampled output from CIC filter with filter overlay and zoomed detail of output spectrum.

This awareness was not obvious because the CIC filter that processed the modulator output had no multiplies. We depart here from the standard solution and ask about a polyphase FIR filter that can perform the same 16-to-1 bandwidth and sample rate reduction as the CIC filter. A polyphase FIR filter that accomplishes the same spectral suppression requires four coefficients per polyphase arm. The data delivered to the polyphase filter from the sigma-delta modulator is a 1-bit sample, the sign-bit, which, conceptually, is delivered to successive filter paths by the input commutator. Thus, the content of each path register is a +1 or –1 in each of the four register positions. The output from any given path is simply the weighted sum of the filter weights where the weighting terms are ±1. The polyphase filter requires four sums for each input sample, but, of course, the four-stage CIC filter also requires four sums for each input sample. That is interesting! The 16-path polyphase filter requires a total of 64 sums to form one output sample at the output rate. The four-stage CIC requires the same number of sums to perform the arithmetic in the four overflowing accumulators. The output of the final accumulator is then passed to the four derivatives in the CIC chain to perform four more sums at the output rate. In terms of hardware resources, the four accumulators in the CIC can be mapped to the four accumulators

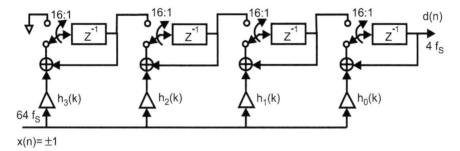

Figure 14.71 Dual form filter showing accumulators in four-tap polyphase filter.

in the dual form of the polyphase filter presented in Figures 5.11 and 5.12. This accumulator variation is shown in Figure 14.71. The advantage of using the multiply free form of the polyphase filter is that we can control the in-band spectral characteristics of the filter, an option not available to the CIC implementation. Another consideration applicable to field programmable gate array (FPGA) implementation is that the content of each polyphase filter path is a four-bit binary word.

There are only 16 possible outputs that any given stage can deliver to its output port. These outputs can be pre-computed and stored in a 16-element array unique to each path that is addressable from the 4-bit word stored in the input register for that path. The sum of the outputs from the 16 paths is the down-sampled and filtered time series. The 16-path filter then only requires 16 table accesses and 16 sums rather than the 64 adds required by the equivalent CIC. Figure 14.72 shows the look-up table implementation of the 16-path polyphase filter operating with binary inputs from the sigma-delta modulator.

Figures 14.73 and 14.74 present the output time series and spectra formed by the polyphase filter performing the first filtering task previously performed by the CIC filter. The two figures correspond to data taken prior to and after the 16-to-1 down-sampling operation. Also shown in Figure 14.73 is an overlaid spectral description of the four-tap 16-arm polyphase filter. Note that the overall filtering effect is the same as that performed by the CIC and presented in Figures 14.68 and 14.69. A housecleaning filter, following the polyphase filter, that finishes the filtering task and reduces the sample rate, will not does but have to correct the, now absent, $\sin x/x$ spectral tilt. Figure 14.75 presents the time response and spectrum of the output from this final 4-to-1 down-sampling filter. Also overlaid on the output spectrum is the frequency response of this 24-tap filter. We also see the spectrum of the final

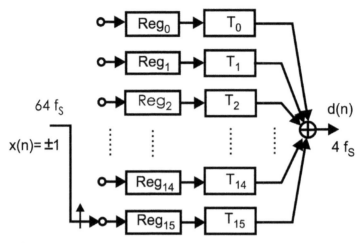

Figure 14.72   Polyphase filter implemented with short look-up tables addressed from binary input data delivered from sigma-delta modulator.

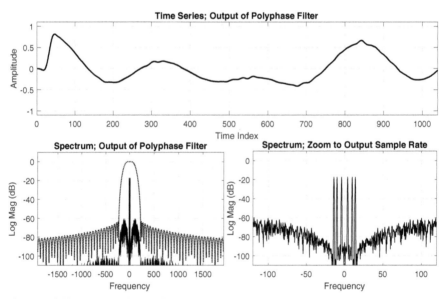

Figure 14.73   Time series and spectrum of polyphase filtered output of sigma-delta modulator with overlaid filter response and zoomed detail of output spectrum.

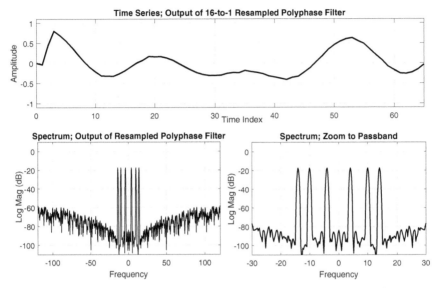

Figure 14.74 Time series and spectrum of polyphase filtered and down-sampled output of sigma-delta modulator and zoomed detail of output spectrum.

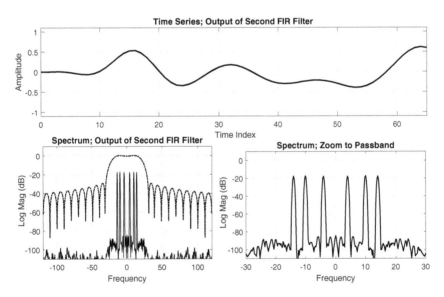

Figure 14.75 Time series and spectrum of polyphase filtered and down-sampled output of sigma-delta modulator and zoomed detail of output spectrum.

down-sampled time series and note that the spectral terms are 90 dB below full-scale response

14.8 FM Receiver and Demodulator

Multirate signal processing has had a significant influence at the physical or hardware layer of modern communication systems. In particular, multirate signal processing is found at the core of communication systems that couple the software defined radio (SDR) and software communications architecture (SCA) to reconfigure system resources for operation over a wide range of modulation formats and waveforms. In this section, we demonstrate one particularly efficient multirate signal processing solution to the task of implement a radio configured to select and extract a single FM channel from the commercial FM band, to down-convert and demodulate that channel, and to perform stereo demodulation and separation of the resulting baseband signal. Sprinkled through this discussion are descriptions of the necessary background material required to understand the various processing requirements.

In the United States, the commercial FM band spans the frequency interval from 88 to 108 MHz with the different FM stations allocated carriers separated by 200 kHz spacing starting at 88.1 MHz. European frequency FM station allocations are separated by 100 kHz intervals. The FM index of the FM signal is 75 kHz, which from Carson's rule leads to a nominal two-sided bandwidth of approximately 180 kHz. The compatible stereo signal, commonly carried by FM stations, is a pair of 15 kHz bandwidth audio signals transmitted as the sum $(L + R)$ and difference $(L - R)$ of the desired audio component signals denoted L and R for left and right, respectively. The $L + R$ signal resides at baseband while the $L - R$ signal is double-sideband suppressed carrier (DSB-SC) modulated to a 38 kHz amplitude modulation (AM) subcarrier. A 19 kHz pilot signal, inserted between the $L - R$ baseband signal and the $L + R$ subcarrier signal, is extracted by the receiver and frequency doubled to obtain a phase coherent reference signal required for the DSB-SC down-conversion. The separated signals are then summed and differenced to form $2L$ and $2R$, the desired stereo components. Figure 14.76 presents a block diagram of a stereo FM receiver as well as the spectral representation of the input and output spectra of the receiver. Also shown in this figure is a second subcarrier, the subsidiary carrier authorization (SCA) signal, carried by many stations.

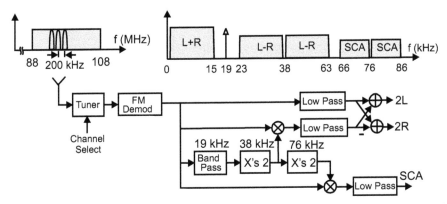

Figure 14.76 Input and output spectra of stereo FM receiver and block diagram of conventional FM receiver and stereo demodulator.

14.8.1 FM Band Channelizer

The tuner section of the FM receiver can be implemented as shown in Figure 14.77. The analog section of the receiver contains an antenna, a bandpass filter with bandwidth matching the 88-to-108 MHz FM band, a gain controllable RF amplifier, and a 10-bit ADC converter operating at an 80 MHz sample rate. The ADC operating at the 80 MHz sample rate performs IF sampling and aliases the FM band centered at 98 to 18 MHz, close to the quarter sample rate of the converter.

The digital section of the receiver, residing in the FPGA, contains a polyphase filter that restricts the signal bandwidth and reduces the sample rate 200-to-1 to 400 kHz. The 400 kHz rate was selected to permit wide transition bandwidth in the filter. The spectral response of the baseband prototype filter designed for the 200-to-1 resampling filter is shown in Figure 14.78. A 1200 tap FIR filter operating at 80 MHz sample rate satisfies the spectral specifications indicated in this figure. When implemented as a 200-path polyphase filter, the length of each path is a reasonable six taps which requires six operations per input sample. The total workload per input sample is increased when we include the complex phase rotators required to extract the desired signal. We now examine a few options that accomplish the channel selection along with the bandwidth and sample rate reduction afforded by the polyphase filter partition.

The first approach to extract the desired channel from the sampled data stream is a standard complex heterodyne and low-pass filter pair with the filters implemented in a polyphase structure that performs simultaneous

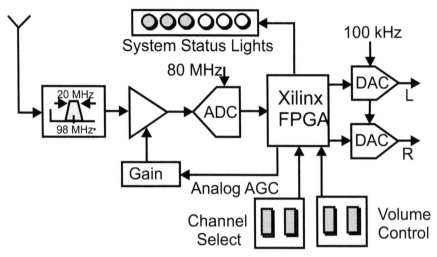

Figure 14.77   Block diagram of DSP-based FM channelizer.

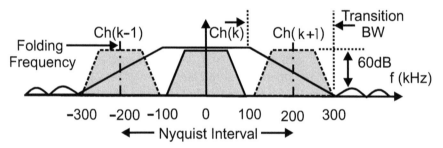

Figure 14.78   Spectral characteristics of prototype low-pass filter in polyphase receiver.

bandwidth reduction and sample rate reduction. This option is shown in Figure 14.79. Here we see that the workload for the real or imaginary output is seven multiplies per input sample, one multiply for the heterodyne, and six multiplies for the polyphase filter. The total workload for this implementation is 14 multiplies per input sample point.

We can apply the equivalency theorem to this channel selection task to obtain a desired reduction in processing load. Rather than down-convert the selected channel to baseband by an on-line heterodyne, we can up-convert the prototype low-pass filter to the selected channel as an off-line heterodyne. The savings we incur is that the on-line post-filtering down-conversion, if necessary, occurs at the reduced output rate rather than at the high input rate. The band-pass version of the channel selection process is shown in

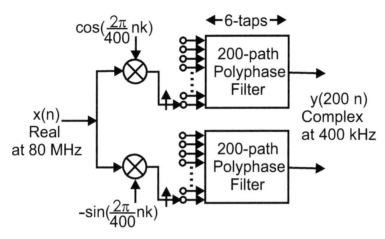

Figure 14.79 Standard heterodyne, filter, and down-sample architecture channel selector.

Figure 14.80. In this system, a signal centered at 8.0 MHz, the image of 88.0 MHz prior to IF sampling, aliases to baseband in the 200-to-1 down-sampling to 400 kHz. Thus, a signal centered at 8.1 MHz, the first FM channel aliases to 100 kHz, the quarter sample rate at the 400 kHz output rate. The required complex heterodyne following the down-sampling is a simple translation from the quarter sample rate to baseband. This operation is performed without actual multiplication since the values of the cosine and sine can be restricted to ± 1 and 0. The workload for this form of the channelizers is six multiplies per input sample for each of the real and imaginary output port; thus, the total workload is 12 multiplies per input sample. By embedding the heterodyne in the filter, we have saved the workload required by the heterodyne process.

A lesson learned early in multirate filtering is that it is wiser to make the data complex as we leave a processing algorithm than it is to make it complex as we enter the algorithm. In the standard channel centered polyphase filter structure, the data is made complex by the output phase rotators that unwrap the desired alias from the multiple aliases that have been baseband by the aliasing caused by resampling process. In the standard polyphase filter structure, all multiples of the output sample rate alias to baseband. In the system we are examining, these frequencies are all multiples of 400 kHz. As noted in the previous paragraph, the original 88 MHz, aliased to 8.0 MHz due to the IF sampling, is a multiple of 400 kHz and folds to zero frequency in the standard polyphase structure. The first FM channel is centered 100 kHz above 88.0 MHz and this frequency will fold to the quarter sample rate at the

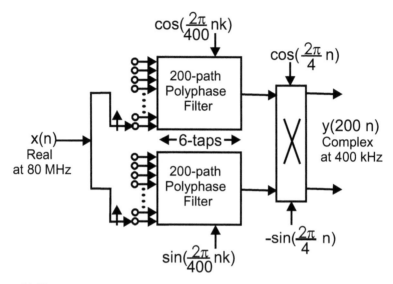

Figure 14.80 Heterodyned filter, down-sample, and down-convert architecture version of channel selector.

output sample rate of the polyphase filter. In fact, all the desired channels will fold to ±100 kHz. We now examine a variant of the standard polyphase filter structure that will fold and translate the FM spectral centers to zero frequency. Equation (14.19) presents the Z-transform of the frequency-translated version of the prototype filter impulse response while Equation (14.20) presents the 1-to-M polyphase partition of the same.

$$H(Z) = \sum_{n=0}^{M-1} h(n) e^{j\frac{2\pi}{M}nk} Z^{-n} \tag{14.19}$$

$$
\begin{aligned}
H(Z) &= \sum_{r=0}^{M-1} \sum_{n=0}^{\frac{N}{M}-1} h(r+nM) e^{j\frac{2\pi}{M}(r+nM)k} Z^{(r+nM)} \\
&= \sum_{r=0}^{M-1} e^{j\frac{2\pi}{M}rk} Z^{-r} \sum_{n=0}^{\frac{N}{M}-1} h(r+nM) e^{j\frac{2\pi}{M}nMk} Z^{nM}.
\end{aligned}
\tag{14.20}
$$

When the frequency index k is an integer, $2\pi nk$ is congruent to 2π, and the selected frequency bin, bin k, aliases to zero in the polyphase partition. A variant of this relationship is to be seen by replacing k with $k + s/4$, for $s = 0$, 1, 2, or 3. This is shown in Equation (14.21) where we see that the inner sum representing the operation of the polyphase stages still has a phase shift that varies with the time index "n." For the example developed here, the residual phase shift term is trivial, simply being powers of j. In operation, when the path coefficients are loaded into the path filters, the coefficients are rotated by the path rotations $\exp(j\,0.5\,\pi\,n)$ for $s = 1$ or $\exp(-j\,0.5\,\pi\,n)$ for $s = 3$. This pre-rotation of the weights results in the successful down-conversion, by the resampling operation, of the frequency components of the FM channels offset ± 100 kHz from the multiples of 400 kHz. Note that the offset is also embedded in the phase rotators on each polyphase arm that are applied in the outer summation of the following equation:

$$H(Z) = \sum_{r=0}^{M-1} e^{j\frac{2\pi}{M}r(k+s/4)} Z^{-r} \sum_{n=0}^{\frac{N}{M}-1} h(r+nM)\, e^{j\frac{2\pi}{M}n M(k+s/4)} Z^{nM}$$

$$= \sum_{r=0}^{M-1} e^{j\frac{2\pi}{M}r\,(k+s/4)} Z^{-r} \sum_{n=0}^{\frac{N}{M}-1} h(r+nM)\, e^{j\frac{2\pi}{4}ns} Z^{nM}.$$

$$(14.21)$$

Embedding the j phase rotator in the path weights has a slight impact on the structure of the polyphase filter arms and the subsequent phase rotator. While no actual complex products are involved in the polyphase arms, the data samples formed by the polyphase arms are now complex rather than real. This means that the formerly complex scalar phase rotators applied at the stage output now requires a full complex product. The structure of the modified polyphase filter is shown in Figure 14.81. The workload for this version of the channel select process is seen to be 6 multiplies per input sample for the polyphase filter and 4 multiplies for the output complex rotator for a total workload of 10 multiplies per input sample. This is the most efficient of the three variants of the channel selection process we have examined.

14.8.2 FM Demodulator

For completeness, we now present a short description of the digital FM demodulator. An FM demodulator performs the task of extracting the modulation signal from the FM modulated waveform. Since frequency is the time derivative of a sinusoid's phase angle, an FM demodulator must form

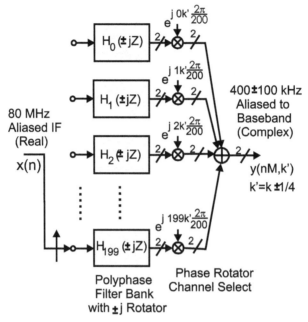

Figure 14.81   Signal flow diagram of modified polyphase filter to permit frequencies offset by quarter of output sample rate to alias to baseband.

the derivative of the received signal's phase. In analog systems, this process is accomplished by a circuit called a discriminator that applies a sequence of operators to the carrier centered FM signal. These operators are a hard-limiter that removes incidental AM from the signal, a derivative circuit that converts the FM signal to an AM signal, and a diode detector that responds to the resulting amplitude variations. A balanced version of this system is used to linearize the derivative process while canceling its DC component.

In the digital receiver, the channel selection process forms a complex baseband signal with the $I(n)$ and $Q(n)$ components defining the complex envelope of the FM signal. The I-Q components define the modulated phase angle of the original FM signal and the digital FM demodulator must form the derivative of that phase angle. One implementation forms the angle $\theta(n)$ of the ordered pair by an arctangent or by a CORDIC routine and then forms its derivative with a FIR filter. In another implementation, the derivative of the arctangent is formed in a single step by the relationship shown in Equation (14.22) and illustrated in Figure 14.82 with the denominator scaling.

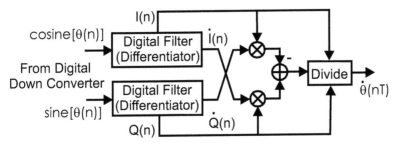

Figure 14.82 Signal flow diagram of digital FM discriminator.

The scaling term serves the same function as the hard-limiter in an analog discriminator, that of removing AM from the input signal.

$$\dot{\theta}(n) = \frac{d}{dn} \arctan\left(\frac{I(n)}{Q(n)}\right) = \frac{I(n)\dot{Q}(n) - \dot{I}(n)Q(n)}{I^2(n) + Q^2(n)} \qquad (14.22)$$

14.8.3 Stereo Decoding

The stereo decoding is straightforward task that closely follows the signal flow suggested in Figure 14.76. The interesting part of this problem is the processing to extract the pilot signal and then frequency double it for use in the demodulation of the DSB-SC modulated *L-R* signal. We first examine the length of the filter required to isolate the pilot signal and then present a clever multirate implementation of the pilot extraction and frequency-doubling task. The brute force filter designed to isolate and extract the pilot signal would have to satisfy the specifications indicated in Figure 14.83. A filter with approximately 273 taps is required to satisfy these specifications. Since the filter bandwidth is a very small fraction of the sample rate, we know that we can use multirate signal processing to implement a more efficient option. We now examine one such option based on down-sampling, filtering, and up-sampling with a delightful little twist unique to this application.

The structure of an alternate pilot extraction filter is shown in Figure 14.84 and the spectra of the signal at various points in the processing scheme can be seen in Figure 14.85. In this structure, we perform a 20-to-1 down-sample with a 20-stage polyphase filter with phase rotators that extract the first Nyquist zone centered at 20 kHz. The prototype low-pass filter used in this partition has a large transition bandwidth equal to 19 kHz, which requires a filter with 60 taps. The output of the phase-rotated polyphase

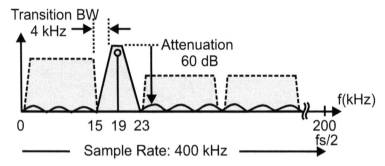

Figure 14.83 Filter specifications for pilot extraction filter.

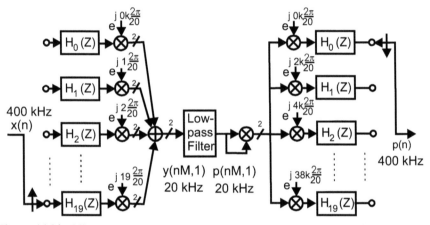

Figure 14.84 Pilot extraction and doubling by polyphase down-sample, filter, frequency doubling, and polyphase up-sample.

filter contains the alias of the 19 kHz pilot now located at −1.0 kHz with a sample rate of 20 kHz. The filter following the polyphase filter limits the bandwidth around the pilot signal and it is designed as a low-pass filter with two-sided bandwidth of 2 kHz and transition bandwidth of 2.0 kHz. This filter requires 30 taps. The complex signal output by the low-pass filter contains only the aliased pilot signal. We now double the frequency of this signal from −1 to −2 kHz by squaring the complex samples. We now up-sample the double frequency aliased pilot by a factor of 1-to-20 in a second polyphase filter. The phase rotators at the input to the second polyphase are selected to output a real signal in the second Nyquist zone centered at 40 kHz. This aliases the −2 kHz baseband signal to 38 kHz, the carrier frequency required for the DSB-SC demodulation of the *L-R* signal component. The

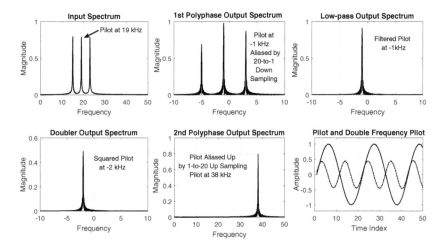

Figure 14.85 Spectrum of input and output of polyphase down-sampler, at output of low-pass filter, at output of squaring circuit, and at output of polyphase up-sampling filter. Also time series of pilot component of input signal and double frequency pilot at output of process.

delight found in this process is that the frequency doubling is performed at baseband between the down-sampling and up-sampling processes and that we have selected different Nyquist zones in the down-sampling and up-sampling process.

The workload required for this technique of pilot extraction and doubling can be determined as follows. The input polyphase filter requires three multiplies and adds for each path filter and two multiplies and adds for the path phase rotators for a workload of five multiplies per input sample. Similarly, the output polyphase filter, the dual of the input filter, also requires five multiplies and adds per output sample. The filtering performed at the reduced 20 kHz sample rate requires 30 multiplies and adds for each of the real and imaginary filter legs and four multiplies and two adds in the complex product that doubles the frequency of the extracted pilot. The total of 64 multiplies is amortized over 20 input or output samples so that the filtering load is 3.2 multiplies per input. The total workload, the sum of these terms, is seen to be approximately 13.2 multiplies and adds per input–output pair. This is a significant improvement relative to the 273 multiplies and adds required of the brute force, non-multirate filter implementation.

References

fred harris, Elettra Venosa and Xiaofei Chen, and Chris Dick, *"Non-Data Aided Symbol and Carrier Synchronization Via 2-samples per Symbol Band Edge Filters"*, 46-th Annual Asilomar Conference on Signals, Systems, and Computers, Pacific Grove, CA, 5-7 November 2012.

fred harris, Elettra Venosa and Xiaofei Chen, and Chris Dick, *"Band Edge Filters Perform Non Data Aided Carrier and Timing Synchronization of Software Defined QAM Receivers"*, WPMC-2012, *(Wireless Personal Multimedia Communications), 24-27 Sept. 2012, Taipei, Taiwan.*

fred harris, *"Band Edge Filters: Characteristics and Performance in Carrier and Symbol Synchronization"*, WPMC-2010, Recife, Brazil, Oct 11-14, 2010.

Aziz, Pervez, Henrik Sorenson, and Jan Van Der Spiegel, "An Overview of Sigma Delta Converters: How a 1-bit ADC Achieves more than 16-bit resolution", IEEE Signal Processing Magazine, , Vol. 13, No. 1, pp. 61-84, January 1996.

Candy, James and Gabor Temes, *Oversampling Delta-Sigma Data Converters: Theory, Design, and Simulation*, Piscataway, NJ, IEEE Press, 1992.

Crochiere, Ronald, and Lawrence Rabiner, *Multirate Signal Processing*, Englewood Cliff, NJ, Prentice-Hall Inc., 1983.

Dick, Chris, fred harris, and Michael Rice, "Synchronization in Software Radios: Carrier and Timing Recovery Using FPGAs", IEEE Symposium On Field-Programmable Custom Computing Machines, Napa Valley CA, April 16-19, 2000.

Fliege, Norbert. *Multirate Digital Signal Processing: Multirate Systems, Filter Banks, Wavelets*, West Sussex, John Wiley & Sons, Ltd, 1994.

harris, fred. "Carrier Synchronization Techniques for DSP-Based Modems" Communication Systems Design Magazine, (Trade Journal), pp 32-34, 36, 38-39, 42, July 2000.

harris, fred. "A Fresh View of Digital Signal Processing for Software Defined Radios, Part I and Part II", International Telemetry Conference (ITC), San Diego, CA, 21-24 October 2002.

harris, fred. "Sigma-Delta Converters in Communication System", Encyclopedia of Communications, John Wiley & Sons, John Proakis, Editor, 2002.

harris, fred "Teaching MODEM Concepts and Design Procedure with MATLAB Simulations", Signal Processing Education, Workshop, Kent, Texas, 15-18 October 2000

harris, fred and Chris Dick, "On Structure and Implementation of Algorithms for Carrier and Symbol Synchronization in Software Defined Radios", URSIPCO-2000, European Association for Signal Processing, Session on Efficient Algorithms for Hardware Implementations of DSP Systems, Tempere, Finland, 5-8 September 2000.

fred harris and Chris Dick, "Performing Simultaneous Arbitrary Spectral Translation and Sample Rate Change in Polyphase Interpolating and Decimating Filters in Transmitters and Receivers", 2002- Software Defined Radio Technical Conference, San Diego, CA, 11-12 November 2002.

fred harris and Michael Rice, "Multirate Digital Filters for Symbol Timing Synchronization in Software Defined Radios", IEEE Journal on Selected Areas in Communications, Vol. 19, pp. 2346-2357, Dec. 2001.

Hentschel, Tim. *Sample Rate Conversion in Software Configurable Radios*, Norwood, MA, Artech House, Inc., 2002.

Jovanovic-Dolecek, Gordana. *Multirate Systems: Design and Applications*, London, Idea Group, 2002. Mitra, Mitra. *Digital Signal Processing: A Computer-Based Approach*, 2nd ed., New York, McGraw-Hill, 2001.

Mitra, Sanjit and James Kaiser, *Handbook for Digital Signal Processing*, New York, John Wiley & Sons, 1993.

Norsworthy, Steven, Richard Schreier, and Gabor Temes, *Delta-Sigma Data Converters: Theory, Design, and Simulation*, Piscataway, NJ, IEEE Press, 1997.

Rice, Michael and fred harris, "Polyphase Filter Banks for Symbol Synchronization in Sampled Data Receivers", MILCOM-2002, Anaheim, CA, 7-10 October 2002

Rice, Michael, fred harris, and Chris Dick, "Maximum Likelihood Carrier Phase Synchronization In FPGA-Based Software Defined Radios", ICASSP-2001, International Conference on Acoustics, Speech and Signal Processing, Salt Lake City, Utah, 7-11 May 2001

Vaidyanathan, P. P. *Multirate Systems and Filter Banks*, Englewood Cliff, NJ, Prentice-Hall Inc., 1993.

Problems

14.1 Equation (14.1) presented the Taylor series for $[\sin(\theta)/\theta]^P$, the main lobe response of the CIC filter. Determine the value of θ for which the droop due to the main lobe curvature is -0.1 dB and then 0.5 dB. What is the signal bandwidth as a fraction of the sample rate corresponds to these values of θ?

14.2 A signal with two-sided bandwidth of 20 kHz is centered on an analog 450 kHz intermediate frequency carrier. Suppose we select a sample rate of 360 kHz to perform IF sampling and alias the signal from the center frequency to a new digital center frequency. Draw a block diagram of this process. Show where the spectra are located and describe the remaining processing required to bring the signal to baseband with filtering and resampling to 90 kHz output sample rate.

14.3 A signal with two-sided bandwidth of 20 kHz is centered on an analog 450 kHz intermediate frequency carrier. Suppose we select a sample rate of 120 kHz to perform IF sampling and alias the signal from the center frequency to a new digital center frequency. Draw a block diagram of this process. Show where the spectra are located and describe the remaining processing required to bring the signal to baseband with filtering and resampling to 40 kHz output sample rate. Be careful of spectral inversion.

14.4 The information a receiver needs to determine correct timing information from a received signal is obtained from the output of a matched filter and the derivative matched filter. We now examine the transfer function of the time error detector. Form a SQRT Nyquist pulse oversampled by 10-to-1. Also form a corresponding matched filter and a derivative matched filter with the same oversampling ratio. Be certain that the derivative filter and the matched filter are time aligned. That is, that the zero crossing of the derivative filter matches the peak of the matched filter. Convolve the impulse response of the shaping filter with impulse response of the matched filter and of the derivative matched filter. Form the dot product of the two responses and plot the results. This function is the timing error detector function. Determine its slope in units of amplitude per time offset. If the amplitude of the input signal is doubled, what happens to the gain of the time offset error detector? How does this impact the time offset error for a QAM signal as opposed to a QPSK signal?

14.5 Determine how to form a band edge filter from a matched filter operating at four samples per symbol. Hint: this will require a four-path polyphase partition, but no resampling, of the matched filter. Where is the band edge located for this process? Can we reduce the sample rate by a factor of 2 as part of separating the two band edges? Why would one try to do this?

14.6 Form a simple band edge filter by operating the matched filter at two samples per symbol and extracting the overlap of the low-pass and high-pass polyphase partition with a Hilbert transform filter. Feed the band edge filters from the output of a shaping filter also running at two samples per symbol and modulated with16-QAM random data. Introduce selectable frequency offsets to the modulated signal and form a plot of the average output level from the energy difference of the two band-edge filters as a function of frequency offset. Be sure to test both positive and negative offsets. Determine the gain of the frequency-offset detector. Is this gain a function of modulation constellation density? Is this gain a function of input signal level?

14.7 A simple digitally controlled, digital delay line can be formed by a cascade of one or more identical all-pass networks with transfer function $H(Z) = (1 - \alpha Z)/(Z - \alpha)$. This is particularly effective when the signal bandwidth is a small fraction of the sample rate as might occur when a signal is oversampled by a factor of 4. Under this condition, the all-pass network exhibits approximately linear phase shift in the signal bandwidth. Determine the expression for phase slope of the all-pass network. You will find assistance in Chapter 10. Then determine group delay as a function of the parameter α.

14.8 Form a delay line using three stages of the recursive one-tap all-pass filter presented in Chapter 10. Form an SQRT Nyquist pulse that is oversampled by a factor 10-to-1 and then pass the SQRT Nyquist pulse through the delay line filter chain with the parameter α set to zero and plot the input and output signal and verify that the delay is three samples. Repeat the experiment for a range of α between –0.5 and +0.5 in increments of 0.1. Examine the relationship between the observed delay and the setting for the parameter α.

Index

μ-dB ripple 357
1/12-Octave
    Proportional Bandwidth 456
1/f decay 65
1/f side-lobe 66
1/f side-lobes 68, 69
1/f stop band 69
2.5-to-1 down sample filter 480
1/3 octave filter bank 452
dB harris filter 93
4-to-1 down sampling of
    quarter-sample rate 480
4-to-1 oversampling 2

A

accumulator bit width 408
alias to baseband 145
alias to DC 145
alias-free partition 246
aliasing 20
all pass filter, single coefficient,
    dual form 335
all-pass filter
    implementation 332
all-pass filter, first order 327
all-pass filter, single
    multiply 334
all-pass filters 327
alternation theorem 56
amortized workload 219
analysis-synthesis filter banks 276

analog down converter 472
analog quadrature filters 142
arbitrary band edge 447
arbitrary resampling 167
arbitrary resampling ratio 185
arbitrary time delay IIR filter 506
Armstrong, Edwin 144
audiometric filter bank 458

B

band edge filter 495
beamforming 156
binary numbers 400
binary numbers,
    2's-complement 401
binary numbers,
    offset-binary 401
binary numbers,
sign modified 401
binary numbers,
    sign-magnitude 401
biquad filter 370
boxcar filter 394
boxcar z-transform 397

C

canceling roots 397
carrier recovery 492
cascade analysis and synthesis 282
cascade channelizer with
    channel masks 284

cascade filter 120
cascade integrator comb 27
cascade recursive half-band 371
CD *See* compact disc player
    channelizer non-overlapping
    bands 244
channelizer even and odd
    indexed bin centers 309
channelizer overlapping
    bands 244
channelizer, aliased polyphase
    filter 538
channelizer, complex
    band-pass filter 537
channelizer, standard complex
    heterodyne 535
CIC See cascade integrator comb
    CIC bifurcate zeros 426
CIC coherent gain 423
CIC down-sampling 417
CIC incoherent gain 423
CIC up-sampling 411
circular buffer 254
circular time shift 249
coefficient quantization 68
coefficient stride 175
comb filter 395
communication system 86
compact disc player 2
compare resampling filter workload
    446
comparison uniform and non- uniform
    phase 410
compensating half-band filter 474
complementary low pass and
    high pass filters 341
complex band-pass filters 142
controlled sampled
    data delay 501

Cooley-Tukey 150
cosine-tapered Nyquist filter 87
coupled channelizers 275

D
decimator 140
derivative matched filter 487
destructive cancellation 155
digital delay line, timing
    recovery example 507
digital down converter 140
digital FM channelizer 460
digital IF 164
digital translation 397
Dirichlet kernel 44
discrete time sampling
    sequence 15
distortionless channel 67
Dolph-Tchebyshev window 61
don't care bands 159
down conversion 419
down converter, multiple
    channels 245
down converter, single
    channel 244
down sampling 14
dual filter 26
dual form FIR filter 111
dual graphs 23
dyadic half-band filter 217

E
early and late gates 411
engineers sampling theorem 33
equal pass band ripple 52
equal stop band ripple 52
equiripple 55
equivalency theorem 130
excess bandwidth 84

even and odd length FFT in
    channelizer bin centers 312, 316

F

failure to acquire 419
Farrow filter 185, 192
FDM See frequency
    division Multiplex
    filter banks, arbitrary
    resampling 245
filter length 52, 60, 63, 99, 131
filter quality 134
filter specifications 43
filter_ten 158
firpmord 59
first generation receiver 472
flexible ruler 29
FM demodulator 539
FM receiver 535
form factor 133
fractional bandwidth 44, 133, 233
frequency derivative matched
    filter 495
frequency division
    multiplex 140, 243
frequency translation 152, 472

G

Gibbs phenomena 47, 498
graphic equalizer 450

H

half-band design trick 225
half-band filter 88, 107, 219
half-band filter workload 232
half-band high pass filter 221
half-band low pass filter 220
harris estimate 61
hearing aid filter bank 463

Herrmann 59
heterodyne narrow band noise 129
Hilbert transform filter 88
Hilbert transform half-band 228
Hogenauer filter 29, 407
Horner's rule 212

I

IFIR comparison 443
IFIR example 439
IF subsampling 477
IF subsampling example 477
IFIR *See* interpolated FIR
Impulses 15, 64
input commutator 150
integrated side-lobes 65, 68, 208
interchange filter and resamp
    147, 151
Interpolated FIR filter 437
interpolated shaping filter
    example 509
interpolation 168
interpolation and up
    conversion 180
interpolator, 3-stage linear phase,
    half-band IIR filters 520
interpolator, 3-stages of
    half-band FIR Filter 515
Interpolator, 8-path polyphase
    FIR filter 512
interpolator, linear-phase 8-path
    IIR filter 517
intersymbol interference 73
ISI *See* intersymbol interference
    iterated half-band recursive
    filter 385

K

$K(\delta_1, \delta_2)$ 132, 134

Kaiser 58
Kaiser-Bessel window 52
Kobyashi Maru Scenario 43

L

Lagrange interpolator 57
lazy S 330, 347
lexicographic 150
linear interpolator 196
linear phase recursive all-pass 281
linear phase recursive IIR 520
linear time invariant 24
linear time varying 29
lineardesign 348
linearly interpolate 202
logarithmic band FIR filters
    for perfect reconstruction 463
lossless all-pass 356
low pass to band-pass
    transformation 361
low pass to low pass
    transformation 357
LTI *See* linear time invariant
    LTV *See* linear time varying
    MAC *See* multiplier accumulator
    main lobe 46, 47

M

masking filter example 445
matched filter 92, 93
MATLAB 52, 58, 61, 63
maximally decimated filter
    bank 244
maximum likelihood frequency
    estimator 494
maximum likelihood timing
    recovery 487
McClellan 57
memory bank data swaps 252

M-fold aliasing 154
minimum mean square 47
Mitra 58
MMS *See* minimum mean square
    move data samples 488
move filter coefficients 488
move the channel 144
move the filter 144
M-path filter 152
M-Path recursive 380
multiple bandwidth
    analysis channelizer 303
multiple bandwidth
    synthesis channelizer 304
multiple bands 242
multiple exchange 57
multiple stage CIC 431
multiple transition bands 382
multiplier accumulator FIR 121
multiply free filters 394, 531
multirate filter 20, 21
multirate filters 2
multirate recursive filter 367
myfrfm 70

N

N/M 134, 135
narrow band beam forming 156
narrow band noise 126
nearest neighbor 186, 189
noble identity, 25, 147
noise feedback quantizer 523
noise generator 127
noise transfer function 525
nonmaximally decimated
    filter bank 254
nyq2 104
Nyquist criterion 32
Nyquist filter 88

Nyquist pulse 88, 282
Nyquist rate 34

O

octave partition 450
odd and even indexed
  polyphase bin center 298, 300
ops/input 134
oversampled ADC 9

P

paired echoes 73
Parks 57
Parks and McClellan 57
partial sum accumulator 121
Pascal triangle 341
pass band and stop band
  responses 355
pass band ripple 55
perfect reconstruction 276
perfect reconstruction
  analysis filter spectrum 279
perfect reconstruction
  synthesis filter spectrum 279
periodic pass band 443
periodically time varying 28
periodically time varying
  filter 129
phase alignment 158
phase coherent summation 155
phase locked loop 493
phase profiles 155
phase response, first order
  all-pass filter 328
phase response, second order
  all-pass filter 330
phase rotators 155
phase shift, 10-path
  FIR filter 505

phase shift, 10-path
  IIR filter 504
phase slope, 10-path
  FIR filter 505
phase slope, 10-path
  IIR filter 504
phased array beam forming 160
pilot extraction filter 542
PLL *See* phase locked loop
  pole-zero cancellation 397
polynomial approximation to
  coefficients 206
polynomial approximation to
  data 210
polynomial order 61
polyphase band pass filter 242
polyphase channelizer 241
polyphase partition 156
predistortion 168
proportional bandwidth filter 450
polyphase down sampling filter 27
polyphase filter 118
polyphase filter bank 243
polyphase half-band filter 110
polyphase synthesizer 246
prune CIC accumulator 410
pruning least significant
  bit (LSB) 418
pruning most significant bit
  (MSB) 410
PTV *See* periodically time varying

Q

quadrature mirror filter 88, 110
quarter sample rate alias 477

R

Rabiner 57
radio receiver architecture 472

rational ratio P/Q 167
rational ratio
    resampling 182, 185
rcosine 113
recursive all-pass filter
    delay line 502
recursive all-pass filters 324
recursive Hilbert Transform 376
register overflow 400
relatively prime 26
Remez algorithm 57
Remez half-band filter 224
Remezfrf 70
remezord See firpmord
    resampling 14
resampling ratio 131
resampling recursive filters 324
residual spectra 189
residual spectrum 188
root, active 350
root, inactive 350

S

sampling function 15
scale factor 44
scaled matched filter 97
scaled shaping filter 97
second-generation receiver 473
separating aliased data 155
serpentine shift 247
shaping filter 92
share feedback and
    feed forward 339
side-lobe 51
side-lobe levels 50
side-lobes 50
sigma-delta CIC filter 527
sigma-delta low pass filter 530
sigma-delta modulator 523

sigma-delta polyphase
    FIR filter 530
sin(x)/x 2, 42, 94
sliding average filter 394
small bandwidth filters 126
spectral mask 98
spectral masking 443
spectral translation 20
sqrt filter 87
sqrt-Nyquist filter 87
sqrt Nyquist harris taper 100
sqrt Nyquist
    harris-Moerder design 103
staircase weight function 68
state machine 184
state machine scheduler 254
stereo decoding 541
stop band ripple 59
stop band zeros windfall 342
subfilter 29
superheterodyne principle 144
synthesized super channels 286
super channel band edges 293
super channel interleave
    wide and narrow bands 297

T

Taylor series, filter weights 205
Taylor series,' time sequence 205
Taylor window 64
Tchebyschev approximation 64
Tibetan hat filter 172, 174
time aligned time series 157
time division multiplex 243
time domain reflectometer 74
time granularity 185
timing jitter 186
timing recovery 411
tony_des2 338

transition bandwidth 59, 90
transmultiplexer 243
TRF *See* tuned radio frequency
    tuned radio frequency 144
two adjacent neighbors 208
two-path all-pass filter 337

U

up conversion 493
up sampling 14
up-converted filter 145
up-sample 1-to-2 231

W

Weaver modulator 228
weight function 69
windowed half-band filter 223
windowing 45
work load 131
workload super channel FIR 289

Z

zero-order hold 2, 187
zero-packed 22
ZOH *See* zero order hold

About the Author

Professor harris is professor of ECE at University of California San Diego where he teaches and conducts research on Digital Signal Processing and Communication Systems. He formerly taught at SDSU, the home of the endowed *fred harris Chair of DSP*. He holds 38 patents on digital receiver and DSP technology and lectures throughout the world on DSP applications. He consults for organizations requiring high performance, cost effective DSP solutions.

He has written over 275 journal and conference papers, the most well-known being his well cited (8500 citations) 1978 paper "On the use of Windows for Harmonic Analysis with the Discrete Fourier Transform". In addition to this textbook, he is co-author with Bernard Sklar, of Digital Communications (3rd edition), and has contributed to a number of other DSP and communication textbooks.

He became a Fellow of the IEEE in 2003, cited for contributions of DSP to communications systems. In 2006 he received the Software Defined Radio Forum's "Industry Achievement Award". He received the DSP-2018 conference's commemorative plaque with the citation: *We wish to recognize and pay tribute to fred harris for his pioneering contributions to digital signal processing algorithmic design and implementation*, and his visionary and distinguished service to the Signal Processing Community.

He was the Technical and General Chair respectively of the 1990 and 1991 Asilomar Conference on Signals, Systems, and Computers, was Technical Chair of the 2003 Software Defined Radio Conference, of the 2006 Wireless Personal Multimedia Conference, of the DSP-2009 and DSP-2013 Conferences and of the SDR-WinnComm 2015 Conference.

The spelling of his name with all lower case letters is a source of distress for typists and spell checkers. A child at heart, he collects toy trains and old slide-rules.

For Product Safety Concerns and Information please contact our EU
representative GPSR@taylorandfrancis.com
Taylor & Francis Verlag GmbH, Kaufingerstraße 24, 80331 München, Germany

www.ingramcontent.com/pod-product-compliance
Ingram Content Group UK Ltd.
Pitfield, Milton Keynes, MK11 3LW, UK
UKHW021111180425
457613UK00001B/26